American Botany

1873-1892

RODGERS, Andrew Denny III. American Botany, 1873–1892; Decades of Transition. Hafner, 1968 (orig. pub. by Princeton, 1944). 340p il bibl 66-29972. 8.75

CHOICE   FEB. '69

Biology

One of a series of books about American botanists by Rodgers, who is well qualified to write these histories. Covering the years in which great strides were made in botanical taxonomy and morphology, it includes the lives of such men as Asa Gray, George Engelmann, and Leo Lesquereux. It is well worth reading by graduate and undergraduate students. Well documented; well indexed. Recommended for all libraries.

American Botany

1873-1892

Decades of Transition

BY

ANDREW DENNY RODGERS III

(Facsimile of the Edition of 1944)

HAFNER PUBLISHING CO.

NEW YORK and LONDON

1968

Reprinted by Arrangement

Printed and published by
HAFNER PUBLISHING COMPANY, INC.
31 East 10th Street
New York, N.Y. 10003

Library of Congress Catalog Card Number: 66-29972

Printed in U.S.A. by
NOBLE OFFSET PRINTERS, INC.
NEW YORK 3, N. Y.

TO CHARLES ALFRED WEATHERBY

PREFACE

THIS is a book conceived for the present generation of botanists, and for posterity. The process of verifying the truth and authenticity of materials has been relentless and extended over more than three years—in all, more than five years—of study. The author's entire time has been given to this and other books of like character involving unceasing investigation in crucial transitional and developmental periods of plant science study, in this instance, the history of American Botany generally from the years 1873 to 1892.

The author avouches the truth of this book's findings as to general trends and the substance of conclusions reached. Moreover, every effort has been made to eliminate errors in every particular. However, in matters of specific character—the detailed information employed in elucidating conclusions—the author, perhaps more than anyone, is conscious of the possibility of inclusion of errors. The analytic and synthetic interpretation of events occurring in the science during the years studied—truthfully and objectively considered—has been the one aim and intention. No partisan effort, no taking sides in any controversial matters still of the domain of science, has been contemplated as part of the enterprise. Enough difficulties in complete and thorough acquisition of materials already exist without adding more. Scarcity of sources in some phases of investigation, abundance of research materials in other phases, and in all, a widely scattered and diverse range of materials, have made the task of organization and verification one not easily performed.

May not such a book be one in which all American botanists share? The history of science is essentially a subject belonging to the humanities group. Yet scientists should, and must, participate as a matter of right in the process of verification, and, indeed, in the interpretation. The author, therefore, invites criticisms and corrections wherever necessary. This book seeks to extend the boundaries of enterprise striving foremost toward arriving at a truthful interpretation of the American scientific scene during two important decades of the last century, confined and limited, however, to a branch of study which has delighted so many of its workers—American botany. May that delight irradiate through its pages!

The authority of the scholarly past is ever with students. What the past has experienced often indicates what the future may bring forward. Nothing is really built except with certainty as to the soundness of foundations. May that confident sense of soundness gratify, and inspire, the mind and heart of every reader!

A. D. R.

CONTENTS

CHAPTER I

Asa Gray–The Great Years Begun

W HEN in the year 1873 John Torrey and William S. Sullivant died, two great collaborations in North American botany ceased. Asa Gray was left to go on apart from Torrey with the completion of their *Flora of North America*. And Charles Leo Lesquereux, apart from Sullivant, was enabled to give more time to the advancement of North American paleobotany.

On New Year's Day of that year, Gray had tendered his resignation to the president and fellows of Harvard College, retiring from his professorship after thirty-one years of service, but requested to be continued as curator of the herbarium. To Alphonse DeCandolle he wrote on January 14: "... this is my last year of university work. I finish in July ... and give my remaining time to the 'Flora of North America.' "[1]

With Torrey gone, Gray was indubitably North America's greatest botanical scholar. A man alive to all new and important movements scientifically in Europe and America, he had already made a "noise" in the scientific world and more than fulfilled the prophecies made for him by his great teacher and friend Torrey. Gray was embarking on new courses unexplored—he was opening that year a botanical laboratory, not only for advanced students but for pupils of the summer school. The dim outlines of a magnificent new era in world botany—a transitional era—were dawning. A new kind of botanical exploration was at hand, an exploration that emphasized work in the large herbaria—taxonomy had not and would not yield its place of supremacy during Gray's life— an exploration that would add to the taxonomic laboratory, and its work with the microscope, the principles of a science with practice was developing. Microscopes were improved and behind the increasingly enlarged study of external structures of plants brought from the field loomed the marvelous new phases of plant morphology, a more intensive study of plants as living, growing, and multiplying things, with consideration given to their internal structures. Long since the famed Lowell Lectures, in 1844 and earlier, Gray had shown interest in physiological botany. But for the most part such studies were still in an *observational* stage. A "pure" science based on *experimentation* had not reached American shores in any magnitude. Even in Europe experimental research was

---

[1] Gray's letters, quoted in this book, are taken mostly from *Letters of Asa Gray*, edited by Jane Loring Gray. In two volumes; Boston and New York: Houghton, Mifflin and Co.; Cambridge: The Riverside Press, 1893.

striving for recognition. Young American botanists, however, would soon return from studies in laboratories there and bring with them the rudiments of a "new botany." Notably among these was William Gilson Farlow, Gray's student and assistant, who on Gray's advice had studied medicine and then gone abroad to study nonvascular plants. He, although concerned primarily with cryptogamic botany at Harvard, was to become an early investigator of plant diseases, publishing in 1875 an article for the Bussey Institute on the potato rot, and a little later, on diseases of olive and orange trees and the American grapevine, on onion smut, and the black knot. Perhaps these studies cannot be said to mark beginnings of plant pathological work in America as most of Farlow's study in these particulars was mycological. Probably not until Thomas J. Burrill of the University of Illinois recognized pear and apple-twig blights as of bacterial origin may it be said that real studies in plant pathology began. But Farlow's researches and the earlier studies of Burrill were comparatively new and inspired by European work. Morphology had to acquire new techniques and methods of research.

At the University of Michigan, indeed at Harvard and a few other places, some classwork in plant morphology had been given for some time. Direct observation under conditions of experimentation, nevertheless, was not prevalent. Gray foresaw the future. In fact, his last lectures in 1872 showed the breadth of his interests, a few topics being, "Reproduction among Algae," "The Sun and Vegetation," "General Principles of Classification," "Reproduction among Fungi," "Reproduction among Mosses," and "Characters of Selected Families of Flowering Plants." Gray showed a growing interest in pathology. But he was first a systematist and of this keenly aware. For "vegetable physiology" he had appointed George Lincoln Goodale, the while watching developments in morphology and physiology. Moreover, working in Gray's laboratory and charmed by his enthusiasm as he lectured, was Charles Edwin Bessey, a young botanist twenty-seven years of age, appointed instructor in January 1870 in botany and horticulture at Iowa Agricultural College. When an American publishing house asked Gray to write an American text similar to an English adaptation of Julius von Sachs's epoch making *Lehrbuch der Botanik,* Gray refused and recommended Bessey. Gray was grooming students for the great task of bringing Europe's "Scientific Botany" to America—a movement which stressed study *of plants* rather than *about plants* and in which objective researches in developmental morphology were advancing. Indeed, when in 1870, Goodale had first written Gray concerning study under him, "our American authority," he said:

I have made myself familiar with your "Structural & Systematic Botany," with Balfour's Class Book, Schleiden (Prin[ciples of] Sc[ientific] Bot[any] and have worked out the microscopy exc[epting] fertilization & embryology). But study by one's self is unsatisfactory, for there is a fear lest some part of the subject may have been misunderstood. . . .

Goodale was not stressing taxonomy more than morphology. He told Gray "I am ready to undertake any work in vegetable physiology," and his cherished wish was to bring "no discredit" to his instructor's learning. Morphology, except as a study of external plant structures largely for taxonomic purposes, was unknown to the older American botanists such as Torrey and Sullivant. Indeed Lesquereux, although born in Europe and associated with able botanists there, had little if any training in developmental morphology. And his astoundingly skillful work establishing paleobotany in North America would two decades later reveal the need for revisions of much of his materials. For development of knowledge of fossil plant structures on the basis of plant affinities and not differences—on the basis of new laboratory and field investigation methods—was to become, as John Merle Coulter later said, "one of the most remarkable chapters in the history" of North American science. Gray and Lesquereux were primarily taxonomists. In 1873 each was left with unpublished works of their collaborators—Gray with Torrey's compilation of the *Phanerogamia of the Pacific Coast of North America* and Lesquereux with Sullivant's and his supplement to the "beautiful 'Icones Muscorum'" and the *Manual of the Mosses of North America* which Sullivant and he had planned and partially begun.

Sereno Watson, Goodale, and Farlow could assist Gray with his many tasks at Harvard. But what aid would be afforded Lesquereux at Columbus, Ohio, would have to be gotten hundreds of miles away at Cambridge. Lesquereux could turn to the Smithsonian Institution in Washington for aid in his paleobotanical studies. For these he could also turn to the Museum of Comparative Zoology of Harvard. The United States Geological and Geographical Survey of the Territories in the West were sending him fossil flora from their explorations. But, at Sullivant's death, Gray had persuaded Louis Agassiz to release Lesquereux temporarily from preparation of a catalogue of the fossil plants of the Museum of Comparative Zoology in order that Sullivant's bryological works might be completed. For most of this work, Lesquereux thought, he would have to labor alone. But he agreed to do them.

One feature was common to both Gray and Lesquereux. Both enjoyed the friendship of George Engelmann of St. Louis. Engelmann by now was taking seriously the suggestion made to him some years before by

Torrey and elucidating various genera of North American plants, which systematizations supplied several valuable investigations to Gray's later editions of his much used *Manual of the Botany of the Northern United States*. Engelmann was particularly interested in certain forest trees—the pines, junipers, and oaks. He was the indisputable authority on Coniferae, on Cactaceae, and other allied western groups. The viburnum and dogwood trees, the yuccas and agaves, the gentians, the North American grapes, sections of Euphorbiaceae, Juncaceae or rush family, Cuscutaceae or dodder family, Isoetes or quillwort, Sparganiaceae or bur reed family, Sagittarias, and Oenothera, the evening primroses, were some of his special interests. On these he was a principal American authority. To him came most of the new materials of these plants from American exploration.

No one surpassed Edward Tuckerman in North American lichenology. In point of age becoming the dean of American botanists, Alvan Wentworth Chapman still retained his place as the most important southern botanist. There were a number of American botanists still living who had shared in greater or less extent the fame of Torrey and Gray's *Flora of North America*[2] by contributions of new plant species for its pages. The number was growing very few. Increase Allen Lapham, Lewis R. Gibbes, Samuel Barnum Mead, Stephen T. Olney, Henry W. Ravenel, and George William Clinton were still living. Those hardy botanical collectors, Charles Pickering, Charles Wright, Ferdinand Lindheimer, Augustus Fendler, Samuel Botsford Buckley, John Milton Bigelow, George Thurber, Arthur Schott, and Charles Christopher Parry were still alive but all except Parry, Fendler, and Thurber were more or less botanically inactive. Daniel Cady Eaton was the authority on ferns. Thomas P. James and Coe F. Austin led the field in mosses and hepatics. Few of these, however, had contributed directly to Torrey and Gray's *Flora*. Gray, Tuckerman, and Chapman were left almost the sole representatives of the very early period of the famous work, conceived from its beginnings on a national scale. Gray's *Synoptical Flora* would begin where the older *Flora* left off and traverse again the ground of that already published, to amplify and bring the original up to date.

Gray, moreover, had another large task. For more than a decade now he had defended against religious and some scientific attack the views of Charles Darwin as to the *Origin of Species*. His great discussions affirming "the general doctrine of the derivation of species" had prevailed over

2 Containing abridged descriptions of all the known indigenous and naturalized plants growing north of Mexico. Volume I: Parts I-II, 1838; Part III, 1840. Volume II: Part I, 1841; Part II, 1842; Part III, 1843.

the Agassizian view "of specific creation," of a special and local creation of forms. But as late as 1873 remnants remained. Illustrative was Sir William Dawson's view:

All that I try to guard against is the hypocritical and unfair attempts which one sees on all hands to confuse Embryology with Geological sequence and frame classifications looking to evolution and not to actual affinity. Much of this, as I am given to have observed, is due to evolutionists. Further I think you must admit that the doctrine of the spontaneous evolution of man from lower animals, which is logically connected with the idea of derivation of species, sweeps away at once not only christianity but natural religion. . . .

Dawson was Canada's foremost early paleobotanist. Lesquereux, on reading Darwin's "great work" in 1860, believed it contradicted by the most "reliable evidence, viz., the geological data." For a long time he struggled with Darwin's thesis reconciling the struggle for life with the "Providential law of development." Both discussed these matters with Gray. Gray, however, held that theism was compatible with even so naturalistic a view as Darwin's. It was important to determine plant origins if possible, their inheritance, purposes, and adaptations. Scientific inquiry was given new impetus. Gray saw that the ascertainment of natural causes need appear no longer inexplicable secrets of the Divine mind.

Indeed over a long period of years Gray had indulged his alert mind in close adducing of many natural laws. The less "close and obvious" observable connection between structure and function in plants than in animals; the infrequency of occurrence of hybrids and plant variations in nature and the more frequent occurrence of such under conditions of cultivation and increased food supply; the return of hybrids to parental forms; the infusion of male and female parental characters and the multifarious inequalities of each induced or happening in the offspring and surviving variably generation to generation; "present" and "anterior" diversity; the perpetuation of individual plant characteristics and their capacities to accumulate and become fixed under agencies of fertilization and selection; the action of foreign pollen on fruits; differences in the life-hold occurring in response to bud propagation and that from seed reproduction; effects of the grafting of variegated varieties on the principal stock itself; plant variations induced by diverse soils, climates, and other environmental and biologic factors were some of many subjects on which he made observations. Gray realized the place of nitrogen and other elements holding complex functions in plant nutrition. As a matter of fact, it is said, Gray's studies in instances, for example, on movements of tendrils of cucurbitaceous plants, led Darwin to initiate investigations.

Yet, Gray was first interested in taxonomy and viewed such matters from implications of systematization. He seemed to favor a lively concept of "the invariability of species." What seemed to impress him was the capacity of species to keep "true in its course by the sum of the heredities which press each individual forward in its actual direction. . . . [A species is like a stream that] has made its bed and lies in it, not escaping from its own valley, it is flexible enough to obstacles, is ever changing its particular course as it flows, and may by its own action send off here and there a bayou (variety) or branch into a delta of channels (derivative species). . . ."

Nor did he preclude the "eddies of atavism (the resumption of dropped characters)." Gray was in a sense America's first great interpreter of results of scientific investigations, not from the taxonomic laboratory only but also from field and garden studies. Over the years the higher plants received most of his attention. Europe looked to him. and only when study extended to the lower plants for more than systematic reasons did his American supremacy in part yield. Furthermore, Gray's efforts concentrated foremost on the plants of nature—the socalled wild plants, not the cultivated.

Sexuality in plants, as we understand it today, was demonstrated by Camerarius in 1694, in *De Sexu plantarum epistola*. The production by man of recognized plant hybrids was made between two species of tobacco by Kolreuter in 1760. Of "the essential meaning of sexuality and as to its operation in respect of fixity" of species, Gray had definite ideas. He said so. The art of plant breeding and a practical knowledge of fertilization and crossing of plants had been known for some time. Gray had kept pace with productions of great European scholars, not only in botany but also horticulture. Although, in 1855, he had said, ". . . we suppose, with Dr. Hooker, that wild plants rarely hybridize," yet he continued, "the possibility and even the probability of the occurrence must not be overlooked in a thorough discussion of the general question of the limitation and permanence of species."

Gray, however, technically was not the professor of horticulture at Harvard. In 1868, after several years of travel in Europe, Charles Sprague Sargent had returned to this country and taken up the practice of horticulture and study of botany. In 1872, becoming director of the Harvard Botanic Garden, he had been appointed professor of horticulture during the years 1872 and 1873; and on November 24, 1873, being also appointed director of the recently created Arnold Arboretum, he was to give up in 1879 his botanic garden directorship and concentrate on developing against great odds a scientific garden dealing largely with woody plants.

Asa Gray

Though not unaware of the immense potentialities in advancement of a horticultural science, Gray saw that this, like agricultural science— agricultural botany, Gray termed it—was not to be achieved on any large scale until later, until expansion of a national experiment station movement, if and when that should develop, until a risen science outgrowing from the arts and business phases of horticultural and agricultural practices should culminate; in other words, until, with a profound transitional development common to all branches of plant science study, a "new horticulture," a "new botany," and a scientific agriculture would sweep lands of many continents. Perhaps it is inferring too much to vouchsafe an opinion that Gray foresaw all these developments. The facts remain, however, that much evidence supports these conclusions and that men trained under him, in some instances his students, made a reality of an envisioned future; and Gray, in a very real sense, prepared the way as a great master and teacher.

At the outset, the province of a book on "American Botany 1873-1892 Decades of Transition" must be limited in its scope by its very title; yet a few relevant observations may be offered for consideration. Lines of division as between the plant sciences at that time—as, even, now—were conventional. So far as the plants were concerned, no one of the three great branches of plant science investigation adhered to any arbitrarily set limits. Botany, generally speaking, adhered to a study of the wild plants of nature; horticulture studied the plants of cultivation; while agriculture as the commonwealth's basic industry studied from the profit and loss standpoint the raising and harvesting of crops of the farm. It seems fair to say that, just because an experimentally scientific program went forward more progressively and effectively in botany than in other branches *during this period,* this fact alone gave botany little cause for any exclusive sense of pride. For, much botanical investigation of a purely scientific nature was done with cultivated and "economic" plants, with "agricultural species," the descriptive terms are numerous. Why progress was attributed to botany, more than horticulture and agriculture in many instances, seems during these years to have rooted in the fact that the most eminent and able investigators of the old and new worlds were botanically trained, were, in largest numbers, botanists. Botany then—as perhaps it remains yet today—was the basic plant science study. The great work of plant introductions from foreign lands or from other regions of the American continent was begun in large part by botanists. Plant breeding in the colleges became important to the theoretical botanist. The work of the horticulturist and the agriculturist —one in the garden, the other in the farm field—was regarded more in the practical than theoretical or experimental spheres, although here and there a progressive voice, pleading for the application of more science in each division, was more than occasionally heard. With the establishment of the Bussey Institution at Harvard, of agricultural departments in various colleges, of experiment stations modeled after those of Europe, and, of course, with the spread of agricultural colleges, American agriculture, and horticulture became more and more institutionalized.

However, it must be remembered that during Gray's life, the United States Department of Agriculture, though established as an independent bureau in the 1860's, remained under a commissioner and was not accorded presidential cabinet recognition and a secretaryship until after his death. The work of the plant sciences generally at that time was taxonomic. Gray was a man firmly set in his task. He, like nearly all of the ablest American botanists and many of the ablest men of European botanical science, was a taxonomist and engulfed in the huge basic task of systematizing the world floras.

He knew that, before experimenters could proceed with experimentation, a knowledge of *what* were the available materials, their relationships and schemes of descent, was basic. That of itself was enough for the last years of the Gray-Sir Joseph Dalton Hooker generation of botanists. The Royal Kew Gardens of England, and also the Harvard Botanic Garden, divided the work of the wild and the cultivated plants, the one for botany, the other for horticulture. No criticism is implied. The tasks were so enormous that there had to be some division of responsibility. But Darwin, Wallace, Sachs, Hofmeister, Agassiz, and numerous other investigators had given or were giving wise teachers, interpreters, and research-minded men much to meditate concerning— certainly their epoch-making studies offered young students of the coming generation a new challenge—a challenge that called for investigation anew of the living plant in all its relationships, all its physical, biological, and environic factors, as well as orderly systematization of herbaria material brought in from world explorations. Students of the next generation, certainly, and not less those of Gray's time, would have to reckon with the great new experimental investigations, most of which were being pursued in Europe—whether from botanists, horticulturists, or agriculturists. An emergence of a transitional period was clearly foreshadowed, although its applications could not be completely foreseen. This does not imply that Gray, Hooker, Engelmann, or any taxonomist of the period—whether of the old world or the new—did not comprehend the importance of studying the living as well as the dried plant.

Nor does it imply they did not understand the importance of studying plants in the garden and the field. Human capabilities, however, are limited. Taxonomists of this generation spent their energies with the task at hand—that begun for them by systematists of Torrey's generation, and earlier. Nevertheless, let us never forget that Gray and Hooker were among the first to take up the challenge and expound with the full force of their strength the new knowledge and conclusions of the evolutionary theory.

In a number of notable reviews after 1855, Gray had added observation after observation bearing on matters relating to breeding and heredity, and when the epoch-making works of Darwin began to appear, he rejoiced that a great leader had arisen who might solve the inscrutable problems of inheritance and elaborate further the little known laws and conditions governing variation. He, as much as anyone, saw the need of a Darwin. On December 8, 1874, appeared his essay, "Do Varieties Wear Out or Tend to Wear Out?" and in this he called attention to "proceedings of pomological societies and the debates of farmers' clubs" in support of the conclusion "it is by no means certain that the nays would win [the argument]. The most they could expect would be the Scotch verdict 'not proven.'" Gray was not in the modern sense a geneticist, but it is not too much to say that he was fully aware of the large mass of data being accumulated and making for formulation of definite physiological laws and the eventual establishment of an exact science in breeding. He must have heard with interest of the investigational breeding work of William James Beal, a former student of his, and others, as it began at Michigan Agricultural College, and elsewhere. Such work was in commodities for the most part. Beal seems to have rebelled against a pedantic teaching of informative facts in science, against the making of "intellectual tramps, and not trained investigators," as he later characterized a prevalent teaching during the period. When Beal had studied under Gray, Gray's correspondence with Darwin was at its height. Gray was enamored of Darwin's evolutionary hypothesis, and with Alfred Russell Wallace's disclosures. Doubtless, Beal caught the Darwinian spirit of investigation. Indeed, even Farlow rebelled against attempting to teach botany to make primarily "botanical specialists" in "descriptive phaenogamy," urging a biological interpretation of botany from the lowest to the highest orders of plants. With all their ingenuity, however, no one seems to have envisioned even partially the more remote future. But each in his chosen field saw the immediate task to be performed. Gray seems not to have envisioned the development of great plant industries. If he did, he must have realized that the systematic work in which he

was engaged was basic to both botanical and horticultural, indeed agricultural, progress. Knowledge of materials was basic. At his age he must,
he knew, merely aid the work of Darwin and work of investigation
called up by his explanation of the role of continued selection in the
modification and origin of species.

The amazing fact is that, although Gray did not regard "the bent of
his mind nor the line of his studies" as fitting him to do justice to writing
on matters of "deductive evolution,"[3] he became one of the world's great
protagonists of the evolutionary view. The more amazing it is when one
realizes that at first few other American botanists responded immediately
to his writings with more than an intellectual interest. Scientific investigation in botany for the most part continued its former course of exploring the North American continent for new materials, manifesting an
interest in the plant-life of other regions, in other words, being vastly
more interested in questions of geographic plant distribution than in
discussions of Agassiz and Gray as to whether plants and animals were
originally created specially and locally by God or had a common origin
with species derived one from another with a "community of descent"
as Gray termed the process. Gray had invoked Maupertius's "principle
of least action"—the Creator did not use more power than was necessary
to originate species generally—and refuted Agassiz.

Gray, with able and most remarkable erudition and foresight, saw it
was important to determine to what purpose or design each kind of plant
was adapted, what use it served. Doubtless influenced by religious adherences, he believed that every species served a purpose with some function
to perform, or some condition to which to adapt. Although Darwin did
not view the world "as a result of chance," he could not look on each
separate thing "as a result of design," as part of a wonderfully perfect
scheme in nature. Life and certain potential powers were originally
"breathed by the Creator into a few forms or into one." Most beautiful
and most wonderful new forms "have been, and are being evolved."
Darwin was not the teleologist Gray was. Design to Gray was like the
concept of Providence, more a philosophical and religious matter within
the province of God's knowledge than scientific within the province of
man's. To Gray, Darwin's great service was in returning teleology to
science—a new teleology based on immense evidence and experimentation, outmoding old fixed concepts of which Gray complained. "[I]n-
stead of Morphology *versus* Teleology," said Gray, "we shall have

[3] See "Preface" in *Darwiniana: Essays and Reviews Pertaining to Darwinism*, by Gray (New
York: D. Appleton and Co., 1876).

Morphology wedded to Teleology." Darwin's new published studies were proof of this.

Faith in an order, the basis of science, was to Gray not possible without faith in an Ordainer, the basis of religion. Believing in unities of type, it was possible to determine the multiform varieties of adaptation to conditions of existence. Gray, following Darwin, believed that varieties of type diverge or were modified into species by natural selection made in the struggle for existence in which all living forms are engaged—a struggle which is the inevitable result of natural causes, some of which may be susceptible of proof but mainly caused by "the high rate at which all organic beings tend to increase." Such a picture did not cause him to shrink from his faith. On the contrary, it strengthened his belief in God. "There is grandeur in this view of life, with its several powers, having been originally breathed by the Creator into a few forms or into one," Darwin said. Like the author of the *Origin of Species* who confessed to a belief in theism, Gray was an incontrovertible theist who believed almightily in a Creator of the universe.

Gray was a forceful and active theorist in more than this one respect. When in 1872 he delivered as retiring president of the *American Association for the Advancement of Science* an address, "Sequoia and Its History,"[4] he elaborated his theories as to geographic plant distribution on the North American continent and the similarities to plant distribution on the northeastern Asiatic continent. In like fashion he found remarkable resemblances between existing North American flora and those of past geologic periods, tracing the descent from a common heritage that once flourished in high altitudes in the Tertiary period, specifically, in the last or Pliocene epoch of the Tertiary before the appearance of man and before the advent of the great glacial advances of the Pleistocene epoch of the Quaternary period. Gray speculated

. . . upon the former glaciation of the northern temperate zone, and the inference of a warmer period preceding and perhaps following. I considered that our own present vegetation, or its proximate ancestry, must have occupied the arctic and subarctic regions in pliocene times, and that it had been gradually pushed southward as the temperature lowered and the glaciation advanced, even beyond its present habitation; that plants of the same stock and kindred, probably ranging round the arctic zone as the present arctic species do, made their forced migration southward upon widely different longitudes, and receded more or less as the climate grew warmer; that the general difference of climate which marks the eastern and the western sides of the continents—the one extreme, the other mean— was doubtless even then established, so that the same species and the same sorts of

[4] *Proc. Amer. Asso. Adv. of Sci.*, XXI, pp. 1 ff. Also *Scientific Papers of Asa Gray*, selected by Charles Sprague Sargent. In two volumes; Boston and New York: Houghton, Mifflin & Co., 1889. Volume II, pp. 142 ff., pp. 156-157.

species would be likely to secure and retain foothold in the similar climates of Japan and the Atlantic United States, but not in intermediate regions of different distribution of heat and moisture; so that different species of the same genus . . . or different genera of the same group . . . or different associations of forest trees, might establish themselves each in the region best suited to their particular requirements, while they would fail to do so in any other. These views implied that the sources of our actual vegetation and the explanation of these peculiarities were to be sought in, and presupposed, an ancestry in pliocene or still earlier times, occupying the higher northern regions. And it was thought that the occurrence of peculiarly North American genera in Europe in the [T]ertiary period . . . might be best explained on the assumption of early interchange and diffusion through north Asia, rather than by that of the fabled Atlantis.

The hypothesis supposed a gradual modification of species in different directions under altering conditions, at least to the extent of producing varieties, sub-species, and representative species, as they may be variously regarded; likewise the single and local origination of each type, which is now almost universally taken for granted.

Continuing, Gray informed his listeners that it seemed to him that

. . . if the high antiquity of our actual vegetation could be rendered probable, not to say certain, and the former habitation of any of our species or of very near relatives of them in high northern regions could be ascertained, my whole case would be made out. The needful facts, of which I was ignorant when my essay was published, have now been for some years made known,—thanks, mainly, to the researches of Heer upon ample collections of arctic fossil plants. These are confirmed and extended by new investigations, by Heer and Lesquereux, the results of which have been indicated to me by the latter.[5]

Oswald Heer, born a Swiss (as was Lesquereux), was the world's "most eminent investigator" of fossil plants and insects of the Tertiary period of the Cenozoic or modern era. Projecting an hypothesis of a lost Atlantic Ocean continent—Atlantis—he sought, as Lesquereux told Gray in 1859, "to show by the fossil plants that it existed between N[orth] Amer[ica] and N[orthern] Eur[ope]: an old continent of which the Canary Islands are the remains." But, said Lesquereux, "The *coal flora* shows that at the coal period this connection between both continents did not exist or rather does not show that it existed." Heer's own researches reduced the theory to a fable. Gray compared by descriptive tables and observations the floral relations between eastern North America and eastern temperate Asia—first in 1859 and later in 1872—commenting on relations of the floras of the United States and Europe, and of Europe and Asia. We have knowledge of lost land masses by marine coverage but little or no factual data support belief of a once Atlantic continent of Atlantis or, for that matter, the mythical land masses of

Pan and Mu.[6] This was realized in Gray's time and is further substantiated today. But how then were they to explain the observed similarities between European, North American, and eastern Asiatic plant life, revealed by fossil and living plant evidence of that day? Even today the study of plant migrations is difficult, botanists will attest. But it was more difficult in the time of Gray, Heer, and Lesquereux. Gray adhered for a number of years to theories advanced by himself and in them for the most part received support from James Dwight Dana of Yale University, conceded North America's most eminent geologist of that time. In 1878, when Gray delivered a lecture, "Forest Geography and Archaeology,"[7] and in 1884, when he lectured on "Characteristics of the North American Flora,"[8] Gray reaffirmed his already announced views, at least in part. In the former address he posed a question, giving the answer as follows:

What would happen if a cold period was to come from the north, and was to carry very slowly the present arctic climate, or something like it, down far into the temperate zone? Why, just what had happened in the Glacial period, when the refrigeration somehow pushed all these plants before it down to southern Europe, to middle Asia, to the middle and southern part of the United States. . . . The clew was seized when the fossil botany of the high arctic regions came to light; when it was demonstrated that in the times next preceding the Glacial period—in the latest Tertiary—from Spitzbergen and Iceland to Greenland and Kamtschatka, a climate like that we now enjoy prevailed, and forests like those of New England and Virginia, and of California, clothed the land. We infer the climate from the trees; and the trees give sure indications of the climate. . . . Wherefore, the high, and not the low, latitudes must be assumed as the birthplace of our present flora; and the present arctic vegetation is best regarded as a derivative of the temperate. . . .

Periodic glaciation occurred in the Quaternary period following the upper Tertiary bracket of the Cenozoic or present era. Combined, the epochs of the Tertiary and Quaternary are, from the base up: Paleocene, Eocene, Oligocene, Miocene, Pliocene, and Pleistocene, in the last of which occurred the origin of man which may have begun as early as the Pliocene epoch.[9] Some periodic glaciation had taken place in the Upper Paleozoic era preceding both the Mesozoic era, the age of reptiles, and the Cenozoic, the age of mammals. But glaciation to which Gray referred must have been chiefly that of the Quaternary. Years before, on June 27, 1859, Lesquereux had reported to Gray: "The quaternar[y] or *sub* or *super* glacial formation of the Mississippi extends itself all along the

[6] See Carey Croneis and William C. Krumbein, *Down to Earth, An Introduction to Geology.* University of Chicago Press (Chicago: 1936), p. 256.

[7] *American Journal of Science and Arts* (3rd ser.), XVI, pp. 85, 183. See also *Sci. Pap. Asa Gray,* II, pp. 226-231.

[8] *Ibid.,* XXVIII, p. 323. See also *Sci. Pap. Asa Gray,* II, pp. 260 ff., pp. 272-273.

[9] Field, Richard M., *Geology Manual* (Princeton University Press, 1941), pp. 4, 5.

Ohio river and its affluents, but much thinner than on the Mississippi. . . ." Above the mouth of Little Sandy River of Kentucky, Lesquereux had found fossil "fruit [seed] of the Papaw" and two nuts similar to our black walnut. The year before he had presented to Gray the value of comparing fossil with living plants:

> You will perhaps say that the identification of fossil leaves of phaenogamous plants is impossible and therefore it is useless to try to make a comparison. In the strata of somewhat different age, especially in the pliocene, miocene and eocene strata along the Mississippi the determination of the leaves, even approximative, is truly important. For example the formation of which the accompanying leaves are taken is supposed contemporaneous with the chalk banks of the borders of the Mississippi bottom which contain only plants of our time and still living on the same latitude.

The great question was not only whence but when came the origins of our modern flora? For years Gray and others sought to account for Europe's endowment of American plant types, for the once possible land connection of Europe and Greenland, and many other such matters. As to America, he observed: "We find the land unbroken and open down to the tropic, and the mountains running north and south. The trees, when touched on the north by the on-coming refrigeration, had only to move their southern border southward, along an open way . . . and there was no impediment to their due return."

Excluding for a time consideration of the plant migration from Mexico to southwestern United States, Gray saw well the migration from the north southward. The puzzling question still, however, was—when? The Pliocene epoch? In his famous address, "Sequoia and Its History," delivered in 1872, Gray rested his conclusions on John Strong Newberry's and Lesquereux's investigations of upper coal deposits west of the Mississippi River, saying: ". . . the facts justify the conclusion which Lesquereux—a scrupulous investigator—has already announced: 'that the essential types of our actual flora are marked in the [C]retaceous period [of the Mesozoic era] and have come to us after passing, without notable changes, through the [T]ertiary formations of our continent. . . .' "

In the Cretaceous period lay the absolute origins. Epochs of the Tertiary—the Eocene, Miocene, and Pliocene—were to be further studied. The important point was that the North American flora's lineal ancestry was being investigated and with the investigation Gray thoroughly sympathized. Other matters of comparatively new and great interest also received Gray's hearty interest.

Gray did not stop with learning theories. Immediately his eager and inquiring mind set about to find means of practical application. He not only watched eagerly for reports of new scientific investigations in paleobotany of western regions but also awaited results of recent explorations in Alaska.

When a letter from the United States Coast Survey reported finding Caulophyllum or blue cohosh on one of the Shumagin Islands of the Alaskan regions, he waited with intense interest to see a specimen. For, Caulophyllum was a characteristic eastern United States plant and had been found by Charles Wright on the North Pacific expedition to the Pacific Islands and Japan. Other examples might serve as illustrations.

Gray had followed closely Darwin's studies on insectivorous plants. On December 2, 1872, he had written Darwin, "Well, it is wonderful, your finding the nervous system of Dionaea!!! Pray take your time next spring, and do up both Drosera and Dionaea. I will endeavor next spring to get hold of *Drosera filiformis* and make the observations." For almost a century the ability of plants to catch insects had been noticed by science. For example Ellis had observed in 1768 the powers of Venus's flytrap. Indeed, Gray regarded the American botanist, John Bartram, Dionaea's "probable discoverer." Investigation, however, reached more or less a climax about 1876. William James Beal reviewed the species and genera of plants catching insects for the *American Naturalist* that year. Casimir DeCandolle wrote a paper on "The Structure and Movements of the Leaves of *Dionaea muscipula*" and Gray reviewed it in *The American Journal of Science and Arts*. From experiments, DeCandolle found that animal matter is not necessary for growth and strength of *Dionaea*. Everything pertinent elicited Gray's interest.

In April 1873, a month after Torrey's death, Gray went to Washington to visit Joseph Henry of the Smithsonian Institution of which Gray was a regent. From there he journeyed to Wilmington, North Carolina, to visit William H. Canby, and found "the spring in all its beauty. . . . I collected a lot of live Dionaeas, etc.," he wrote DeCandolle. Returning to Cambridge, he wrote Canby, in June, "My Dionaeas grow finely, and are the delight of my heart. *Drosera longifolia*, also cultivated, is almost as good a fly-catcher. Now and then I see a little exudation inside base of hood of *Sarracenia flava.* . . ." Again in July he wrote, ". . . Conundrum? Why does the Dionaea trap close only part way, so as to cross the bristles of edge only, at first, and afterwards close fully? Darwin has hit it. . . ."

Gray's studies were evoked by Darwin's studies—from about 1860 to 1875—when the latter published a work on *Insectivorous Plants* and one on *The Movements and Habits of Climbing Plants*, both of which Gray reviewed together in *The Nation* in 1876. Indeed, Gray's interest, like Dr. Joseph Dalton Hooker's, had been aroused much earlier and he himself had written two years before an article entitled, "Insectivorous Plants"—the same year Farlow published results of researches commenced in Europe on asexual growth from the prothallus of *Pteris cretica* and *Pteris serrulata*. Gray traced historically American study in "carnivorous plants," including Canby's and his analyses, together with some of Mrs. Treat's of New Jersey, another botanist, friend of Gray, and to be one of his early biographers. Darwin had begun with the round leaved sundew (*Drosera rotundifolia*) and studying nearly all plants of the family and other such plants, had recorded an amazing amount of information concerning stimuli and responses, noticing a parallelism between digestive powers of Droseraceae and animal gastric juices. Gray wrote Canby:

> ... I have also seen here that water is secreted in the pitcher of *Sarr*[*acenia*] *flava* before the lid is open. But I have also seen some time ago, when the weather got rather warm, very minute globules like finest dew on the erect part of the lid, near base, inside. And, lately, during the very warm days, I found in some this increased, and the droplets running together into a clammy exudation. But I want to see more of it. I shall watch, as I get a chance, and the weather gets hot. Look at yours. See if there is anything of the sort in *S*[*arracenia*] *purpurea*; I think not.

On August 12 Canby replied he had been in New Jersey, had examined leaves of *Sarracenia purpurea* for moisture in the lids, but "could find nothing of the kind, tho' many leaves were examined." For several years, following work of Moses Ashley Curtis, in which Gray was also interested, Canby had studied the subject of fluid poured around captured insects, and published on it.

Insectivorous plants was a subject to which people responded with interest. It was curious to learn that some plants fed on animals as well as animals on plants. To Darwin and especially Gray it had a greater significance. Gray observed in his article on "Insectivorous Plants," published in 1874:[10]

> Why should these plants take to organic food more than others? If we cannot answer the question, we may make a probable step toward it. For plants that are not parasitic, these, especially the sundews, have much less than the ordinary amount of chlorophyll—that is, of the universal leaf-green upon which the formation of organic matter out of inorganic materials depends. These take it instead of making it, to a certain extent.
>
> What is the bearing of these remarkable adaptations and operations upon doctrines of evolution? There seems to be a field on which the specific creationist, the evolutionist with design, and the necessary evolutionist, may fight out an interesting, if not decisive, "triangular duel."

Later in his review on Darwin's works he said:[11]

10 *Darwiniana, op. cit.*, pp. 306-307.   11 *Ibid.*, p. 331.

Whether these carnivorous propensities of higher plants which so excite our wonder be regarded as survivals of ancestral habits, or as comparatively late acquirements, or even as special endowments, in any case what we have now learned of them goes to strengthen the conclusion that the whole organic world is akin.

Gray sought the teleological explanation always. With regard to "climbing plants," he observed:[12]

Climbing plants "feel" as well as "grow and live"; and they also manifest an automatism which is perhaps more wonderful than a response by visible movement to an external irritation. . . .

Most leaves make no regular sweeps; but when the stalks of a leaf-climbing species come into prolonged contact with any fitting extraneous body, they slowly incurve and make a turn around it, and then commonly thicken and harden until they attain a strength which may equal that of the stem itself. Here we have the faculty of movement to a definite end, upon external irritation, of the same nature with that displayed by Dionaea and Drosera, although slower for the most part than even in the latter. . . .

In revolving tendrils perhaps the most wonderful adaptation is that by which they avoid attachment to, or winding themselves upon, the ascending summit of the stem that bears them. This they would inevitably do if they continued their sweep horizontally. But when in its course it nears the parent stem the tendril moves slowly, as if to gather strength, then stiffens and rises into an erect position parallel with it, and so passes by the dangerous point; after which it comes rapidly down to the horizontal position, in which it moves until it again approaches and again avoids the impending obstacle.

Climbing plants are distributed throughout almost all the natural orders. In some orders climbing is the rule, in most it is the exception, occurring only in certain genera. The tendency of stems to move in circuits—upon which climbing more commonly depends, and out of which it is conceived to have been educed—is manifested incipiently by many a plant which does not climb. Of those that do there are all degrees, from the feeblest to the most efficient, from those which have no special adaptation to those which have exquisitely-endowed special organs for climbing. The conclusion reached is, that the power "is inherent, though undeveloped, in almost every plant"; "that climbing plants have utilized *and perfected* a widely-distributed and incipient capacity, which, as far as we can see, is of no service to ordinary plants."

Inherent powers and incipient manifestations, useless to their possessors but useful to their successors—this, doubtless, is *according to the order of Nature; but it seems to need something more than natural selection to account for it.*[13]

In other words, Gray found in the evolutionary process more than only natural selection. He found in plant processes an adaptation of means to ends—selections made on more than merely adaptations of *necessity* for the ends of preservation and survival. He found evidence, admittedly inconclusive, but *probable*, of a wisdom, skill, and power directing the process. Belief in this was just as acceptable with Darwin's

---

[12] *Ibid.*, pp. 332, 335, 336-337.   [13] Italics are mine.

theory, Gray said, as it was before without the theory. Theism was thus vindicated as a probable, and not merely possible, part of the process. After all, something cannot come out of nothing. If the universe manifests order and not chaos, an Ordainer, of whatever nature, is implied. The world no longer entertained the notion of the fixity of the earth. So it might change from its past belief in the absolute fixity of species which inhabit it. Species originated by modifications or derivations one from another.

Gray, however, was not bigoted. He allowed to others the right to disagree with him. He admitted his argument for design or purpose was as to design more metaphysical than scientific. We might say today that his argument for design was more an argument of persuasion than one susceptible of absolute scientific proof. Gray, like almost every one of the ablest of his science, was a very religious man.

On July 20, 1874, William M. Canby, one of the North American botanists whose imagination was much stimulated by Gray's speculations, wrote Gray, saying:

I just wonder that after all your coaching and my asking myself "What harm if the leaf of Dionaea did take a small insect? that I was such a boob that I couldn't 'see it.' . . ."

Well the *discovery* is "real Darwinism"—The power I like to think of as originating higher; and these revelations don't shake my faith, but cause me to wonder and glory all the more in the Great Power, *all* whose works praise Him—whether directly created or "evolved."

I cannot remember exactly what Darwin wrote about climbing plants, except the swinging round of a vine until it reached a support on which to climb. But a year ago I had common "Cypress vines" planted in a vase in front of my porch. They did not revolve or swing round their stems to catch on something but every one *made directly* for the first post behind the vase . . . in spite of all I could do to persuade them not to. This they did repeatedly. How did they know it was there and evince such determination to reach it? The main sunlight was all in the other direction too.

Are we coming to *Instinct* in plants?

Gray's belief, however, was more than this. Perhaps he too conjectured on the possibility of "coming to *Instinct* in plants." Perhaps this observation of an action of a climbing plant did not surprise him. He, too, experimented similarly, studying and observing all the powers of movements in plants. Where he certainly must have disagreed was in *liking* to think of the power "originating higher." Gray did believe the power originated higher. But he maintained a scientific standard. He admitted such was not *provable* by actual scientific measurements. The evidence was there, however, sufficient to convince him if on no other than philosophical and religious grounds—there was a Higher Wisdom, a Divine

Intelligence. He accepted that on faith and his faith always held firm.

Louis Agassiz's death stilled the great discussion and debates of Gray and Agassiz on Darwinism. The origin of species, as Darwin described processes, was gaining in favor although clear minded expositors of Darwin's views such as Gray were still much needed. "Gray," said Lesquereux, "is somewhat aristocratic of character like Agassiz. He well knows that he is a prince of science and if he does not openly despise small men and poor things, he keeps his whole regard for high standing subjects. . . . [H]e is certainly frank and honest and despises every kind of duplicity. I see it in that way at least."

Nevertheless, for years after its publication Darwin's *Origin of Species* was not fully understood by all American scientists. To illustrate, Lesquereux read the work and as late as 1864, neglecting its essential points in science, emphasized religious implications, saying:

> I have studied and studied again Darwin's & the origin of species and the more I read it the better I am pleased with it. This system explains to me some of the mysteries of our Christian revelation which had been obscured to my mind till now. For to tell you the truth I read the Bible every day and constantly find it new light and new life.

Philosophical comprehension of the province of Darwin's theory was gradual. By 1873 Lesquereux foresaw science pointing the way to a more truthful theology, and religion encouraging scientific research. He wrote:

> No scientific system not even that of Darwin has done anything against true Christian dogmas. Those dogmas are essentially for the food of the spiritual in man which teach faith, love, and force constant improvement toward immortality or future life. If ever man should discover the secret of immortality and that may happen, why not? We would find then Christus still proving to us that without the application of his principles: love to the omnipotent Ruler, love to our fellow men, we can not have here peace, content, happiness, indeed and then immortality would be a hell. . . . Man's happiness is in immediate relation to his moral development, a state which has nothing to do with dogmatic & church influence. I believe that scientific adepts never would fight against Christianism if they were acquainted with it by their own study. But they receive it by transmission.

One might argue that, because of Lesquereux's close friendship with Agassiz, he was inclined toward Agassiz's opposition to Darwinism. Quite the contrary was apparently true. Lesquereux's beliefs evidently were more in harmony with Gray's. Following Agassiz's death, Lesquereux expressed himself in a letter dated February 1, 1874:

> The death of Prof[essor] Agassiz has caused me to feel as if I was myself about to leave this world; as if my task which compared to his has been nothing should too be laid aside; this, not only by the deep regret, by the sadness of his loss, but by the more palpable evidence of the variety of all my efforts, trials, aspirations for

something of this life . . . I have spent with Agassiz in Switzerland, in our rambles in the peat bogs; in America too, for I was always his guest when at Cambridge, very pleasant time in intimate association. At Cambridge, we have, time
and again talked over his own [Agassiz's] and Darwin's opinions, going up or
descending (as you like) in philosophical considerations or ideal religion etc, and
I think that I have seen deeper in him and knew him better & his so-called spiritual
disposition than any friend of his. On many points our opinions were quite at
a variance; that never in the slightest degree changed our quiet manner of considering things. He admitted personality or rather individuality of mind, intellectual
and moral as well as of body and therefore he never became troubled or angry in
any discussion of this Kind. . . .

Lesquereux was of the same age as Agassiz and also Arnold Guyot,
the three who constituted a most important triumvirate given to American science by the French cantons of Switzerland during the Sonder-
bund War. With Guyot, like Agassiz, a friendship persisted, Lesquereux
accompanying Guyot in 1862 to the Adirondack Mountains and often
visiting him at Princeton, New Jersey, where Guyot taught in the college, a science building of which today bears Guyot's name. Lesquereux
also maintained a friendship with James Dwight Dana of Yale, who
eagerly watched the progress of Lesquereux's studies in North American
paleobotany. Dr. Leidy of Philadelphia corresponded with Lesquereux.
Nearly all the great centers of scientific interest in the East employed
Lesquereux at one time or another to study or arrange their paleobotanical specimens. The totally deaf and financially poor Swiss scientist, who,
moving to America, went to Columbus, Ohio, to aid William Starling
Sullivant in bryological study, had, therefore, the best opportunities to
keep pace with scientific progress in Europe and America. That he at
least recognized the importance of Darwin's *Origin of Species* and was
moved to philosophical discussion with Agassiz, Gray, and probably
Sullivant—the three whom he regarded as his closest early American
friends—redounds immensely to his credit. Not all American scientists
were so quick to respond to Gray's words, both spoken and written, in
furtherance of Darwinism and scientific experimentation which followed closely the work of the great Englishman.

CHAPTER II

Government Surveys and Explorations. Paleobotany Included

WITHIN the past half century, geologists had displaced the once held theories of a ready-made or specially created earth. The origin of the earth had proved a difficult and complex problem—a matter first for scientific hypotheses. No simple explanation was complete. Moreover, the history of the development of the earth was equally difficult of complete ascertainment but, seemingly, it could be determined gradually with hope of someday realizing perfection of a geological calendar. For example, the earth's surface had been found altered, not by catastrophic changes only—earthquakes, volcanoes, and other once inexplicable agencies of configuration—but by many other causes now subjects of scientific investigation and study. Over the grand sweep of numerous geologic ages, covering millions and millions of years and divided into many periods and epochs,[1] the earth's "crust" manifested a history which enlisted the avid and eager interest of every scientist of consequence in the entire world. Mountains, valleys, rivers, forests, plains, and all geographic phenomena—their origins and development—were objects of intensive study in almost every part of the world and, in many instances, their investigation employed government and privately supported exploring expeditions to gather materials on the ground for students of the laboratory. The early eras of rock formation, of widespread igneous action; the early periods of marine submergence, of uplifting and sinking of land surfaces, of glacial activity, of erosion and concomitant sedimentation; indeed, periods of the origins and development of invertebrate and vertebrate life—man's origin as well as that of the other animals—were all included in the regimen of increasingly important and learned specialization in various branches of science. To study origins required studies of development. For these tasks, therefore, studies now turned to fossil remains of plants and animals.

North American paleobotanical studies were being directed in great part toward the several million years period of late Cretaceous time, before the appearance of man, in which great changing conditions of topography and climate had proved besetting problems to not only geologists and paleobotanists but also zoologists, biologists, entomolo-

[1] This entire subject is ably presented in Carey Croneis and William C. Krumbein, *Down to Earth: An Introduction to Geology* (Chicago: University of Chicago Press, 1936), Chapters 30-47.

gists, and other men of science seeking to determine the nature of land surfaces; if any, their plant and animal life; and a multitude of environmental factors millions and millions of years before our own epoch in time and space. State geological surveys in the eastern and central United States had produced much knowledge of the ancient upper coal bearing periods; the great beds of lignite similar to the ancient and more recent coal beds; the early vegetations—their climate, temperature, schemes of plant distribution—but the more knowledge acquired, the more complex and difficult became the problems. With development of the knowledge of stratification and the presence of oil reservoirs came an "oil craze." As already indicated, much investigation in fossil remains of epochs of the Tertiary period had taken place in Europe and far northern North America, especially in Greenland and the Arctic regions. Now studies had begun to extend to the Western Territories of the United States. On December 16, 1866, Lesquereux had written:

That question of the distribution of plants of types still in our Flora has taken some more consequence from the discovery of Magnolia in the Cretaceous both of Europe and of America and from the ascertaining that the strata where H[a]yden plants were collected in Nebraska and which were considered by Heer and myself as tertiary are true Cretaceous. This show[s] either that the relation of the vegetation between the Cretaceous and the Tertiary of Europe is very little known and that it is useless for us to compare our old vegetations with those of an ascertained age of Europe. For the more we ascend above the coal formations the greater is the difference between the European and the American types of vegetation. It is here then that we have to look for points of comparison. And of course the first thing to be done is to study plants of known formations, to publish them and to put in that way the first sticks which have to be planted for showing the direction of the roads. We are in America for fossil plants as in every other branch of palaeontology p[a]rticularly more favoured than in Europe. Our formations are of wide extent because they are scarcely disturbed and we can therefore study not only the stratigraphical variations and changes of vegetation but those also which are due to geographical stations. Ah! if I could! Now there are many places in Tennessee and Mississippi state where strata are exposed and where plants are found from the lower [C]retaceous to the bluff or [Q]uatenary formations along the Ohio and the Mississippi [Rivers]. Buckley writes me from Texas that he has a great deal of fine specimens of fossil plants which he would like me to examine: a true series from the Cretaceous to the Upper Tertiary. And Prof. Whitney would like to have me in California for the study of plants of the same recent formation. All this is very fine, very attractive, but I am afraid that I shall not be able to do any thing more in the way of Palaeontology or at least of recent fossil plants. If I may get some body to send me to the Rocky M[oun]t[ain]s on a tour of exploration for the coal strata there, I will go and at the same time or while on my way push to California. . . .

Seven years earlier Lesquereux had been delighted to tell that, "Fossil plants of the tertiary are now sent to me from every corner . . ."—enough to fill his small house and shed at Mound and Fourth streets in Columbus, Ohio. Lamarck and Darwin, from abundant fossil and modern flora materials, had given precise and concrete expression to the evolutionary theory. Theory had led to discussion and discussion to wider and further study. American botanical exploration had by no means finished its task of learning our flora. Fossil study, consequently, went along therewith.

During the last four decades of the nineteenth century, exploration pushed further into the American continent interior—not only in the United States but also Canada, Alaska, Mexico, Central America, and South America. Many rivers and trails gave way as paths to the railroads. New trails pushed to unknown rivers and mountain areas. Scientists sought to enlarge, limit, or negate the many new theories or hypotheses being enunciated by men of both the laboratory and the field. Huge areas remained scientifically unexplored. The great world explorations of the British, French, and American governments continued. But not on their former grand scale. Settlement and cultivation of already explored lands occupied governments. Yet much geologic and geographic information had to be obtained. The United States government sent out four major surveys for these purposes, commonly known as the King survey, the Hayden survey, the Powell survey, and the Wheeler survey. Men learned in science accompanied, in some instances were in charge, of these surveys. In the field, their duties were not merely to collect natural history specimens as formerly in most instances but also to study the geology, geography, and stratigraphy *in situ*. Their collections still went to the laboratories but their studies went into books. And they went West.

In the United States the great Interior Basin of the West, including a great section of desert land, was explored. During the last three years of the 1860's, the United States geological exploration of the fortieth parallel of north latitude under Clarence King went over regions of northern Nevada and northwestern Utah covering "what was at first designated as the 'Great Basin,' the high plateau, without outlet for its waters, separated on the north by low divides from the valley of the Snake River and continuing southward until it merges into the desert of the Lower Colorado." In 1867, they explored a large area in western Nevada, at the east base of the Sierras, including vicinities of Pyramid Lake, Trinity Mountains, Hot Springs Mountain, West Humboldt Mountains, Soda Lake, and Truckee River. The winter of 1867-1868 was spent at Virginia City, Nevada, studying the Comstock Lode mines. In 1868, beginning at Carson City, Nevada, they passed over a wide region of the Great Basin in central and northern Nevada to northern portions of Utah with Ogden as their destination, exploring with thoroughness many of the mountain ranges, valleys, and arid lands, along their winding and intricate course of journey. The last year, 1869, was spent around Great Salt Lake, and regions east and south, the Utah desert ranges and Wasatch Mountains, going in the course of their explorations to Provo City and the Uinta Mountains. The whole embraced an exploration of the fortieth parallel from the Sierra Nevada to the western slope of the Rocky Mountains. Later the survey was extended to the northwest and eastward toward the Great Plains.

Sereno Watson was the botanist of this difficult but valuable exploration and his report on the botany was published in Volume V of the final government report, although it was, moreover, published separately before the final report of 1871.[2] So remarkable was Watson's ability as a botanist shown in plant determinations of the collection that, after study with Daniel Cady Eaton and Gray, Watson became permanently associated with the Gray Herbarium at Cambridge where he remained the balance of his life.

On December 18, 1867, W. W. Bailey, later an instructor of botany at Brown University but then associated with Watson, wrote Gray from Carson City:

Now that I have settled in winter quarters, I will give you an account of my summer's work. It was much interrupted by sickness chiefly fever and ague from which nearly our whole party suffered. The pain is too recent, and my recollection of it too vivid for me to speak much of it now. Luckily my associate in this department Mr Watson was well all the time—very energetic and industrious—and his herbarium probably contains twice the number which I have collected. I cannot speak in terms of too high praise of this gentleman—always genial and kind—and ever persevering. His botanical work was in addition to that of topography. He works early and late and seems never tired or ruffled. In writing my report—which is a separate document from the above named gentleman's, I have divided the flora into sections which are in a measure natural, but partly arbitrary. . . . The summer work began at Hunter's Station on the Truckee, and the examination extended from that place to the Big Bend of the river; in the mountains about Camp 12 situated at that place and from thence up the valley of the Humboldt to Oreana. . . . [A] camp was established for sanitary reasons in Wrights cañon in the West Humboldt range, and afterwards during nearly two months stay at Unionville the opposite side of the same range was explored. . . .

Much else, among which were matters of a topographic and geographic nature, was explained by Bailey to Gray. Plants were arranged

---

2 Ably and interestingly reviewed by Gray in *The American Journal of Science and Arts* (3rd ser.), III, pp. 62, 148, where Watson's observations concerning turning the plants to "some profitable account under the necessities of a future population" including horticultural species for the climate are considered. See Sargent's *Sci. Pap. Asa Gray*, I, p. 183.

as found on river bottoms and margins of sloughs, on the desert plains or valleys far or near from water, on the mountains, on alkali flats, meadow tracts, and numerous other subdivisions; and many times speculations where prominent fields of erosion and other agencies of geologic history had occurred were included. The importance of the botany of the fortieth parallel can scarcely be overestimated. Not only techniques pursued in the conduct of the work but its forms of publications served as models for other great surveys which investigated other regions for the United States government during the next decade. The geological survey of the fortieth parallel is not noted for its paleobotany but for its botany. There were, however, three other surveys of great prominence conducted by the United States government between the years 1867 and 1879.

Paleobotany came into its own through the agency of the great United States Geological and Geographical Survey of the Territories under Ferdinand Vandeveer Hayden. In the course of this survey's progress, techniques and methods of exploration, survey plans, and even some standardization of forms for publications were perfected, and their materials became "a storehouse of geographic, geologic, ethnologic, and archaeologic information"[3] concerning the western United States and Territories, sharing as the survey did a, if not *the*, most prominent place of the four government surveys. This survey had begun in Nebraska in the spring of 1867 under modest circumstances with a Congressional appropriation of $5,000 and extended over the great Western Territories—at first, regions west of the Missouri River, Wyoming, Colorado, New Mexico, Montana, Idaho, and Utah—until by 1873 the survey's appropriations had increased to more than $75,000 with an allowance of $20,000 for engraving. For twelve years Hayden's survey was to be of much importance scientifically.

Hayden knew the West. Not only had he spent a year in each of the Territories of Kansas and Nebraska surveying or exploring but as early as 1853, a graduate of Oberlin and Albany medical colleges, he had gone to the Bad Lands of South Dakota on White River and gathered remarkable evidences of extinct animals, vertebrate fossils, and other natural history objects. Again and again, sometimes at his own expense, Hayden had journeyed to the upper Missouri and Bad Lands regions, exploring rivers, aided in blazing new trails, and served several famous reconnaissances, among them, Lieutenant G. K. Warren's Northwest

[3] For a list of all the publications of whatever nature of these surveys see L. F. Schmeckebier, "Catalogue and Index of the Publications of the Hayden, King, Powell, and Wheeler Surveys," *Bulletin* 222, Department of the Interior, United States Geological Survey (Washington: Government Printing Office, 1904), pp. 207 ff.

reconnaissance over wide areas of Montana and Dakota and Captain W. F. Raynold's exploration of the Yellowstone, Gallatin, and Madison rivers and mountains. So impressed with his collections were the St. Louis Academy of Sciences, the Smithsonian Institution, and the Academy of Philadelphia, these institutions had either financed him or aided materially in securing his appointments as geologist, for in that capacity he usually served. As a consequence Hayden's explorations had taken him over large areas from Kansas to Montana. For a time, however, the Civil War had required his services as a medical officer and an election as professor of geology at the University of Pennsylvania had kept him occupied until 1872 when he resigned his position owing to increasing duties directing the Geological and Geographic Survey. Hayden liked the West. His interests were not confined to geology but, like many scientists of his time, extended to all branches of science not the least of which were botany and paleobotany. Among his early finds were fossil leaves similar to those of our present forest trees.

The importance to North American botanical herbaria of collections made by the government exploring expeditions of these years can scarcely be overestimated. In the year 1873, for example, the commissioner of agriculture reported that over 5,000 specimens had been added to the United States National Herbarium. So energetically was the government's agricultural work being prosecuted that a microscopic division studying plant diseases and parasitic fungi as well as plant structures and organic growth had been established, along with a widely operating statistical division and a well equipped scientific library. Of importance, therefore, were many new government scientific publications, among which were Hayden's *Preliminary and Annual Reports of the Survey*, published by the government and containing reports on botany and paleobotany.

In the report of 1871 made by Hayden, appeared catalogues of plants by C. C. Parry and Thomas C. Porter. Parry's catalogue consisted of a list of plants collected by one Thomas in eastern Colorado and northeastern New Mexico during the survey of 1869. And Porter's list was described thus:

This catalogue embraces the plants collected in Wyoming Territory by Dr. F. V. Hayden during the geological survey of 1870—at Camp Carlin, from July 25 to 30, on the route from Fort D. A. Russell via Fort Fetterman, Sweetwater, South Pass, Wind River Mountains and Green River, to Fort Bridger, from August 1 to September 13; in the Uinta Mountains, south of Henry's Fork of Green River in the latter half of September, and on Henry's Fork in the month of October. To these are added his collection in the North Park, Colorado Territory, August, 1868, and another, made by B. H. Smith, in the region around the city of Denver, during the summer of 1869.[4]

In 1872 Porter published another plant catalogue, this time enumerating plants collected by Hayden on an expedition to the headwaters of the Yellowstone River in 1871 and a few plants gathered by George Smith on Grays Peak near Georgetown, Colorado. Aid in the determinations was given by Torrey, Gray, Engelmann, Olney, Thurber, Lesquereux, and Tuckerman.

Since about 1850, as indicated, North American paleobotany had had Lesquereux in its service. From Rhode Island to California Lesquereux's very interesting researches were to go. One of the most unselfish lives ever devoted to American science, his collections today still constitute the foundation of American paleobotanical study. As a student of glaciation and of the formation and reproduction of peat in Europe, he had bound fast his tie with Agassiz.[5] They were both interested in European bryology. Both migrated to America though they were separated when Lesquereux went to Columbus, Ohio, in 1848 to study "muscology" with Sullivant. After Sullivant's death in 1873, Lesquereux's work principally in paleobotany won him recognition of an LL.D. degree conferred in 1875 by Marietta College. But his long labors were difficult.

On March 15, 1849, Sullivant reported to Gray: "Lesquereux started today on an excursion to the southeastern portion of the state, the coal & iron regions—He will be engaged collecting most of the season." Early in 1850 Sullivant proposed Lesquereux's going south "to explore the pine barrens of Alabama and after [that] the mountains of Georgia and Carolina." The learned young European decided to go and "examine the country carefully." At least three months may have been spent collecting fossil plants, a work "so much in harmony with what [he] had worked before in Europe." A trip was made in 1851, the same year Professor H. D. Rogers employed him for work on the first Pennsylvania geological survey to study Pennsylvania's coal plants. Thus, incident to moss and hepatic collecting for Sullivant and by individual employments, Lesquereux began studying American fossil plants.

About 1854 he published 110 "New Species of Fossil Plants from the Anthracite and Bituminous Coal-fields of Pennsylvania,"[6] including carboniferous flora of "the adjacent coal-fields of Ohio and Virginia,"

[4] Pages 472-483.
[5] The subject of Lesquereux's studies with Agassiz in Europe is developed in Andrew D. Rodgers, III, *"Noble Fellow" William Starling Sullivant* (New York: G. P. Putnam's Sons, 1940), Chapter XIII.
[6] *Boston Journal of Natural History*, VI (August 1854), Number 4, Article XXV, pp. 409-414 ff.

and noted an impressive similarity between North American and European species which seemed to establish "a close accordance, if not identity, in the geographical and climatal conditions prevailing at their formation." This work was later revised as survey study and publications involving plants not only of Pennsylvania but the United States increased. The subject, as Lesquereux told the Pottsville Scientific Association in 1858, was "comparatively new and uncultivated."

Rogers was not the only eminent American scientist who early employed him. David Dale Owen did also, for examinations in Kentucky and later, Arkansas. The results of the Kentucky coal studies were published between the years 1856-1861 and those of the geological reconnaissance of middle and southern Arkansas during years 1859-1860 were published in 1860. Of the Arkansas survey Lesquereux wrote, "I was charged with the examination of palaeontological and stratigraphical distribution of the coal of North Arkansas and had moreover to explore the botanical distribution in relation to agriculture, formation &c. Of course the examination of the beds of lignites and of their age was also in the program of my explorations. . . ." A. H. Worthen of the Illinois geological survey soon sought him for systematization of the carboniferous flora of that state,[7] offering him a "year round" position in 1860; about the same time he went into Indiana "for Owen" and, becoming associated with the geological survey there, with E. T. Cox and Richard Owen as assistants, he was to study fossil marine species from carboniferous measures, the results being published in 1875 when Cox was state geologist. Mississippi, Tennessee, and specimens "from every corner" were added. Indeed, investigations for various objectives—coal, petroleum, minerals, etc.—had been or were in progress in many American localities.

Lesquereux's careful and thorough workmanship quickly won the admiration of J. Peter Lesley of Pennsylvania with whom a lifelong friendship was established. In 1851 Lesley had written from Pottsville, where they became acquainted, an interesting description of their work on the anthracite coal tips searching for plant impressions. He said:

Lesquereux was hammering away an hour in a good locality, while D[esor] returned, picked up a rock, cracked it open, and there was a cabbage or young suckling palm, as big as a child's head, and when broken open full of young leaves all branching from an axis or stalk or root. It is a splendid discovery. I am sitting in the midst of palm branches, ferns, algae, slime plants, and turtle tracks, all dis-

[7] In Illinois alone Lesquereux determined 109 new species of fossil plants. See Raymond E. Janssen's *Some Fossil Plant Types of Illinois* (University of Chicago Press, March 1940), introduction. An elaborate study of Lesquereux's collections and type specimens is published in *Isis*, XXXIV (1942), pp. 104-106. By W. C. Darrah.

covered in the red shales—and—to think! how many years our whole corps were crossing and recrossing these strata, and never *knew* what we looked at.

A reputation was soon won by the young European scientist. Although severely handicapped by deafness and inability to make himself understood in English, Lesquereux forged quickly to the foreground as North America's ablest paleobotanist. When, in 1863, Joseph Henry, secretary of the Smithsonian Institution, proposed compilation of a fossil flora of the United States (especially that of the coal fields), and turned to Lesquereux, the latter wrote him on November 10:

I have begun the study of the fossil plants of this country in 1850 and ever since, have pursued my researches in the whole extent of our coal fields. The object[s] of these researches were two fold 1st. To get as good as possible an acquaintance with the Fossil Flora of the United States coal fields by comparison with that of Europe and of other countries 2dly. To try to obtain some positive Data concerning the distribution of the fossil plants in connection with the stratification and thus to find, if possible, some characteristic species to aid in the identification of the different beds of coal. The results of the last part of these researches have been, I think, most satisfactory, and have solved a geological problem of great practical importance. Some of them have been at different times published in the geological reports of the State surveys of Pennsylvania,[8] Kentucky, Arkansas, Indiana, and Illinois. (This last report is still unedited.) The result of the first part of the researches have been also partly given to science and embodied in the same reports as also in various scientific periodicals. But I consider what has been already published as only a meagre part of a whole which according to my plan should contain: 1st A *sectional* examination of the United States coal measures, in all the essential parts which have been surveyed, together with a comparison of the distribution of the coal plants with the stratification as far as it is known till now. 2d Description with figures of all the fossil plants of our coal fields, even of the species already published either by European or American authors. . . .

Lesquereux supposed that a work on the fossil flora of the United States coal fields would contain four to five hundred pages of quarto print and at least 150 plates. This would facilitate, he believed, the study of fossil floras of the coal, rendering it attractive and a reliable guide for future exploration "in the field containing our greatest and most valuable riches, our coal measures." It had been an ambitious and interesting project, involving study of a number of public and private collections of fossil plants in the East, besides his own numerous collections. Moreover, completion of the task would involve numerous publications and many years. Lesquereux said, ". . . there is not a single complete [American] work on the fossil flora of the coal, nothing indeed. . . ."

Mention of Lesquereux's participation in the Ohio geological survey

---

[8] There were two Pennsylvania geological surveys, commonly referred to as the first and second. Rogers directed one and Lesley the other.

was absent in Lesquereux's letter. Probably for a definite reason. When he had moved to Ohio he had begun immediately to study the phanerogams "in the Ohio C[ountr]y," by 1850 having "determined and ascertained more than seven hundred species." Along with this work he had commenced a collection of fossil plants and by 1852 his number was so noteworthy that he received a visit from Dr. John Strong Newberry, then of Cleveland. Pennsylvania, Kentucky, Arkansas, Illinois, and Indiana explorations had followed in order; and these, together with tours in Ohio and "All the best collections of fossil plants of the U[nited] S[tates]" sent him for examination and classification so increased his collection that by 1863 he had seen "an immense number of specimens." In 1859, when an Ohio geological survey was considered, Lesquereux was offered "the department of botany and palaeontology." However, what he wanted was a legislative appropriation for a survey of Ohio coal fields. Sullivant had no political influence. But Newberry had, Lesquereux thought. And so when the survey did come into existence and Newberry was placed in charge, Lesquereux's work was confined to a study at his own expense of certain southern Ohio oil and coal lands. In 1872 he regarded the Ohio survey as "a failure. It was a money-enterprise," he said. "[N]either the directors nor the assistants had any plan, or any other purpose but to get as much money (as possible) out of the state for doing nothing at all. Newberry could not leave New York and could not give any direction for work to be done." Nevertheless, Lesquereux became an authority on "oil geology" and the mode of formation of "oil reservoirs," and of oil itself. He said:

. . . I do not say that oil has been formed by the decomposition of bituminous coal but for the oil of the coal measures by the decomposition of coal plants and for the oil of other formations by the decomposition of plants of other formation. I have seen at Breckenridge [Kentucky?] the sand stone under the cannel coal and especially where the coal is absent and thus under the place which the coal should occupy entirely impregnated with oil. At other places when the coal was still present, the underlying sandstone has no oil. . . . I believe that at *divers epochs* the decomposition of plants (maybe with animals) has been made under certain atmospheric circumstances and under the influence of water and has produced oil. I say . . . that oil is produced or has been produced at every epoch of our Geological formation. I do not think that because the coral mounds contain oil it is a proof that the matter is of *animal* origin. . . .

Theoretical observations on the origins of oil, coal, prairies, and other formations of geologic and geographic interest were indulged in by Lesquereux, but at first only incidentally. He realized the necessity of assembling and systematizing the North American paleobotanic flora and to this task he bent his energies. North American paleobotany was not, nor did Lesquereux regard it as, a recognized science. During his life it would become tremendously important in solving problems of classification of our modern flora, and even more significant, problems shedding light on evolution. On July 28, 1867, Lesquereux wrote Lesley:

Botanical palaeontology has never been studied in America and even in Europe it is still a new study, for I will not call it a science. . . . [We] can not be blind to the immense advantage that Geology has derived from the application of the palaeontological science (animal) by such men as Hall, Meek, Worthen, and how many others. If they have not constructed the scheme of the rocks, they have at least patiently identified them at such distance where stratigraphy and topography would have failed to do it. . . .

The study of "botanical palaeontology" was in largest part then taxonomic. The search for relationships and distribution schemes in fossil plants, as in other branches of science—for example, in geology with rocks—had not really begun. By 1875 he was to say, "Palaeontological botany has made of late a great way." And by 1881-1882, after publishing one of his greatest works, Lesquereux was to tell Lesley: "The publication of the coal flora has awakened such an interest in fossil botany that I have quite as much as I can do, to examine and determine the specimens sent to me from all the states of America (about)." Four months later he said: "I am now overcrowded with fossil coal plants sent to me from everywhere coal is worked[:] Arkansas, Tennessee, Georgia, Alabama for the subconglomerate; Ohio for the high coal; Illinois for No. 1, especially from Morris and Mazon Creek. . . ."

Furthermore, during practically all of Lesquereux's active years, North American paleobotany remained taxonomic, notwithstanding the fact that in the first year or so of his employment by Rogers he prepared for him a paper "on the morphology of the coal plants." Lesquereux sought specimens in which "the leaves [were] attached to the stem and the stem to the roots." Surprising as it may be to American botanists, he examined "the nature and internal structure" of plants and he wished with all his heart to establish "by the great amount of materials" at hand in 1872 "the continuity, identity and relation of all the coal strata of our American measures." Lesquereux anticipated the future and the restriction of his labors was deliberately conceived, born of the necessities of experience. As proof of the latter point may be cited his letter to Lesley dated July 28, 1867:

I am still a poor beginner in the study of Palaeontology . . . it would have been far better if knowing this I should have been satisfied with collecting and recording facts before asserting anything on the distribution of fossil plants. But for my excuse I can say this which is certainly true. This application of botanical palaeontology ought to be made. How then to begin it? Collecting fossil plants and describing and naming them would not have helped this beginning in any way. I have risked deductions. Some have been wrong: others may be true. In any way I do not think that this study has put any confusion into the part of geology to which it had application. Though it may be, I intend now, if I can still give my time to palaeontology, to study it in itself to collect Data which may be used in the future and to leave for more enlightened times the care of drawing deductions or of applying to geology the collected materials if it can be done with advantage. . . .

On the point of his anticipations of the future of paleobotanical studies may be cited another letter to Lesley written in 1886:

I should . . . like to make a clear exposition of the vegetation of the coal period in the succession and modification of its essential types, from the Devonian or even the Silurian up to the Permian, Permo-carboniferous or rather Upper Carboniferous, all expressions representing the same thing, viz. the gradual effacing and disappearance of the Carboniferous types in their passage to those of [the] Trias[sic]; the Vegetation of the Paleoz[o]ic time. . . .

Indeed, this was not all. The great branch of botany known today as ecology had its real beginnings and rise after Lesquereux's time. However, in respect to both living and fossil plants, Lesquereux appreciated the value of studying their ecological factors. In a letter to Gray written in February 1850, he said, "It is no[t] possible to study one['s] plant[s] when one has not seen it growing in its natural place: when one was not able to compare it with the many variety of forms to which every one is liable."

As late as 1871, however, Lesquereux characterized the results of his researches in paleobotany as "meagre." The early history of such work in North America had had little significance. In New York and some eastern states, studies of Amos Eaton, James Hall, Ebenezer Emmons, and a very few others had laid foundations. Nevertheless, not until Desor persuaded Lesquereux about 1851 to commence extensively fossil plant study "so much in harmony with what [he] had worked before in Europe" did North American paleobotany receive a real impetus.

Lesquereux resented the all too prevalent jealousies among American men of science. In fact, he himself was not altogether amiable, becoming involved in controversies with nearly all of the men with whom he dealt. He was impressed with neither their knowledge nor ability, especially he despised the flair for publicity in which some of the state surveys indulged at the expense of thorough workmanship. Lesquereux conceived his theory of testing "the value of the palaeontology for the identification of the coal." Some of the most prominent scientists "sniffed" at him but in 1857 while exploring in Kentucky he became confident. "I was certain from my anterior observation that the coal slates were perfectly reliable in their carracters and that I would find in the fossil shells the same analogy as in the fossil plants," he told Lesley, "and I was not mistake[n]." To the view that coal beds could be identified by fossil floras he clung for many years. Returning to New Harmony, Indiana, from his Kentucky "rambles" in 1857, he wrote Lesley:

. . . those researches offered me the first opportunity to test the value of the paleontology for the identification of the coal. . . . Though I have seen the low coal at more than one hundred places, I have never missed one of its true either animal or vegetable c[h]aracteristic. The examination of the Kentucky coal has fixed for me the true place of many beds of coal that I had seen in Ohio and Pennsylvania and of which I did not know the true geological level. . . .

Seventeen years later he still adhered to his view, saying:

I still believe that in a limited area or for identification of the coal beds of a basin of small extent, the distinction of the species of the flora of each bed of coal can be positively made. I am now differently posted [than] I was twenty five years ago when I had to study the fossil plants in their characters therefore to make descriptions or identification of them; to clear the rub[b]ish of the ground and find the stones and the material for the building. Now I have made an intimate acquaintance with the flora of the coal and now only could the work of application be pursued with some chance of success. If there is no possibility to find clear lines of demarkation between the different beds of a coal basin, there are however peculiar stages which are clearly characterized by their vegetation. . . .

When Lesquereux believed he was right, he was fearless. In 1875, when told that his theory of the age of the great lignitic coal formation of the Western Territories was opposed by all geologists and paleontologists of the Rocky Mountain areas, he simply replied, "All right." Throughout his life, he valiantly fought his handicaps. "Poor, deaf, without books, without friends, without any occasion whatever of making my labors profitable for myself and useful for others," he told Lesley in 1857, "I am going along in the path of science like a blind man on the streets of a large town." Nevertheless, he was often philosophical. Paleobotany, he said, "is often a Jack o' Lantern which brings me to some muddy ditch. But what of that? Good enough if it helps others."

Having studied paleobotany in Europe, he could not resist speculating on possible relations of the North American flora with the European— "to show from the beginning of the vegetation on earth the remarkable similitude of American with European types always broken by characters of dissimilarity as difficult to appreciate now as they were at the epoch of the coal." Gray's announced theories evidently interested him. But, as years passed, the conviction grew "that our American paleontology has to be based on what we know from our specimens." About 1860, he wrote for *The American Journal of Science and Arts*[9] on fossil

[9] XXVII (2nd ser.), pp. 359, 366. Some species were evidently also of the Pleistocene period or "Post-pliocene."

leaves, in part, at least, of the Pliocene epoch found at Chalk Bluffs on the lower Ohio River near its entrance to the Mississippi not far from Paducah, Kentucky; on fossil leaves collected by Lesquereux and Owen in the chalky banks of the Mississippi River near Columbus, Kentucky; also on species of fossil plants collected by J. M. Safford near Somerville, Tennessee. In 1871 he wrote Mrs. Lesley saying, "As yet we know little of our Pliocene Flora," and added:

. . . these materials are very scanty and their isolation from animal remains prevents a reliable acquaintance with the period which these leaves represent as our arborescent vegetation has already its essential typical characters clearly marked in our lower Cretaceous; [A]s these characters become still more defined in our [T]ertiary we want only a knowledge of the vegetation of the Pliocene to complete the chain of evidence, demonstrating that from its origin, the flora of our[s] is [A]merican, only more isogenetic (if this word is right) and far older than that of Europe which in the tertiary especially is mixed with [O]riental, Australian and American types. It is probably the oldest flora of the world. . . .

Lesquereux was officially chosen about 1870 by Hayden's survey to study evidence of fossil floras found in the Western Territories of the United States. He was to assemble data on the ages and characteristics of past geologic periods of the North American continent. Specifically, he was to study the age of the Cretaceous Dakota group, and, also, the Lignitic formations of the West, believed of Tertiary origin in which were coal beds seeming to have a value comparable to the already recognized valuable coal measures of the East from the Mississippi River basin to Massachusetts. Foundations were being laid to infer conditions of climate, temperature, rainfall, glaciation, land texture, stream composition, and like matters of each geologic period. For example, palms, especially Sabal, were found in fossil flora of the Lower Lignitic, indicating a warm, moist climate like that of Florida today had once prevailed in the West. A tropical climate may have extended northward to Vancouver Island and beyond. Indeed, as early as 1860, Lesquereux said, "Palm trees, figs, Cinnamomum, and [Proteaceae] are now generally distributed at least 30° lower than they were then." Such findings were important for historical and *economic* reasons, reflecting not only plant origins and development but also origins and existence of coal, petroleum, and other commodities.

Beginning in the West, in Kansas and Nebraska, Lesquereux's investigations continued over a wide area from New Mexico through Colorado into Wyoming, Utah, Minnesota, Montana, Dakota, and California.

On March 19, 1868, Lesquereux had completed and sent to *The American Journal of Science and Arts* for publication "the first paper published in America describing fossil plants of the Dakota group," entitled "On Some Cretaceous Fossil Plants from Nebraska." This paper enumerated forty-one species regarded as new and marked the real beginnings in America of an intensive investigation of the Great Central Plains and Rocky Mountain areas for fossil plants of the two imposing geological periods—the Cretaceous and Tertiary—whence originate the development of our modern flora, at least our flowering plants as we know them today. During the Cretaceous period, covering as it did, millions of years, the great areas where now are our Great Central Plains and Rocky Mountains underwent a period of initial sedimentation caused by aggrading streams which preceded inundations of the inland sea that then bisected the North American continent. The geological formation which resulted—a great depositional area—has been known as the Dakota group. Other beds formed during the Cretaceous period were the Colorado group—beds largely representing sedimentary deposits of widespread seas; the Montana group, also largely marine, but showing evidence of withdrawing seas; and the Laramie group formed after the seas had withdrawn and showing today some of the great coal areas of the Rocky Mountains. Gatherings of waste materials in rock strata produced great beds of lignite resembling coal measures of earlier geological periods. To these evidently Lesquereux applied the term Lignitic.

About 1867 Dr. John L. LeConte, connected with the Union Pacific Railway survey, had found near Fort Harker "a number of fine specimens of fossil leaves from red shales of the Dakota group" and sent them to Lesquereux who, combining them with "some splendid specimens of Cretaceous fossil plants" sent by F. B. Mudge, state geologist of Kansas, and a large number sent by Hayden as director of the geological survey of Nebraska, prepared his now famous paper on the Cretaceous fossil flora of America.[10] Apparently, however, all the plants enumerated by Lesquereux came from the Cretaceous formation north of Fort Ellsworth, Nebraska, or its vicinity. Mudge's plants were included just before publication, and Lesquereux, giving an account of all literature on American Cretaceous fossil plants to that date, was enabled to increase the total to seventy-three known species. Lesquereux's comments are one of the classics of North American botanical literature:

. . . the more we know of the floras of the geological ages of America, the more we recognize in them peculiar types which in their grouping constitute what may be called an *American facies*, which by successive transitions has passed to our present flora and assigned to it its general character. Is it not remarkable, for ex-

[10] *The American Journal of Science and Arts*, XLVI (2nd ser.), pp. 91 ff. In all, 53 species were described.

ample, that our Cretaceous fossil plants should have a more evident relation with our present flora than with that of any stage of the Tertiary of Europe. Some of the Cretaceous species are undistinguishable from predominant species of our time. And when we consider merely the general facies of our present arborescent vegetation, we can but recognize it in the Cretaceous. Liquidambar, Populus, Betula, Fagus, Quercus, Platanus, Credneria (closely allied to Coccoloba of which we have two species in Florida), Laurus, Sassafras, Lyriodendron, Magnolia, Acer, Paliurus, Rhamnus, Juglans, Prunus, &c., all genera of ours and this in seventy species discovered! Is it possible to point out a more evident characteristic affinity!

From this we may at once admit, that we do not have to look for the origin of our actual vegetation to some far distant country, and to account for its nature by peculiar and cataclysmic transportations. Its origin is not Australian as it has been sometimes admitted, nor Asiatic, still less European: but it is born, has been cradled, and has grown up on this continent. The preservation of peculiar types, present at divers geological epochs, indicates a succession and slow development of formations without such great disturbances as are recognizable in other countries; and it proves also that the climatic conditions of our North American continent have continued about the same as they are now from the Cretaceous through the Tertiary. No species found in these formations of ours indicates a warmer temperature than that of the Southern States.

We know very little yet of the vegetation of our Tertiary formations, and it is impossible to attempt now a comparison of the floras of the Tertiary and of the Cretaceous in America. Nevertheless, from the species already published, even from those . . . of Nebraska, obtained from Dr. Hayden and Dr. LeConte, the generic affinity is striking and therefore the general American facies is equally represented in both.

Vegetable remains are the records of the natural phenomena which have governed the surface of our earth at different epochs. Nowhere else can the successive development of a long series of vegetable cycles, without cataclysmic interruptions, be followed as well as in America. When, then, the fossil plants of our country have been thoroughly studied, they will unfold to us the history of nature's proceedings during the geological times. Questions of a high order are therefore intimately allied to the study of those remains of fossil plants, so little valued among us even now.

Lesquereux's activities were not confined to Cretaceous fossil flora. There were also the Tertiary. Concerning "Fossil Fruits found in connection with the Lignites of Brandon [Vermont]," he stated his belief in a *Report of the Geology* of that state published in 1861 that: ". . . I have no doubt that the Brandon lignites belong to the same epoch as the upper bed of the lignite of the tertiary . . . they cannot belong to the Pliocene. There is no living species among them."[11]

As early as 1867 he had also studied fossil flora of lignitic measures of the Western Territories. That year LeConte had collected around the base of the Raton Mountains near Trinidad, and in Colorado, specimens

[11] *Report of the Geology of Vermont* (two volumes; Claremont, New Hampshire, 1861), pp. 240, 712.

of these deposits, and, after briefly referring to them in his *Notes on the Geology of the Survey for the Extension of the Union [Kansas?] Pacific Railway, from the Smoky Hill River, Kansas, to the Rio Grande*, sent them to Lesquereux. When soon Hayden sent more fossil leaves from Laramie Plains, and Golden, Colorado, though many specimens were fragmentary, Lesquereux had begun to describe them in 1869. The lignitic flora of a vast then believed Tertiary formation,[12] rich in combustible material, occupied an immense area in the territories—Colorado, Utah, Wyoming, and others. As years passed and materials were brought to view, profusions of ferns and palms from Colorado and New Mexico vied in interest with fossil remains of Viburnum and Ficus from Black Butte and Point of Rocks, Wyoming. Species of Cinnamomum, Sequoia, and even Ottelia, a tropical type, were discovered. Many discoveries evoked discussion and wide interest. By 1882 seven Cycadaceae from the Dakota group were believed known.

Lesquereux's investigations continued. In Hayden's 1871 report was published his study "On the Fossil Plants of the Cretaceous and Tertiary Formations of Kansas and Nebraska," and in a preliminary report of the survey of 1871, published 1872, appeared two other considerable studies by him on "Fossil Flora."

I. Enumeration and description of the fossil plants from the specimens obtained in the explorations of Dr. F. V. Hayden, 1870 and 1871.
II. Remarks on the Cretaceous species.
III. Tertiary flora of North America.

To the third branch of this report was also attached a supplement concerned with Tertiary plants from Hayden's explorations in 1870.

In 1872 Lesquereux, with his son, went to many localities in the West to obtain specimens and ascertain the geologic position of various coal beds. For specimens from the plant bearing Cretaceous strata of the Dakota group, they visited the Saline River valley and the Smoky Hill Fork of the Kansas River. A wide area of lignitic formation of the Rocky Mountains was then visited to study the formation and its relations, its scheme of plant distribution, and probable age of its measures: Raton Mountains, along the base of the Spanish Range and Greenhorn Mountains to Canon City, the Colorado basin around Colorado Springs and Denver to Golden, Coal, and Boulder creeks; and in Wyoming along the route of the new and famous Union Pacific Railroad at Carbon, Black Butte, Evanston, and other places. Officials sought to persuade Lesque-

---

[12] Subsequent investigation has altered findings in regard to epochs and periods of much paleobotanic materials. References here are made in terms of what Lesquereux believed—the majority view at that time.

reux to stay and devise means of smelting ore with their coal. But he declined because of his lack of knowledge of practical "enginery." "Every day I feel the need, the want of a mathematical mind and regret more and more my ignorance of topography," he told Lesley. Again in 1873 he returned west and "by myself alone," he said, visited former and new localities, adding to fossil collections of Hayden's survey. Although places in Kansas and Nebraska were visited, Lesquereux's special area in western paleobotany at this time was between Denver on the east; Santa Fe on the south; and Cheyenne, Wyoming, and Ogden, Utah, on the north and west. His showing of an intimate relation between our present flora and that of the Tertiary period, especially that shown by fossil plants of Green River station, won an interesting comment from *The American Journal of Science and Arts*, which said:[13]

Among species of *Salix*, *Myrica*, *Ilex*, and *Rhus*, whose representatives are intimately related to species of our time, the fossil flora of Green River has an *Ampelopsis*, and a *Morus*, which by their marked affinity indicate in the Tertiary the origin of our now so predominant and widely distributed Virginian Creeper and Red Mulberry.

But the Rockies did not impress Lesquereux. "My Jura M[oun]t[ain]s are one thousand times more beautiful even in their ordinary aspects than the finest places of the base of the Rocky [Mountains]. No water, no trees, all dry: barren, desolate," he commented.

"The Tertiary coal of the Rocky Mountains," Lesquereux observed, "precisely indicates the same degrees and difference of transformation in their matter as those of the Carboniferous measures from Rhode Island to the Susquehanna and to the Allegheny Mts. Near the primitive m[oun]t[ain]s especially where the upheaval is marked, the coal is hard compact sometimes semi-anthracitic." In 1873 he was invited to join an expedition west of Fort Rice, to the Yellowstone lands. Urgently Hayden solicited him to accept and he did "conditionally to the acceptance of Prof. Leidy and Prof. Porter" for whom he later did systematic work at Lafayette College about the time he compared for Hayden materials in the Torrey Herbarium at Columbia University. Lesquereux had to decline the expedition invitation but on May 18, 1873 he wrote:

. . . If I live some years more, I shall have by my researches contributed something to the history of our Coal formations, lignitic and carboniferous. And therefore I am not at liberty to refuse the opportunity of exploring the only field which I have as yet not seen, of the whole extent of our Carboniferous and Lignitic formations. I forget the Richmond Coal. But that too may pass under my view sometime. . . . Had I not been called to the West I should have spent two months with Prof. Porter in the Anthracite field for examination and collection of specimens.

[13] IV (3rd ser., 1872), p. 494; see Hayden's *Supplemental Report for the Survey of 1871-2*.

He sought acquaintance with all vegetations in black shales, the coal, the rocks, and all formations. But he could not do all he wished to do. Cutting fossils and carrying their weight great distances proved often too much for Lesquereux's strength. The time of appearance or as he termed "the apparition? of the Dicotyledonous plants" interested him. He planned a work on American coal floras from Pennsylvania to Kansas "about like that made by Heer of the floras of Switzerland in his Urwelt"; but before that he had to prepare several monographs for Hayden.

Paleobotanical reports, however, were not the only reports made by the survey. There were also reports on botany which covered an even wider area. Apparently the survey's explorations for botany began in northern Colorado and spread north and west into parts of Wyoming, Utah, Idaho, and Montana, seesawing during various years north and south of the Union Pacific tracks. Generally and briefly stated, in 1868, including northern Colorado, the survey explored from Cheyenne and Fort Laramie on the east westward to Fort Bridger and South Pass. In 1869 the field of operation again went south, more deeply into Colorado and New Mexico. But in 1870 it shifted to the north of the railroad. Nevertheless, of Wyoming only the southern and western parts of the territory were for several years the subjects of the study. For a number of years after the survey began operations, no portion of eastern Wyoming or of Dakota was formally included. The line of march proceeded into portions of northwestern Wyoming, Utah, Idaho, and Montana. On September 1, 1872, Hayden wrote James D. Dana, editor of *The American Journal of Science and Arts*:[14]

Two large and well-equipped parties have been in the field at work since about the first of July. The largest party made Ogden the point of departure. It was under the direction of Mr. James Stevenson, my principal assistant. There are attached to the party a geologist, topographer, astronomer, and meteorologist, with the necessary assistants for each. There is also a botanist, who has already collected over 1200 species of plants through that new and interesting region, the Snake River valley. . . . The party surveyed a route from Ogden to Fort Hall, Idaho, where full preparations were made for a pack train for a given time. The party passed up the west side of the Snake River valley, forced their way across the mountains, made a careful survey of the Teton range, then passed up the valley of Henry's Fork, entered the Madison valley through the Targee pass, and reached the Geyser Basin of the Madison August 14th.

The party under my charge traveled by stage to Fort Ellis, and there spent about three weeks preparing the outfit, then started up the Yellowstone valley, over about the same route as last year. The party consisted of about thirty persons, among them a chief topographer, astronomer, meteorologist, mineralogist, with their

14 Found in the *Journal*, IV (3rd ser.), pp. 313-314.

assistants, and a number of others who acted as collectors. A careful examination of the Yellowstone Valley was made, and a map in contour lines of 100 feet each constructed.

Both parties met in the Geyser Basin on the same day within a few hours of each other. . . . The opening up of that great Snake River valley will prove one of the most important events in American explorations for the year 1872. . . .

In 1872 John Merle Coulter, a young man twenty-one years of age, born in Ningpo, China, of missionary parents, and a recent graduate of Hanover College, was an assistant geologist of the survey. Early that year, while awaiting the arrival of Hayden to take up the year's work, Coulter, finding himself in an interesting floral mountain region, occupied himself collecting plants instead of playing cards as other members of the party did. When Hayden arrived, Coulter's collection so impressed him that he appointed the young man botanist of the expedition and, when they arrived at the hot springs and geysers of Yellowstone Park, meted out a task much to Coulter's liking.

There had been explorations in the Yellowstone area before the arrival of the Hayden party but none of a strictly scientific character. For years rumors had been rife that a great geyser and hot springs country existed in this region unsurpassed for natural beauty and geographic interest. But the stories were principally those of traders and trappers. And it was not until 1871 that any confirmation of these reports was had. At that time, Hayden with the survey of that year explored in Montana and parts of near by territories and included the Yellowstone vicinity in their route. The examinations, however, were not made with thoroughness and so Hayden's 1872 expedition returned there.

As they approached the great steaming regions, men of the survey climbed trees to establish their route to the objectives. Once there, they made camp and the scientists went about their various tasks. Coulter, although originally appointed assistant geologist, was assigned the task of gathering botany when the originally appointed botanist, it is said, proved too fond of the alcohol provided for the animals. Coulter, while at Salt Lake City, had interested himself gathering the flora of that region but he did not confine himself merely to assembling plants. He studied the physical and environmental factors of the plants collected and in the Yellowstone region became greatly interested in studying the effect of the geyser and hot spring water on the plant floor. Perhaps here was an anticipation of his later pronounced ecological interest. At any rate it brought him some amusing adventures which for many years he delighted to relate. In Yellowstone Park today, Coulter Creek remains as a memorial to a joke. Coulter, the "boy" of the expedition, fell from his horse into this creek through the fault of an animal preceding his. The men of the survey named the creek for Coulter.

Hayden, seeing Coulter's interest in the sixty minute geyser of the park, assigned Old Faithful geyser to the young scientist for investigation.[15] In the survey's sixth annual report, embracing portions of Montana, Idaho, Wyoming, and Utah, appeared Coulter's report on botany. Forwarded by him to Hayden on April 15, 1873, it contained also an elaboration of the Phanerogamia found in the explored area on both the eastern and the western slopes of the continent. The Cyperaceae were done by S. T. Olney; Graminaceae, by George Vasey; Musci, by Leo Lesquereux; Lichens, by Henry Willey; and Fungi, by Charles H. Peck. Coulter's collections north of the Union Pacific Railroad would undoubtedly have been more thorough had the survey remained the next year north of the railroad. But after 1872 for three seasons it went south again into southern and southwestern Colorado, going into northern New Mexico and eastern Utah, and within four years completing the mountain parts of Colorado.

Coulter did not remain all four years with the survey. But during the first year in Colorado he collected plants and with Thomas C. Porter prepared a *Synopsis of the Flora of Colorado*, published as a miscellaneous report, Number 4, by the survey on March 20, 1874. Porter wrote of their *Synopsis*:

> The work is based chiefly on collections made, in 1861 and succeeding years, by Dr. C. C. Parry, whose indefatigable labors have added so much to our knowledge of the flora of the region; in 1862, by Messrs. Hall and Harbour; in 1867, by Dr. W. A. Bell, of Manitou Springs; in 1868, by Dr. F. V. Hayden; in 1869, by B. H. Smith, Esq. of Denver; in 1871, by Dr. George Smith and W. M. Canby, Esq.; in 1871 and 1873, by Messrs. Meehan and Hooper; in 1872, by J. H. Redfield, Esq.; in 1872 and 1873, by T. S. Brandegee, Esq., of Cañon City, Rev. E. L. Greene, of Pueblo, and T. C. Porter; and in 1873, by J. M. Coulter.

The mosses and hepatics were elaborated by Lesquereux; the lichens by Willey; and the fungi by Peck. To Brandegee and Greene the authors acknowledged special indebtedness—to the former for southern Colorado collections and to the latter for lists and specimens of rare species.

Coulter evidently spent from May till late summer collecting in Colorado. Exploring plains of the Platte River he went west of Denver into the foothills and beyond to the mountains, visiting such famous localities as Longs Peak, Grays Peak, North and South Parks, Mt. Lincoln, Mt. La Plata, White House Mt., Mt. of the Holy Cross, James Peak, Horse-

---

[15] See *The Dictionary of American Biography* (New York: Charles Scribner's Sons, 1930), IV, pp. 467, 468.

shoe Mountains, Buffalo Peaks, Sierra Madre range, St. Vrain's Canyon, Clear Creek Canyon, Boulder Canyon, California Gulch, Bear Creek Canyon, Monument Park, Pleasant Park, Eagle River, Twin Lakes, Weston's Pass, Chicago Lakes, Ute Pass, and the Arkansas River.[16] In the preparation of his materials from the Colorado and Wyoming surveys he was compelled to go to Washington where he met Asa Gray. This meeting resulted in the beginning of a correspondence and collaboration which continued to the end of Gray's life and had much to do with establishing the sound taxonomic foundation on which Coulter built creatively in later years in botany.

Porter, a professor of botany, zoology, and geology at Lafayette College, in Pennsylvania, also did much for Coulter. Porter had had the benefit of collaborations with John Torrey and had already done much botanical exploration in Pennsylvania, Georgia, and other eastern states, and besides, had preceded Coulter by at least one year in serious study of the Colorado flora. To him Coulter owed a debt and this he generously acknowledged ten years later when publishing his subsequent works on the Rocky Mountain flora. Porter and Coulter had used for their *Synopsis* the plan of Watson's *Botany of the 40th Parallel*. And their work announced the beginning of the real systematization of Colorado botany.

The botany of southern Colorado, Arizona, New Mexico, Wyoming, and Utah, was more completely investigated before the end of 1874. That of western Wyoming was more thoroughly explored by studies of Charles Christopher Parry as botanist in the reconnaissance of 1873 from Fort Bridger to Yellowstone National Park under Captain Jones's command. That of southern Utah was explored by Parry's continued travel of the next year. And that of southern Colorado, Arizona, and New Mexico, was more amply revealed by botanical searches of Joseph Trimble Rothrock and his party as part of the United States geographical surveys west of the one hundredth meridian under First Lieutenant George M. Wheeler.

Rothrock's work was of much interest. In areas where his party went, botanical collections had already been made by Ferd. Bischoff and others of the Wheeler survey of 1871 and 1872, the collection extending from the southern portion of the Great Basin in northwestern Arizona to the adjacent desert section of California. These, with a small collection made near Kanab, southern Utah, by Mrs. Ellen P. Thompson [in 1869?] while accompanying her brother, Major Powell, in his survey of the Colorado River of the West—the United States Geographical and

16 Based on an analysis of Coulter's plant collections as shown in the *Synopsis of the Flora of Colorado.*

Geological Survey of the Rocky Mountains—were published in Volume VII of *The American Naturalist* by Sereno Watson to secure priorities for several new species and in anticipation of a fuller report which appeared in 1874. *Whipplea utahensis, Peucedanum Newberryi, Angelica Wheeleri, Chaetadelpha Wheeleri,* and *Salix nevadensis* were among the species enumerated.

Powell conducted a few explorations in Colorado Territory before his daring exploration of the Colorado River from a point where the Union Pacific Railroad crosses Green River to his emergence from the Grand Canyon and arrival at Kanab. Botanists, it is said, accompanied these explorations but, apparently, no published results in this regard were made; why, in at least one instance, is explainable. In 1867, Powell, then appointee to the chair of natural history and geology at Illinois Industrial University (now the University of Illinois), selected Thomas Jonathan Burrill, a superintendent of the Urbana public schools and who became professor of natural history at the University the following year, to serve as botanist of an expedition to the Colorado Rockies. Only a small part of the botanical collection of this journey reached Urbana, the larger part being lost when a burro, loaded with dried specimens, was drowned when fording a swollen mountain stream. This small collection, however, together with a collection Burrill had made of the flora of Champaign County, Illinois, became the nucleus of the present University of Illinois herbarium.

Joseph Trimble Rothrock was born in McVeytown, Pennsylvania, in 1839. "In 1860 I first went to Cambridge to study Natural Science,"[17] he wrote, "I had previous to this given it all my *spare hours* for two years. During the time I was there I spent all my time in botany and labored to my utmost capacity. I expended all the means I could obtain for books, instruments and the facilities for study. After graduating seeing no opening by which I could support myself I came home and most reluctantly took up the study of medicine. Before graduation I was asked by Kennicott to accompany him to the N[orth] W[est].[18] I at first declined, and only accepted after you [Dr. Gray] had written me that you thought the chance a good one—which I ought to accept. The hardships I endured whilst there and the complete failure I made so far as scientific work was concerned are only too familiar to my friends. This failure was due to no fault of mine. It did still sting and mortify me, and when,

[17] See *Letters of Asa Gray,* II, p. 530, where Gray says he had been a student of his in 1865 "three or four years . . . that botany [was] in him, and [would] probably come out. . . ." Gray characterized him as "a bright lively pupil. . . ."

[18] See *Letters of Asa Gray,* II, p. 531, where Gray describes the exploration for a telegraph route along the Arctic Circle.

anxious to return and redeem my reputation, I offered to go back and collect if only my expenses were paid, I found the science of botany could not give me the means of doing so. Medicine was the only thing left. After graduation came my chair at the P[ennsylvania] A[gricultural] College."

Rothrock's "failure" was the loss of his plant collections in the Fraser River while with the Kennicott and Pope expedition to British Columbia and Alaska from which issued his *Sketch of the Flora of Alaska*.[19] On June 22, 1865, he had written Gray:

Well here I am at last on the Fraser River in Latitude 54° N[orth]. Circumstances have conspired to prevent my doing any scientific work this year at all. You perhaps know before this what those circumstances were and Mr. Kennicott and myself were involved. Please let your opinion in the matter remain in statu quo until I see you. Or if you don't know about the affair I will explain on my return to the States. Next year or as soon as I get back from this overland trip to Nortons Sound. I have the promise of being sent to the Kurile Islands or wherever else I desire. Don't think I shall return to the States for at least four years. Our next move will be to Lake Fraser thence to Babini from which place we set off into the Terra incognita between there and the head of the Steekine River. . . . I shall have time to get to Lake Francis before snow sets in. Once at this Lake we shall have down stream a good river—Pelly to our Ultima Thule. I have walked the party 280 miles during the last 17 days. This was done to test who possessed the "grit" for explorers. Some have failed to come up to the standard and ergo were dismissed. The rest I think would prefer a month on trips. . . . Not more than six of us will go on from Steekine River to Fort Yukon. I don't shut my eyes to the fact that we run a risk in attempting it at the late season we will have to. Yet if a fair share of will and an honest devotion to the Cause will be of any avail we have at least the first earnest of success. Once through with my connection with the Company I will go home or at least try to by way of Peace and Red Rivers. . . . The scenery along the Fraser is grand, for at least one half its way the river runs at a fearful rate through a cañon at least 1000-1500 feet deep which it has worn through the solid rock. The character of the forest vegetation is rather monotonous, which with the exception of cottonwood is made up of evergreens. In a few days we expect to strike the region of Birches. You see I dont presume to think of Botany this year. If I did I should forget the work which it became necessary for me to enter. But next and the succeeding years I hope to prove I have not forsaken my first Love. . . .

Rothrock returned and on May 7, 1868, sent Gray a Polygonum from Alaska and other plants which would "wind up [his] Alaskan troubles, Trillium & Asplenium." Though he was dissatisfied with his teaching position, wanted to get married, to get a new microscope, and to get his herbarium in order, he continued teaching, keeping botany first if pos-

19 A catalogue of Alaskan plants was part of the article, published for year 1867 by the Smithsonian Institution. See *Ann. Rep. of Board of Regents of Smithsonian Inst.* (Washington: Government Printing Office, 1872), pp. 433-463.

sible. By December 1869 he was married, had left the college in disgust, but had been induced to return where he said, "Here I profess—Botany, Anatomy, Physiology, Hygiene, Zoology, Geology. Practice for the College Carrying our coal, curry my own horse.—for $1500.00 per annum, which I don't get punctually. I make neither money nor reputation." He planned to go to Philadelphia, study surgery, and later settle in Harrisburg. But Wilkes-Barre became his home and, practicing medicine, he aided in the establishment of a hospital there.[20]

Rothrock's health had, from young manhood, never been good. Once to regain his health, it is said, he took a position as axman on the Philadelphia and Erie Railroad, a position which did much to quicken his interest in botany and forestry. In the early 1870's his health again became impaired and after some years of relinquishment, he turned again to botany. A position as botanist on the United States geographical surveys west of the one hundredth meridian was obtained and on June 3, 1873, he wrote Gray:

We are now camped on Cherry Creek just outside Denver, awaiting our outfit. Mules are there and aside from graver scientific work our chief amusement is attempting to break them and getting "bucked off" over their heads. Wolf[21] is with us as botanical collector, and I am therefore directed to expend my energies as Zoological collector & Surgeon. However between us you may I think look for a good collection this summer. It would be a great assistance to us if you could forward by mail any of your lists of the Rocky Mountain collections such as published on Parry's Collection. Our route lies through southern Colorado & Northern New Mexico. Among the mountains the peaks tower up grandly and reveal their snow clad outlines beautifully against the wonderfully clear sky. Astragali are out in force on the plains. Two species of Ranunculus & Plantago, Senecio, a Cactus, etc.

I can't tell you how overjoyed I am at getting back to this life. My only care now is being absent from my dear family. Consider me as returning with a new lease on life in December unless I meet with some accident meanwhile such as there is always a slight possibility of on such trips.

We have a splendid set of men. Genial and earnest—no boys. I feel well as I ever did. And look forward now to a permanent return to science.

Love to Mrs. Gray. Tell her if we ascend Grays Peak I'll press & send to her the flower nearest the snow line.

Rothrock's party evidently went west across the plains from Denver to the foothills and on to the region near Georgetown. Whether they climbed the mountains in the vicinity of Torreys Peak and Grays Peak is not certain. Plant discoveries by Wolf on Grays Peak are listed in Rothrock's *Report*.[22] In any event, the party moved on across the flat parts of

---

[20] See *The Dictionary of American Biography*, XVI, p. 188.   [21] John Wolf, 1820-1897.

[22] See Volume VI, "Botany," *Report upon United States Geographical Surveys West of the 100th Meridian in Charge of First Lieutenant George M. Wheeler* (Washington: Government Printing Office, 1878), pp. 3-37.

South Park not far from Fair Play and then, presumably, to the region near Twin Lakes and Granite. "The valley of the Upper Arkansas, as we first saw it," wrote Rothrock, "twelve miles above Twin Lakes, certainly looked like anything but a land of promise." From South Park they had descended several hundred feet into the Arkansas River valley to explore, the indications are, as far as Poncha Pass and as much of the valley as possible as far as Pueblo. Whether or not they went east to Canon City and Pueblo, they gathered information concerning their regions botanically and went through the Pass where before them lay the interesting San Luis valley. Southward they proceeded to the Rio Grande del Norte, passing evidently near Fort Garland, examining in their explorations several river tributaries. A point called Loma on the headwaters of the Rio Grande was the western limit of examinations in this area. And then they proceeded into northern New Mexico. Interestingly, Rothrock studied and reported on the floral, timber, and agricultural resources of the regions.

Rothrock observed that "a marked change in the flora appears about the headwaters of the Arkansas and runs east out into the western edge of the Great Plains at Pueblo, whence it shades off gradually more markedly into the flora of the warmer and more arid regions as we go toward the south." North of there, the "Piñon Pine" seldom appeared in Colorado, and about Pueblo "not less than ten species of Cactaceae" were shown. But south of a line somewhat north of the Colorado-New Mexico boundary, there seemed to be a natural floral division, notwithstanding that on mountains and peaks even as far as the Mexican boundary they found "characteristic Northern plants to suggest the inquiry as to whether the influences of the Glacial period may have extended so far south, and driven these plants before it, as it did those of Labrador to the latitude of New Jersey and Pennsylvania on the Eastern coast."

The Wheeler geological survey was a large outfit. In 1874 when it left Pueblo for work in New Mexico, the engineer survey was divided into eight parties, of which six were primarily topographical and two organized for geological and biological investigation.

Rothrock's party explored for botany and zoology and went into New Mexico and southern Arizona—still later, into California. Going to Santa Fe where they explored the surrounding hills, they "moved toward the Rio Grande, which we struck," he related, "at the Indian town of San Felipe." After studying the Rio Grande valley, which he observed had "an abounding fertility where irrigation is possible," they began exploration of a wide area extending westward across New Mexico and into southern Arizona, and principally studied there during the summer of

1874. In June they are reported to have been at Algodones and their investigations west of there including Fort Wingate, the Zuñi Mountains, Zuñi, and other of the more important localities must have consumed about a month. For July and August were given mainly to Arizona work. Many important, some hitherto unvisited, areas were explored in Arizona: Willow Spring, Camp [Fort?] Apache, Gila River east of San Carlos, Camp [Fort?] Grant, Cottonwood Creek, the Mogollon Mesa, Camp [Fort?] Bowie, the San Francisco Mountains, the Sanoita Valley, Tucson, Camp Crittenden, and others.

Always Rothrock observed a southward slope with vegetation changing in interesting but noticeable particulars; intimately connected with which was the geographical distribution of the forest growth. "Indeed," he said, "we may ... consider the entire country from South Park to the Mexican line as a series of continental swells and depressions, illustrating still this southward slope." Comments on climate, rainfall, soil, and many other factors much more thoroughly investigated since his time—ecologically—were contained in his *Report*.

Rothrock's *Catalogue of Plants*, collected in Nevada, Utah, Colorado, New Mexico, and Arizona, with descriptions of those not contained in Gray's *Manual of the Northern United States*, and Volume V, *Geological Exploration of the Fortieth Parallel*, was first published as Chapter IV of Wheeler's *Report of the United States Geographical Surveys West of the 100th Meridian*, Chapter III of which by Rothrock, "Notes on Economic Botany," also was a significant special study on western botany. Rothrock was most noted for his *Catalogue*, however, which, together with Watson's study of the fortieth parallel botany, became the authoritative works on botany from the Great Basin to Mexico in the interior southwestern states. Nor was this all that Rothrock's explorations accomplished.

Compilation of California's flora was in process, and nearing completion. This work, like that of a study of western Texas botany, was much needed. Rothrock did not participate directly in publication of either but materials from his explorations aided much. In 1875 he extended his study of southwestern United States botany to California—taking another summer for this purpose. Once again a list of all his localities is more than space permits. However, excepting early summer collecting around Santa Barbara and the Coast Range, most of his work was in central California to the Sierra Nevada Mountains: in the Santa Clara valley, Ojai Creek valley, Olancha Mountains, Cassitas Pass, and Mount Piños in July; around Fort Tejon in August; and in September along the north and south forks and headwaters of Kern River, Walkers Basin, and Mount Whitney; among others. His work was published as an appendix to the *Report*, and, briefly stated, concerned central California regions.

Rothrock's health regained, he returned east and became a professor at the University of Pennsylvania, his alma mater in medicine. His collections were placed with the Smithsonian Institution and finally the herbarium of the United States Department of Agriculture. More than fifty new species, it is said, were found in the course of his explorations —two distinct new genera—and many duplicates for North American herbaria were obtained. The genus Rothrockia and a number of species perpetuate his fame as an American collector.

Probably the greatest botanical collector of this period was Charles Christopher Parry. When John Torrey died in 1873, Parry turned to Asa Gray as his principal correspondent. He, of course, also corresponded with his great friend George Engelmann, supplying him with specimens of his favorite plant groups. Parry collected for many great American botanists. And his career was an interesting one.

During his incumbency as botanist of the United States Department of Agriculture, Parry went to the North Carolina Mountains where, although he found himself "too early for the best flora," he collected "some nice plants" on Roan Mountain and in other localities. Parry's genius as a plant collector, however, had been shown in western collecting. In the department's monthly report for January 1871, he had written on "North American Desert Flora between 32° and 42° North Latitude." And in this he specialized. Nevertheless, the government's growing interest in foreign or unexplored lands occasioned investigations in other than western territory. In 1871 Parry served as member of a commission of inquiry to Santo Domingo and returned to report on the botanical features, agricultural products, and timber growth of the peninsula of Samana. On this expedition, trouble, engendered evidently by a gardener's incompetence, occurred which did much to shape Parry's future career. After the expedition had returned, and while Parry was away from Washington on another collecting trip, representations were made to the commissioner of agriculture and, although Parry was given no opportunity to remonstrate, he was dismissed from his position. The scientific world was shocked. Torrey, Gray, Joseph Henry, Edward Palmer, and others came quickly to Parry's defense. *The American Naturalist* and *The American Journal of Science and Arts* carried strong protests. Parry was not returned to his position, however. Fortunately for American science today, he once again became a collector—in the West.

On April 1, 1872, Dr. George Vasey of Illinois was appointed to Parry's position. Vasey had accompanied Major John Wesley Powell in 1868 on an exploring journey to Colorado, had served as curator of the natural history museum of the State Normal School of Illinois, and since 1870 had been associated in the editorship of the *American Entomologist and Botanist*. He was born an Englishman, had been brought to this country when a child, and had become interested in botany through Dr. P. D. Knieskern while living in New York State. Moving to Illinois he continued his correspondence with eastern botanists such as Gray, Stephen T. Olney, and John Carey. No one challenged Vasey's ability or fitness for the position as botanist of the Department of Agriculture and its herbarium. But few approved of his appointment at the time.

Parry, however, turned again to western collecting. On February 7, 1873, Engelmann wrote him:

You have reason to complain of me—as you and other correspondents often have—the example of Torrey and Gray—the great Rocky Mountain Peaks—is too seductive for a lazy man! . . .

So you go to Hayden's Park! Was it you or somebody else who wrote to me about Hayden's grand services to Science, "still he ought not to consider all he surveys his own." Pretty good.

I expect some nice things from you, though "more suo" littlebits, but to say the truth, that country is botanically well explored. You may get good seeds e.g. Lewisia—and trace the conifers.

Hall[23] sent me some of my families from Texas. . . . Nothing new, it was an unfortunate trip. . . .

If you visit to 49° Parallel see that I get good instructive sp[ecies] of Pines—Young cones or female aments—same to you!

Dr. Anderson sends me an Arceuth[24] from Monterey, the old thing, least, a wonder, a date to it! So you see instructive instructions sometimes do good!

No, have not heard from Lapham. . . .

Hall says he can make more in raising taters than in drying plants.

Engelmann's allusion to Gray's ascension of Grays Peak, 14,341 feet high, and Torrey's visit to the region later in the year 1872 must have amused Parry. He had entertained Dr. and Mrs. Gray in his cabin in the Colorado Rockies. Never once did Parry's love of this and other western regions show any sign of diminishing. And so when in 1873 he was asked to accompany a reconnaissance of northwestern Wyoming[25] under Captain William A. Jones of the United States engineers, Parry accepted with alacrity. The reconnaissance covered the headwaters of the Snake, Bighorn, Greybull, Clarks Fork, and Yellowstone rivers, and in large

[23] Elihu Hall, 1822-1882.　　　　[24] Arceuthobium.
[25] See William A. Jones, *Report upon the Reconnaissance of Northwestern Wyoming* (Washington: Government Printing Office, 1874). "Botanical Report," by Parry, p. 308.

part examined regions already passed over by Hayden and other explorers. But it afforded T. B. Comstock and Parry opportunity to study more thoroughly the regions and when the results were announced, they expressed much satisfaction.

On April 20 Parry wrote Gray that he was not surprised to learn that he might be called to the curatorship of the Torrey herbarium. "But," he commented, "my sense of personal loss is too keen to cause any feeling of exultation—indeed I should shrink from the responsibility of occupying a position that has been so long filled by one whom we all delight to honor. Under the circumstances I could do nothing personally, whatever is needed, in view of your exceedingly friendly and almost too strong recommendation. I have no direct knowledge that Dr. Torrey had any such intention towards me but I know to the last he was full of kindest wishes and some of his allusions seemed to point that way." Parry indicated that he would accept but also intimated that he believed it was not best that he be called to the place. He planned to leave for the Yellowstone expedition May 20. He was ordered to report at Fort Bridger before June 9 and this trip he preferred as, he said, ". . . we go direct into the high mountains and have no time in the plains & 'bad lands'—good for fossils[26] but not so good for live things. I wish Watson could go [as] much is out there *somewhere*." He had spent a day with Engelmann whose brother had told him that at Salt Lake City the earliest spring flower was already in bloom. Engelmann himself was finishing reading proofs on his Yucca paper.

Parry wrote Gray on June 4 from Denver, telling him:

I leave tomorrow on route to F[or]t Bridger and next week I suppose we will be on our way to the Wind River and Big Horn country. . . . Mr. Greene promises to meet me (in September). I also have some hopes of seeing Dr. Engelmann then.

I have seen Gardner, Hayden's topographer who is now commencing a thorough triangulation of the mountains—they are now measuring a base & putting up signal towers. He has agreed to adopt *Torreys* name for the West peak, not simply on my representations but as a matter of scientific justice & proper respect to the memory of a worthy distinguished American. This will of course *fix* the matter in spite of Whitney's opposition. . . .

On June 10 at Fort Bridger the order was issued to be ready to start on the march the next day for Camp Brown on Wind River. Parry accordingly packed up the few plants he had collected to be stored till his return. At first their march was slow and tedious. Their heavily laden wagon trains found difficulty getting through the rain soaked earth but

---

[26] Among the several important explorations for fossil plants in Yellowstone National Park, that of Lester F. Ward and F. H. Knowlton in 1887, the latter collecting particularly fossil woods, was especially important. See *Botanical Gazette*, XII, Number 9 (September 1887), p. 234.

to the botanist the verdure was immediately interesting. "Even the repulsive 'sage plains' and 'grease wood flats' so monotonous and forbidding to the ordinary traveller," wrote Parry, "yielded up unexpected treasures of rare plants."[27] The evanescent annuals were in great profusion. Though there was a prevalent growth of Artemisia, Tetradymia, and Linosyris and many "equally forbidding Chenopodiaceous shrubs," several species of Astragalus appeared. Along Green River they found many rediscoveries—plants early found and determined by Thomas Nuttall. Bright yellow and blue flowers in mats of dark green and silvery foliage made neat contrasts. *Aplopappus acaulis* Gray, *Astragalus simplicifolius* Gray, and "a showy asteroid plant with large white flowers . . . a new species *Aster Parryi*," and many others excited Parry. As they reached the higher ground of Green River valley, the desert growth gave way to vegetation of a subalpine character. On June 20 Parry wrote Gray from a camp on Sandy Creek twenty miles west of South Pass:

Here is another installment of *scraps* from our route to Wind River. [T]omorrow we move to South Pass and will leave the desert region of Green River basin. I am glad of the opportunity to gather up some of the evanescent forms brought out by recent rains. You will be glad to see *Astragalus flavus* Nutt. Several other species new to me. I am collecting carefully in fl[ower] & fruit. We will reach Camp Brown next week, and then a short delay before starting for Yellowstone. [A]long the Eastern edge of Wind River M[oun]t[ain]s we will ascend Fremonts Peak. . . . I hope to make a pretty nice general collection but will collect only the rarer species. . . .

On July 9 Parry wrote from Camp Brown:

We are to leave tomorrow with a pack train for Yellowstone. Expect to be about 20 days on the route and will go over some interesting country. I have just returned from my 1st M[oun]t[ain] trip near the foot of Fremonts Peak. I enclose you some scraps, part of the fruits. You will be pleased with the Aster or whatever it may be —the Oxytropis also. . . . As far as I have seen the M[oun]t[ain] flora here is much less rich than in Colorado, a few new things. . . .

Parry noticed the prevalence of *Lewisia rediviva* in the Wind River valley. On the smooth tables of the summits of the Wind River range, the grasses interspersed with brightly colored flowers were a pleasant sight to him. In some regions the pine growth was scant; in others more abundant. Instead of following down the valley a great distance, they crossed Wind River and a low spur of Owl Creek range and, going near the base of high mountains, passed Owl Creek, Greybull, and several other tributaries of the Stinking Water which they ascended to its source in the high divide separating it from those of the Yellowstone basin. The

[27] See "Botanical Observations in Western Wyoming," by Parry, *The American Naturalist*, VIII (1874), pp. 9, 10, 11, 12, 13, 102, 103, 104, 105, 106, 175, 176, 177, 178, 179, 211.

Owl Creek range was comparatively unexplored. And near the main Wind River bottoms, they found a new species of Astragalus, "sending up a loose spike of white flowers," which Gray later named *A. ventorum*, and other new species. On the crests of the dividing range at 9,000 feet elevation in rock crevices was found "the charming dwarf columbine, which, in compliment to the enterprising commander of the expedition, and its first actual discoverer," Parry wrote, "I have named *Aquilegia Jonesii* n. sp." On the northeastern slope of the Owl Creek range which formed the western edge of the Big Horn basin, Parry found "*Stanleya tomentosa*," a new species. This plant, Parry related, was "in the full glory of its dense spike of cream-colored flowers. . . ." Another discovery was *Astragalus Grayi*.

The detached topographical party which went to the peak named by Jones "Washakie's Needle" found a more distinct alpine flora. As the party went up the valley of the Stinking Water dense woods and grassy parks with "attractive and varied flora" alternated. "The occurrence of so many peculiar Californian forms in such an isolated locality on the Atlantic slope is suggestive," commented Parry.

On August 5 Parry wrote Gray from a camp on the Yellowstone saying they had had a "successful but rather hard trip"; however, now were in a very attractive locality amply described by Hayden in his 1871 report. Later Parry's opinion changed. On August 14 from a camp below Big Canyon on the Yellowstone, he wrote Gray:

I find very little of botanical interest in the Park. [T]he high M[oun]t[ain]s are not high enough to reach the true alpine vegetation and the lowlands and pine forests (mostly *P. contorta*) are mostly common, the lake spring & mud volcano district of which this region is full offers nothing new except perhaps in microscopic forms. I shall be able to add considerable to Hayden's list however and will get up a pretty good *Park list*. We have yet to see the great sight of the *Geysers*. The falls are more pretty than grand and the Cañon is "no great shakes" being accessible everywhere, and Hayden's "immutable colors" are mere red and yellow daubs. [H]is description has caused much amusement to our Company when read on the spot. I have seen one patch of Porterella, very Lobelia looking. . . . I send specimens of *Aster Haydeni* Porter which is the curious high M[oun]t[ain] Aster and I judge hardly a distinct species. . . .

I hardly expect to do much more on my return route as by the time we reach the high M[oun]t[ain]s the alpine plants will have retired for the season. I shall however do all I can & collect mainly of choice desirable things. As soon as possible after my return I want to get up a paper for The (*American*) *Naturalist* to contain description of new or little known species collected on the trip, as otherwise my report may not be printed for a year or more, if at all. The collection sent via F[or]t Ellis ought to reach Omaha before us, and I shall have all the balance with me so that I can quickly sort up on reaching Davenport & send you a complete set. . . . So must again close, after seeing the Geysers shall be anxious to push for home.

We find the atmosphere of the Sulphur Spring district *quite debilitating.* Not good for invalids! The Lake & Falls are good in their way, but the Park as a whole I regard *a failure....* I am getting Allium for Watson.

When, however, Parry wrote his article for the *American Naturalist*[28] on arriving home, he was more lenient with Hayden's reports. "The very full botanical list contained in Hayden's Reports for 1871-2," he said, "includes most of the plants met with in the Upper Yellowstone basin, being comprised within the limits of the Yellowstone National Park." Near the falls of the Yellowstone, he discovered a dense sub-aquatic growth in a muddy pond which, upon Engelmann's later study, proved to be an Isoetes—*Isoetes Bolanderi var. Parryi.* The azure blue blossoms of *Gentiana detonsa* Fries and the brilliant colors of *Pentstemon secundiflorius* Benth. also came in for more praise than Parry described in his letter to Gray.

After the expedition left the head of Yellowstone Lake on the south, they passed by "an almost insensible" grade to a branch of the Snake River. Going along the irregular mountain ranges, they came into view of the Grand Tetons, from which they proceeded to a low divide at the head of Wind River. Near the summit of the peak overlooking Snake and Wind River valleys, Parry discovered *Draba ventosa,* another new species. But one phenomenon interested Captain Jones more—the discovery of "Two Ocean Water" or "Two Ocean Pass." Jones wrote:

I ... for a moment could scarcely believe my eyes. It seemed as if the stream was running up over this divide and down into the Yellowstone behind us. A hasty examination in the face of the driving storm revealed a phenomenon less startling perhaps but still of remarkable interest. A small stream coming down from the mountains to our left I found separating its waters in the meadow where we stood, sending one portion into the stream ahead of us, and the other into the one behind us—the one following its destiny through the Snake and Columbia Rivers back to its home in the Pacific; the other, through the Yellowstone and Missouri seeking the foreign waters of the Atlantic by one of the longest voyages known to running water. On the Snake River side of the divide the stream becomes comparatively large at once being fed by many springs, and a great deal of marsh.

The expedition went rapidly down the valley of Wind River, reaching Camp Brown on September 12 after two months absence. Parry's catalogue of plants published both in *The American Naturalist* and in Captain Jones's report in 1874 enumerated 311 plants.

Parry immediately wrote Gray of his intention to proceed to Empire, Colorado, where he expected to join Mrs. Parry and do some collecting. "On the whole," he said, "I am well pleased with the season's work. Of all the choice & desirable things I shall have plenty for distribution...."

[28] VIII (1874), pp. 9-211, Appendix; Description of New Species.

On September 20 he reached Denver, met Mrs. Parry, and wrote Gray again, saying, "I shall stay as long as the season warrants." He sent on many roots, seeds, and plants, some for Backhouse, an Englishman and a friend of Gray. But whether he met Edward Lee Greene is not definitely known.

Greene had been collecting that year near Pueblo. During the winter of 1871 and 1872 he had made what he termed a "botanical ramble" to western Wyoming, but spent much of his time in the vicinity of Cheyenne. In May, June, and July he was in the field searching for plants and making observations concerning the comparative floras of Wyoming and Colorado, some of which were published in an article in *The American Naturalist*.[29] Greene was one of the most interesting personalities of the first great period of North American botany and, at the same time, one of the most difficult to comprehend. Debate will probably go on for years as to whether he was a truly great man. Certainly he was fearless. And he made people think during his time. Beyond doubt he hastened the settlement of the rules of nomenclature. In North America Greene was among the first to insist on the study of plants *in the field*. He definitely stood for scientific independence and freedom to work in science as one saw fit.

He was born in Hopkinsville, Rhode Island, but when twelve years of age his parents moved to Illinois and then to Wisconsin. At the age of twenty-three he was graduated from Albion College and four years later after a residence at Decatur, Illinois, moved to Colorado. In Volume III of *The American Naturalist*,[30] there appeared a short article by him on "The Botany of Central Illinois," an article of no especial promise and exhibiting little but an inquiring and observing mind concerning the science.

Greene[31] was for two years—1862-1864—a private in the Thirteenth Wisconsin Infantry. While on guard and doing patrol duties of one sort or another, as opportunity permitted, he studied Alphonso Wood's *Class Book of Botany*, and succeeded in botanizing in Kentucky, Tennessee, and Alabama. His early knowledge in botany, moreover, had been amplified while a student at Albion Academy, said to have been an institution of collegiate standing then. Not far from Albion, near Lake Koshkonong, lived Thure Kumlien, "a naturalist of the old school," with whom Greene studied and was to maintain a correspondence lasting for twenty-six years.[32] He taught for a few years after graduation but

29 VIII, pp. 31, 210.   30 Page 6.
31 *The Dictionary of American Biography*, VII, pp. 564-565.
32 See Mrs. Angelia Kumlien Main, "Life and Letters of Edward Lee Greene," *Transactions of Wisconsin Academy of Arts and Sciences*, XXIV, pp. 147-185, particularly from p. 179.

kept up an interest in botany which brought him an instructorship in botany during the years 1871-1872 at Jarvis Hall, an Episcopal seminary at Golden City, Colorado. In 1873 he was ordained and his first congregations were at Pueblo, Colorado; Vallejo, California; Yreka, Siskiyou County, California, and Georgetown, Colorado. The missionary activities incident to his work sent him over areas in Wyoming and Colorado, and later were to send him into New Mexico and Arizona. On September 3, 1874, from Vallejo, California, he was to write Ludwig Kumlien, son of Thure:

What a time I have had since I saw you, ranging over the wonderful plains of Wyoming and northern Colorado, and the cactus desert, even away down to the borders of New Mexico; and climbing about the perpetual snows of the Rocky Mountains, up to the altitude of 14,245 feet; and now at last have been six months on the Pacific coast. I have naturally added somewhat to my knowledge of botany and have a splendid collection of western plants; and have been so fortunate as to discover a few new species, and have rediscovered a number of long lost ones, that had been found only by such early botanists as Nuttall, James, etc.

Greene had faults which increased in intensity as he grew older. He had a rather egocentric nature which manifested itself in tendencies toward quarreling. Intrepidly he was unafraid to stand alone. In 1875 he quarreled with the bishop to whose diocese he was going—northern California. And with Asa Gray he soon developed a controversial spirit, accusing the latter of loose and careless handling of western genera and species. But Greene acquired during his life an unrivaled familiarity with west North American plants. And for by far the most part a hearty attitude of friendship toward botanists was maintained by him. On September 26, 1872, he had written Thure Kumlien[33] concerning his meeting of Gray and Torrey in Colorado:

Yes, I had the honor of meeting the great Prof[essor] Gray; and certainly he is one of the most delightful men I ever saw. I went to the mountains ninety miles away to Empire City near Grays Peak. . . . I met him at Empire City on Saturday evening Aug[ust] 10th just at dark. Was presented to him by the celebrated Dr. Parry, of fame as a collector in Mexico, California, and Colorado. . . . Well, on Monday the 12th Dr. Gray, Dr. Parry, and myself, with some unscientific gentlemen, made the trip to the top of Parrys Peak, and made our first tramp together, collecting no end of fine things. . . .

Wednesday and Thursday were occupied in making the ascent and descent of Grays Peak, the highest mountain this side of the Parks. Mt. Gray is 14,245 feet high, and the view from the top is probably one of the most magnificent which this world affords. . . .

Dr. Parry is one of those quiet, diffident men who knows a great deal more than one would think. Gray regards him as a great botanist though he has never published anything.

[33] Sometimes spelled Kumlein.

Well, I have seen more than Prof[essor] Gray, for last week I was down to Denver for a few days; at the depot one morning I noticed on the platform waiting for the train a very aged gentleman, whom I took to be a botanist when I saw among his hand baggage an unmistakable bundle of specimens in press. I stepped up and begged his pardon for asking his name. Judge of my delight when he replied, "I am called Dr. Torrey." . . . He was on his way to the mountains to visit Grays Peak. I was not able to go with him though he urged me. I had never corresponded with him but he knew me and we were as familiar friends from the moment I gave him my name.

. . . Yes I have certainly had a rich delight in seeing these three great botanists.[34]

Parry, with his good nature and never ceasing interest in botany, never had troubles of any magnitude with Greene. They botanized together in the Rocky Mountains and a friendship subsisting to the end of Parry's life was established. Doubtless, when Parry returned from his northwestern Wyoming reconnaissance, he met Greene for at least a while and they probably did some botanizing together then also. On October 14, 1873, Engelmann inquired of Parry, "Have you seen Greene? He is doing well in Botany—zeal and intelligence." Engelmann usually thought well of people. Aside from his family, however, his greatest personal interest was in Parry. Especially he admired Parry's accomplishments in Colorado. March 15 of that year he had written him:

Fremont at least will have to knock under (as he, poor rascal, seems to do every where now—no, rascal, is too severe—poor weak humanity!) Benton said of him that he had made so many more hypsometrical observations than Humboldt—and you will have made so many more than Fremont—so the conclusion is evident, and I am too modest to heap encomiums on you—In fact I follow Benton's example, who left his hearer, Dr. Wislizenus, to form his own conclusions.

Parry had measured the height of several mountains of Colorado and planned others he would figure scientifically. Engelmann realized that Parry was becoming a great pathfinder and scientific observer. Some of Greene's specimens sent from Canon City, with queries, made Engelmann realize the more that collaborations of Parry and Greene would contribute much to North American botany. But Parry had other plans.

[34] See an able and interesting address on "The Botanical Work of Edward Lee Greene," by Harley Harris Bartlett, *Torreya*, XVI (July 1916), pp. 151-175, in which early correspondence with Gray, Olney, Engelmann, and other leaders is given; also Willis Linn Jepson's "Edward Lee Greene, the Man and Botanist," *Newman Hall Review*, October 1918.

CHAPTER III

Other Western Explorations in
North America. Canada

P ARRY planned to go in 1874 to St. George in southwestern Utah. This was a district visited by few naturalists and, with the exception of Dr. Edward Palmer, by no botanical collector. Palmer in 1870, while in the joint service of the Department of Agriculture and the Smithsonian Institution, and touring from Salt Lake City to the mouth of the Colorado River and the Pacific coast, had spent three weeks near St. George and collected a number of new species of plants which Watson had described in his *Botany of the 40th Parallel*. The Wheeler surveys of 1871-1872 and that of Major Powell, ending at Kanab, after exploring the Colorado River, had touched portions of the district and their discoveries had also been published by Watson in *The American Naturalist*.[1] Two collectors of lesser importance, A. L. Siler and J. E. Johnson, had visited the region. But no thorough investigation had been made of this important botanical area.

On December 7, 1873, Engelmann wrote Parry that his planned trip "to St. George is tempting indeed—and therefore a tramp among the Apaches (with or without scalp) and Yuccas and Agaves and Cacti. But—!" Engelmann could not accompany him. He asked Parry to keep an eye on junipers for him and observed that St. George was in the southwestern corner of the Territory, twelve miles from the southern and twenty-five miles from the western lines, and not far from the Rio Virgin, an area having "names familiar through Fremont's tramp 31 years ago." Engelmann hoped that Parry would find nice things and other agaves in the canyons of the Colorado. He wondered if *Quercus Emoryi* grew in Utah. Engelmann told Parry he himself might go to Colorado in June or July but "Colorado without Parry as guide" disturbed him. However, he admonished Parry, "I shall go with my wife, you must know, if I go at all." He wanted full particulars on specimens from Parry—"o, I get mad at you," he exclaimed, "you do not tell me any thing, must pump! pump!! pump!!!" for information. "And what does the great traveller Dr. C. C. Parry say? Echo—what?" He cautioned Parry that the "unbotanical collectors sometimes hit things that our 'botanists' have missed vide Mrs. Millington, Dr. Wislizenus, Dr. Gregg and others. They do not take things for granted—for old and

[1] VII, pp. 299-303.

stale—*everything* is new and interesting to them." Engelmann had found another good friend in Arizona at Camp Apache, Dr. Girard, who had made his "thirsty soul happy . . . by sending another fresh, mature Juniper, which," he wrote, "was not unknown to me in Utah, Nevada and Arizona, but not well understood. . . . I am inclined to take it for a new species. Look out for it in St. George. . . ."

Parry arrived at Salt Lake City in March to find a snowfall of nearly two feet which would make the 350 mile journey to St. George difficult. By April, however, after he had passed over the rim of the Great Basin and was in the valley of the Virgin, "the whole floral aspect assumed a change almost magical; orchards in full bloom including peach, almond, and apricot, marked at a distance by a perfect blaze of blossoms." The "lucerne fields with their deep green foliage were nearly ready for a first forage crop" and over "the intervening desert table-land the aspects of advanced spring were evidenced in rainbow-colored patches of *Phacelia Fremontii* Torr. and bright yellow clusters of *Eunanus Bigelovii* Gray."[2]

At first there were many species of Phacelia and Gilia in evidence and many dwarf forms of spring flora. Later a different class of annuals, largely represented by Eriogonum and Borraginaceae "came forward to continue the series of evanescent forms." Parry's attention was called to the bushy shrub known to the inhabitants as "wild almond."

On April 12 from St. George, Utah, Parry wrote Gray:

I have been more than a week here and have made a fair reconnaissance of the adjoining hills & ravines. I shall soon be ready to move *in force* and take possession of the many nice things coming to view. Every ramble turns up something I have not seen before. . . . I am now giving special attention to the evanescent annuals including a number of Hydrophyllaceae. . . . I hope this week to make a trip 20 miles to the 1st outcrop of *Yucca brevifolia* Engl. called by the Mormons "*Joshua*" [I]t is now in flower. . . . I am quite comfortably fixed, have a working room & bunk in Mr. Johnson's library, and an excellent neat boarding house near by. The Mormons are inclined to be *suspicious*, but will get over that when they find I am not going to trouble their wives or churches. I heard Brigham Young hold forth to the faithful. [H]e has now left for Salt Lake, this being his winter quarters.

I want you to look sharp for a nice *new genus* for Canby, as nice (if possible) as the man himself. . . . The weather is delightfully pleasant now, will be hot enough soon, great prospect for fruit including everything that grows from *Apple* to *fig*. What a charming tree the *Almond* is.

On May 23 he wrote again, telling of the increased heat, but saying of the discovery of a new Gilia and several new species of Oenothera which bothered him. Among the latter was a large flowered specimen

2 C. C. Parry, "Botanical Observations in Southern Utah," *The American Naturalist*, IX, pp. 14-346.

which Parry dedicated to his friend, J. E. Johnson, *Oenothera John-sonii*. Another with small yellow flowers was named by Watson *Oenothera Parryi*. Moreover, cacti—perfect masses of delicate pink rosettes, set in beds of spines—were a season's attraction.

Early in May Parry had gone to the mountain range which separates the valleys of the Santa Clara and Muddy rivers—the Beaver Dam Mountains. The desert flowering willow was abundant and over *Cowania mexicana* was "a profusion of pure white flowers, almost hiding from view [Cowania's] finely divided varnished leaves. A pleasant balsamic fragrance, exhaled in the clear atmosphere from this charming shrub, lent," Parry said, "additional attractions which seemed to be appreciated by a swarm of hovering insects." Near the close of day, Parry sighted one of the principal objects of the journey— *Yucca brevifolia*.

Engelmann had asked Parry concerning this plant. "The only question to solve about the Joshua," he wrote, "is the fruit. Is it erect or pendulous? how does it appear before maturity—how when fully ripe?" When Parry sent on specimens, Engelmann jestingly replied, "A few days ago your letter and the box purporting to contain 'Yucca angustifolia . . .' arrived—safely—i.e. letter and box safely, but contents were fully transformed—whether this Yucca is another insect feeding vegetable or whether the insect has eaten up the Yucca is uncertain—as there is nothing but a stinking putrid slime. You have seen or heard of the insect eating Dionaea, Drosera, and Sarracenia, which Gray is popularizing now in the N[ew] Y[ork] Nation and N[ew] Y[ork] Tribune—publishing Dr. Mellichamp's observations—So you see I am now somewhat Insectivorous."

In June Parry wrote Gray that "the Flora has seemed to have *dropped out* in this section, very few new things coming on. I have therefore in a measure dropped the *portfolio* and taken to my *seed bags*. I expect next week to make a trip to the M[oun]t[ain]s on my return here in case the prospect does not improve I shall pack up my duds & move northwards stopping by the way at any desirable points. I have so far about 200 good species. . . . Dr. Engelmann reports on the Echinocactus? sent as *Mamillaria Arizonica* var. *Chloranthe* a n[ew] sp[ecies]. . . ." During the month Parry found an Eriogonum, later named by Gray *E. Parryi*, an Asclepiad, named by Engelmann, *Astephanus Utahensis, Petalonyx Parryi*, Gray, and after leaving St. George, *Gaillardia acaulis*, and other new species. Southwestern Utah explorations were concluded with a visit to Pine Valley and the adjoining mountain districts around Pine Mountain to add alpine floral collections.

Parry went north the last of June to Cedar City where the flora proved scanty and so he turned his attention "to the high Mountain range of the Wahsatch, rising abruptly to the East, and overlooking the southern extension of the great interior basin." There the botany proved "similar to other elevated pastoral districts in the interior West" and after spending a few enjoyable days in the crude homes of the hospitable sheepherders there, he returned to Davenport where soon his collections came through safely and in excellent condition. Within two weeks he went west again—to Empire City, Colorado—and joined Dr. and Mrs. Engelmann who were there. On September 6 he wrote Gray:

I have been here now over 2 weeks & have commenced collecting seeds. [F]rosts hold off longer than usual, but I shall soon have to put into root digging. I have been on the range twice but it has been too stormy to collect much. In company with Dr E[ngelmann] & wife & Mrs. Parry we made an *un*successful attempt on Parry's Peak, were caught in a snow storm in Berthouds Pass, & had to beat a retreat. [O]n the way home Mrs. E[ngelmann] had the misfortune to fall from her horse and strained her left hand, is otherwise uninjured. Fine fall weather has now set in and we still hope to make the peak. The accident to Mrs. E[ngelmann] will delay their leaving for a week or more. A wagon road is now in process of construction via Berthoud Pass to Middle Park. [I]t is nearly finished to the pass, a log cabin will be built in the pass for the accommodation of travellers, so that Parrys Peak can be easily reached. Dr. E[ngelmann] keeps busy with *pines*, and with his singular taste for ugly spiney things has tackled Cirsium, making dissections & fresh drawings. . . . I shall be anxious to get back to my Utah collection. . . . I shall be able to make up near 20 sets of Utah & Wyoming plants, excluding yours. [O]f these are already bespoke, viz D. C. Eaton, Canby, Redfield, Torrey Herb[arium], Crooke, myself. So there will be about 12 more to be disposed of . . . at the usual price $10.

Parry again wrote on September 26, from Denver, sending plants for the botanic garden at Cambridge, seeds of *Abies Engelmanni* and *Abies Menziesii*, and informing Gray that Dr. and Mrs. Engelmann were at Colorado Springs but would join the Parrys for the trip east the next day. Parry wanted Gray to send him a Raspail or student's microscope for his fall and winter work. Two days later Engelmann wrote, telling Gray of the "beautiful gorge or cañon of Glen Eyrie near Colorado Springs" where he met his old friend *Abies concolor*, and saw it again in Ute Pass. He had spent three pleasant days at Manitou and the canyons near but "Alas the happy days of wild mountain and pleasant valley life are passed," he wrote, "and eastward the foot is turned." Engelmann had spent three days with Brandegee, a

Yale College man, and pupil of William H. Brewer. ". . . a good and very obliging fellow, to whom I believe my 3 days company did a great deal of good and gave much encouragement," commented Engelmann. This was Thomas S. Brandegee, afterward an able botanist of western regions but then a county engineer. His *Flora of Southwestern Colorado*, concerned with vegetation of the San Juan, Mesa Verde, and near by mountainous regions, was published in 1876.[3]

Engelmann's respect for Parry was not impaired. Parry "seems to hold a sort of supervision over that part of the Rocky Mountains" where they visited, he said. Engelmann's studies of junipers, oaks, and pines were much accelerated by his visit to the regions. With Parry's Utah yuccas, agaves, and Cactaceae, in addition, he had enough to do for the winter. Always he studied geographic distributions of the plant families. When Parry went to Wyoming he asked Parry to study the pines there. However, after his return, he wrote Parry, "More doubtful than your plants are specimens from the British Boundary in Patterson's Exp[edition], and sp[ecimens] from still farther North—Arctic, which may be real *alba*."

Sometimes Engelmann let plants lie for years before examining or naming them. He had some of Lindheimer's 1849-1851 collections still undisturbed. He enjoyed Mexican plants but even these in some instances went unexamined and unnamed—for example, some plants collected by Gregg. Engelmann welcomed the Alaska plants which arrived from Gray with a set of Hall's Oregon plants. Soon Hall's catalogue of Texas plants was also received. These all stirred Engelmann. Would that he had had more time! Not unusual with industrious men, the more Engelmann had to do the happier he seemed to be. His botanical correspondence was dwindling. Bolander was "mum," Lindheimer old, Torrey and Sullivant gone. George, his son, also a doctor, had begun to relieve him of some of his medical practice and responsibilities. But there was so much that he saw yet to be done. He was working some with grapes again. For years Edward Palmer had been sending him yuccas, agaves, and Cactaceae, collected in western regions, particularly, in Arizona, southern Utah, Lower California, and California.

On February 19, 1875, Engelmann wrote Parry, "I have got Rothrock's things—nice—some Pines and Oaks, or rather 2 pines and 1 oak that none of you ever got in Arizona! And settling some Agaves—

[3] Department of the Interior. *Bulletin of the United States Geological and Geographical Survey of the Territories*. F. V. Hayden, U.S. Geologist-in-charge. Volume II, Number 3 (Washington: Government Printing Office).

Parryi in bloom and fruit, and leaf! Also *Gentiana Wislizeni*, a *Pinus Chihuahuaensis*, but that has already been found by Wright in Arizona. . . ." At first Engelmann was inclined to criticize Rothrock's collections severely but Rothrock took his criticisms good naturedly, making "excuses for *haste*." Professor Cope believed the collections, however, "fine," and by March Engelmann wrote Gray, "Rothrock has done well. . . . I must admit it under protest. He got Oaks and Pines new to our Flora. Rothrock has also settled some of my Agave muddle. . . ."

Porter had written Engelmann that he had found a locality for *Abies concolor* discovered since Engelmann's "sojourn" in Colorado. This, he regarded, as "connecting the Pikes Peak with the New Mexican localities—no doubt a common plant!!" Engelmann added, "But as to *grandis?* The whole thing must be revised. . . ." When Rothrock went on to California, Engelmann wrote Parry on June 12, 1875: "Rothrock, whose headquarters are at Los Angeles this season, must try to trace it, *concolor*, from the San Francisco Mountains, where I know it to grow farther west—and Prof[essor] Dawson of Montreal will have to trace *A. balsamea* on the 49° lat[itude] westward to find how it becomes *A. grandis*—as well as *alba* & *Engelmanni*. I have written to him about it." So continued Engelmann's exhaustive researches in the geographical distributions of the plant families in which he took an interest. He refused more work many times—Watson wanted him to work up the poplars of the West—but while he would not admit he was too old, he argued a lack of time and books, and the abundance of materials he had on hand. He told Parry that he would have to run away from yuccas, "if they continue to disturb me thus! I fear I shall dream of them: just think of their bayonets attacking me from every side, and their big pulpy fruits choking me! Ugh!"

In the autumn of 1873 Engelmann and Parry had been surprised to learn news concerning Edward Palmer. The rumor circulated that Palmer had been dismissed from the Smithsonian Institution's service, although evidently he retained an employment with the Department of Agriculture. Probably the story was unfounded. Professional botanical collectors were none too plentiful and the demand for their services was extensive. In every branch of science, institutions were still gathering much valuable fundamental taxonomic data and materials. Although Palmer was not in the field as many years as Parry, in many particulars his contributions were great. Palmer's passion was collecting. Like Parry, a medical doctor by profession, he appreciated the

value of his work and took an almost unimaginatively serious pride
in it. His health required out-of-doors activity much of the time. He
had a love of plants and he gave much time to their study in the field
and in cultivation.

Palmer's most important work had begun in 1869 when the United
States commissioner of agriculture had sent him to New Mexico and
Arizona to report on agricultural resources, commercial products, and
matters of climate and soil fertility in those regions. During that year
and the next he had collected plants, gathering also valuable information concerning the lives and habits of various Indian tribes, and before
his return to Washington he had explored wide areas of what is now
southwestern United States, Lower California, and northern Mexico.
Sometimes accompanied by military escort, he had made long and
difficult journeys across deserts and mountainous regions—exploring
some localities botanically for the first time—and although in some
instances his specimens were found scanty, his plant collections were
of necessity light since food, arms, blankets, and medicines weighed
down their packs.

In 1871—the year of Parry's dismissal as botanist of the Department
of Agriculture—Palmer spent much time in Washington. He regretted
Parry's leaving, visited Torrey and Gray on his behalf, but, when
Vasey was appointed, went to work with him and they got along "on
the best of terms." The west, however, appealed to Palmer and he
yearned to return to Arizona. Apparently he sought an appointment
as an Indian agent there and, being deprived of this, went to Woods
Hole where he assisted the commissioner of fish and fisheries in the
collection of marine invertebrates and algae along the New England
coast. Parts of these years were spent with his plants in Washington.
But during collecting seasons he pursued his work along the Atlantic
coast until, at Gray's suggestion, he went to Florida and the Bahama
Islands to collect. Daniel Cady Eaton published in 1873 a *List of
Marine Algae Collected Near Eastport, Maine, in August and September, 1872, in Connection with the Work of the United States Fish
Commission* under Professor S. F. Baird, and in 1875 *A List of the
Marine Algae Collected by Dr. Edward Palmer on the Coast of
Florida and at Nassau, Bahama Islands, March-August, 1874.* That the
rumor of Palmer's dismissal from the Smithsonian Institution's service
in 1873 was probably unfounded is borne out by a letter written October
12 of that year to Vasey: "I shall not return to Washington this winter,
as Prof. Agassiz has offered me a better position. . . ."

Palmer furnished Engelmann valuable materials from his Southwest

explorations.[4] Parry thought well of Palmer. Indeed, Torrey compli-
mented him highly, saying he had "done great service to North
American botany." Torrey, Gray, and Engelmann, however, looked
primarily to Parry who wrote May 10, 1875:

> My present plans are not materially changed. Go out in June to the Wahsatch
> range in Utah. Then in the fall to St George to winter. Early next spring to the
> Lower Colorado. . . .
> I found the Torr[ey] Herb[arium] in a very unsatisfactory condition. [N]o
> conveniences for study, and I was there three times on rainy days to find it *locked
> up* & Leroy not there!

While east Parry conferred with George Thurber, an authority on
grasses and erstwhile associate on the Mexican Boundary Survey; with
Eaton at New Haven; and with Gray at Cambridge. By July 3, however, Parry was at Spring Lake, Utah, and he wrote Gray:

> I have at last reached the expected field of my season's work, and [am] ready to
> take hold in earnest. I think the prospects are reasonably fair of doing at least
> something in the way of botanical discovery at least in the high M[oun]t[ain]
> districts.
> I have made one considerable climb, far enough to reach a snow bank, but the
> M[oun]t[ain] flora is meagre compared with Colorado. Among other things I
> found the enclosed Viola. . . .
> I have made some enquiries as to the chances of securing Indian implements,
> Mound relics &c&c and may possibly be able to. . . .
> Please say to Mr. Sargent that the Abies question for this season is easily, if not
> satisfactorily solved. Passing the other day through a forest of *Abies concolor, not
> a single cone could be seen*. So that my surmizes were correct that this would be the
> *Off year* for *conifers*, at least as far as the Abies section is concerned. I now feel some
> curiosity to *settle* whether (as I have some time suspected) the *Abies grandis* of the
> Cal[iforni]a botanists is not our *A. concolor* of the Wahsatch? I find here again the
> true *A. grandis* of Colorado, but confined to the highest ridges. Please keep me
> posted on Botanical matters. . . .
> This is rather a pleasant locality. [W]e have our headquarters in an extensive
> fruit orchard & garden—with trees loaded down with apricots, peaches, apples &
> plums &c&c Duncan Putnam is with me, but a man *wreck* passing rapidly through
> the early stages of consumption.

Again Parry wrote Gray in July, saying that he was "pitching in"
to botany in a quiet way and had found some new and more old
things. He expected to make a trip to Salt Lake, returning by way of
the American Fork Canyon. Though he had found what he believed
a new species of "a pinnate Aspidium" and some eastern aquatic

---

[4] See William H. Safford's study of Palmer, *Popular Science Monthly*, LXXVIII (April 1911),
pp. 341-354. For an exemplary study of parts of Palmer's collections, see Rogers McVaugh's
and Thomas H. Kearney's article, "Edward Palmer's Collections in Arizona in 1869, 1876, and
1877," *American Midland Naturalist*, XXIX (May 1943), 3, pp. 768-778.

plants not credited to the section and though on the whole he was disappointed in the mountain flora, he had decided to stay till October
and then "put out for California." He planned to write an article on
"Summer Botanizing in the Wahsatch."[5]

By August he had returned from Salt Lake, had gone to Mount
Nebo, had met Captain F. M. Bishop, the first collector in southern
Utah, and had visited the Utah Museum where he found "some
scrappy things." The American Fork Canyon had proved "a wild
rough country but inferior in botanical interest to Mt. Nebo." Near the
foot of the latter mountain he planned to camp till Greene arrived
from Colorado en route to California. There he remained getting up
a collection of dried plants and investigating the alpine slopes of Mount
Nebo. "The most remarkable thing," he wrote, "was to find great
patches of *Primula Parryi* growing on a *dry* rocky slope," and in seed.
Mrs. Parry was to join him about September 1 he later discovered.
But it was not till the last of the month that he decided to go to San
Francisco with his plants, numbering about 100 species, and a collection of Indian things for the ethnological museum. On September 23,
1875, he informed Gray:

Dr. Palmer now in San Diego, talks of coming on here and go down to St
George to winter to dig Indian mounds & go over to Moquis. I have encouraged
his plan, and agreed to put the Indian matters into his hands. I have also suggested that he rig up a conveyance next March & meet me on the Lower Colorado
for a joint botanical collection, I working from California Eastward, then go to
Bill Williams &c. I expect to stay perhaps most of winter in vicinity of San Francisco. [M]ay take quarters at Oakland to have the benefit of University library.
Would it be worth while for you to give me a letter to Pres[iden]t LeConte? No
late word from Greene as to whether he goes to Cal[iforni]a this fall. Of course I
shall be glad to see advanced sheets of [the] Botany of California. . . .

While in Utah, Parry had studied again Watson's *Botany of the 40th
Parallel.*

These,[6] and the *Bibliographical Index to North American Botany;
or Citations of Authorities for All the Recorded Indigenous and Naturalized Species of the Flora of North America, with a Chronological
Arrangement of the Synonymy,* were the great works of Sereno Watson of this period.

In 1860 William Henry Brewer had gone as assistant in the botanical
department of Josiah D. Whitney's geological survey of California and
in its service during a period of four years explored many California

[5] *Proceedings of the Davenport Academy of Science* (1876), pp. 145-152.
[6] Watson's *Botany of California* (at this time, volume I in preparation) and "Botany," in King,
Report Geol. Explor. of 40th Parallel.

regions: Southern California, Los Angeles, Santa Barbara, the Coast Road, Salinas Valley and Monterey, the Mount Diablo Range, Napa Valley, the Sacramento River, Mount Shasta, Fort Tejon, Tehachapi, Walkers Pass, Yosemite, Mono Lake, Aurora, Sonora Pass, Lake Tahoe, Lassen Peak, Siskiyou, the San Joaquin Valley, the High Sierras, and many other localities. With his materials he had returned east by way of Nicaragua to become professor of agriculture in the Sheffield Scientific School at Yale, to share with Samuel William Johnson (honored as one, if not the, "father of scientific agriculture in America") in developing "the first American institution" which had recognized as early as 1846 "the claims of agricultural science," an institution furthermore which had participated notably in the nationwide agricultural movement ostensibly made possible by land-grant funds of the historic Morrill Law of Congress (dated 1862) which had established agricultural instruction in States of the Union. Brewer typified in a sense the early and persistent fundamental relation of botanical study to agricultural instruction, certainly the relation of the botanical taxonomist who enlarged the horizon of plant science study to include the whole of agricultural study. With the presence of Johnson—professor of theoretical and agricultural chemistry—agricultural instruction at Yale had men skilled in laboratory study, both from the standpoint of plant classification and the standpoint of "pure" scientific advancement. Other names loom large in plant study progress at Yale—Daniel Cady Eaton—and, as far as the Connecticut experiment station's history is concerned, Wilbur Olin Atwater. But we must be concerned with Brewer's accomplishments as a botanist.[7]

Collaborating with Watson and Gray, Brewer had begun preparations for the publication of a flora of California. When finally published as the *Botany of California* in two volumes, the first by Watson, Gray, and Brewer appearing in 1876 and the second by Watson in 1880, the work, containing nearly twelve hundred pages, became the standard authority on the botany of the Pacific slope. The work appeared not without leaving a long trail of trials and investigations. At one time the California survey seemed financially "bushed," to use Watson's term. Long years of studied examinations of materials; continuous explorations and searches

[7] See *Up and Down California* (in 1860-1864), *The Journal of William H. Brewer*, as edited by Francis P. Farquhar. New Haven: Yale University Press, 1930, 1931. The quotations are taken from U. P. Hedrick's *A History of Agriculture in the State of New York*, printed for the New York Agricultural Society, 1933, p. 413, and from A. C. True's article, "Origin and Development of Agricultural Experiment Stations in the United States," *Report of the Commissioner of Agriculture* for 1888, Washington, Gov't Print. Off., 1889, p. 541. See also *Letters of Asa Gray, op. cit.*, II, p. 532.

to determine comparative geographical distributions of the plants, especially the trees; large expenditures of money and effort, and many other factors made the results of the work a real achievement. Where possible, all North American botanists aided with the preparations of the work. Engelmann gave years of study to Coniferae and other families involved. Henry N. Bolander, though connected with the state department of instruction, found time to visit unexplored regions such as "Yolo Bolo, a snow-covered mountain of the Coast Ranges near Red Bluff," going also to southern California regions and into the Sierras. But Bolander's usefulness as a botanist was diminishing. He had had to fight politics which, he said, if one is "forced into it, as I was, are sufficient to tire out and crush out all nobler impulses in man." Bolander compiled a *Catalogue of the Plants Growing in the Vicinity of San Francisco* in 1870 and an article on the "Genus Stipa in California" in 1872, published in the *Proceedings of the California Academy of Sciences* and these were among his last important works in botany. When Watson published the second volume of the *Botany of California*, there was contained a "List of persons who have made botanical collections in California" compiled by Brewer. It showed the considerable number who in one way or another had contributed to their great work by explorations and collections. However, when in 1875 Parry went into California for the first time in several years, much botany remained to be done and an able group of botanists arose. Botanical exploration in the main was moving from the interior Territories—Colorado and Utah —to the boundary lands, though instead of going principally to the southwestern parts of the United States and up the Pacific coast, as it had gone years earlier, exploration now was moving northward, northwestward, and into more remote interior western areas.

In 1872 Edward Lee Greene published in *The American Naturalist* articles on "The Alpine Flora of Colorado" and "Irrigation and the Flora of the Plains of Colorado"[8] a subject speculated on by Sereno Watson with respect to the botany of the fortieth parallel. Greene's Colorado works, together with publications of Gray, Engelmann, and Parry on Parry's Colorado collections during the 1860's (including collections made by Hall and Harbour about the same time), the smaller catalogues by Parry and Porter on collections made by Hayden's Colorado surveys, Wolf's Colorado collections published by Rothrock in the Wheeler survey publication, Porter and Coulter's *Synopsis of the Flora of Colorado*, and Brandegee's later *Flora of Southwestern Colorado*[9] constituted for that time a rather thorough system-

---

[8] Pages 734-738; 76-78.   [9] Published by the survey, *op. cit.*

atization of that Territory's botany. Hayden's surveys had contributed much toward a knowledge of Colorado's botany. In 1875 Brandegee went into the mountainous regions of southwestern Colorado with Hayden's party, visiting areas such as Mesa Verde and the San Juan River. Future exploration would bring further knowledge.

What botany of larger areas needed was assimilation in floras of wide, defined range. For instance, no flora of the great Rocky Mountain regions had yet been published. By 1877 Colorado's important land localities, especially the mountainous, were explored by Hayden's survey and on their completion it was determined that the United States geological and geographical surveys should move northward again into Wyoming and Idaho north of the survey of the fortieth parallel—from Fort Steele, Wyoming Territory, to Ogden, Utah, and north to Yellowstone National Park. Gradually, Montana, the Dakota country, Minnesota, Washington Territory, Oregon, and California—all United States boundary lands—became centers of great scientific interest. Hostile Indians were still occasionally being encountered. As late as 1875, exploration in localities remote from settlement had brought forth attacks from the savage Ute Indians. Two Hayden parties had been so encountered. The Jones Wyoming expedition had also met unfriendly Indians, some of whom had stolen horses and equipment. Exploration, therefore, had to move slowly and cautiously when going into areas of no settlement and little or no exploration.

For almost two decades the botany of the upper Missouri River regions had engaged interest. Minnesota was not regarded as part of the domain of Hayden's surveys of the Western Territories although during the year 1865 Hayden as a part of his explorations went to mine regions of Lake Superior. Minnesota had its own geological and natural history survey, supervised in large part by the University of Minnesota. In 1893-1895, as part of the survey's work, Lesquereux's report on Minnesota fossil plants was published as "Cretaceous Fossil Plants from Minnesota."[10] The principal areas of Hayden's surveys were the Territories to the west—the Dakota country and Montana—both Territories south of the British-United States boundary line.

One. of the most interesting facts concerning development of knowledge of western American botany was that early exploration went, in large part, to the far west, the southwest, the far northwest, and even arctic regions before it came thoroughly to the interior west of North America. Since the beginning of the century, some botany of remote

---

[10] Minn. Geol. and Nat. Hist. Surv. 1872-1901, *Geol. of Minn.*, III, pp. 1-22. It is possible these determinations, at least part of them, were made by Lesquereux as early as 1875 and were published posthumously.

interior regions had been studied.[11] The Lewis and Clarke expedition returned with some plants; and famed explorers such as Thomas Nuttall and others occasionally had made extensive collections. Transportation difficulties and lack of settlers kept scientific investigation on a large scale to the south in Rocky Mountain regions. True, J. N. Nicollet's expedition up the Missouri in 1839 to Fort Pierre had collected fossils from Cretaceous formations. Even Prince Maximilian of Neuwied had told of certain well known Cretaceous fossils observed on his journey. Early explorers, however, though reporting presence of lignites and other formations on the Missouri above Fort Clarke, failed to recognize variations in land deposits, and proof of great lignitic and fresh water beds along the upper Missouri was long coming, especially from the interior. It was a while before the great North American paleobotanists, Lesquereux and John Strong Newberry, received the rare evidence of American geological history revealed by abundant stores found in regions of this famous river and its tributaries. Very interesting is this as the regions since have proved fertile.

Lesquereux never went to the interior northern territorial regions although, it is said, in 1856 he explored in the southwestern portions of Minnesota near New Ulm and must have investigated extensively along the Minnesota River.

Newberry was much more of an explorer although for botany even more than paleobotany. In 1866 he had been appointed professor of geology and paleontology in the then recently founded school of mines of Columbia College. He had served as surgeon and naturalist of the expedition of Lieutenant Joseph Christmas Ives which explored a substantial part of the Colorado River of the West, had studied its geology, and later in 1859 accompanied the exploring party of Captain J. N. Macomb which went over the region from Santa Fe to the junction of the Grand and Green rivers of the Colorado of the West. Included in his report were descriptions of "a large number of Triassic plants."[12] Earlier, on one of the notable Pacific Railroad surveys he had served as botanist on the Williamson and Abbott expedition which explored northern California and the Oregon Territory from San Francisco to the Columbia River. In 1869, the same year he accepted the directorship of the

---

[11] Account based in part on introduction to *Report on the Geology and Resources of the Black Hills of Dakota*, by Henry Newton and Walter P. Jenney (Washington: Government Printing Office, 1880), pp. 5 ff.

[12] See an able biographical sketch of Newberry by Nathaniel Lord Britton in *Bulletin of the Torrey Botanical Club*, XX, Number 3 (March 1893), pp. 89-98. See also a recent article by A. E. Waller, "The Breadth of Vision of Dr. John Strong Newberry," *Ohio State Archaeological and Historical Quarterly* (October 1943), pp. 324-346.

geological survey of Ohio, his once fellow citizen of Ohio, Lesquereux, sold a collection of more than 4,000 fossil specimens to the museum of comparative zoology of Harvard—a collection which Lesquereux described as the "typical specimens of most of our American species" and "some of them of great value." Lesquereux and Newberry were never competitors. Had they been, there is much reason for believing that Newberry would have become America's great paleobotanist as at first most materials were sent to him rather than Lesquereux. Nevertheless, they differed in scientific findings involving western materials mostly and never so far as is known actively collaborated. Lesquereux was poor; Newberry was influential; an unfortunate controversy developed and at length Lesquereux possessed most of the field.

But Newberry did much for early American paleobotany. Not all, but most, of early fossil plant discoveries in northern territorial regions was sent to him—including plants from Raynolds and Hayden's exploration of the Missouri and Yellowstone rivers; from the northwestern boundary commission's collection made by George Gibbs on Vancouver's Island, Orcas Island, and on the coast of Washington Territory at Bellingham Bay and other places; from the important Fort Union flora of Montana and an area extending into Canada, Wyoming, and the Dakotas; from Oregon, Colorado, and other regions. James D. Dana of the United States Exploring Expedition had found some plants in Washington Territory and described them. Dr. John Evans, United States geologist of the Territory of Oregon, had sent a collection to Lesquereux who published his conclusions as to its materials in *The American Journal of Science and Arts*,[13] "Species of Fossil Plants collected . . . at Nanaimo (Vancouver Island) and at Bellingham [B]ay, Washington Territory," along with fossil plant species collected near Somerville, Fayette County, Tennessee, by J. M. Safford, and "Fossil Leaves collected in the Chalky banks of the Mississippi River near Columbus, Kentucky, by Dr. D. Dale Owen and L. Lesquereux."

Nevertheless, Newberry's studies of northern Territory materials and his several years of "study of the geology of the interior of the continent exploring a large area . . . in Kansas, Colorado, Arizona, New Mexico, and Utah," established his work as authoritative. His early studies, "Notes on the Later Extinct Floras of North America,"[14] "The Ancient Lakes of Western America, Their Deposits and Drainage," his publication of the Gibbs collection in the *Boston Journal of Natural History*,[15]

---

[13] XXVII (2nd ser., 1859), p. 359.
[14] *Annals of the Lyceum of Natural History*, IX (April 1868) and reprinted.
[15] See *Proc. Boston Soc. of Nat. Hist.*, IX, p. 160 (1862, 1863, and reprint, 1863).

caught Gray's admiring interest as works of a real American scholar. Later Newberry turned his energies to the important industrial subjects of mining and metallurgy but, notwithstanding, found time to do studies on the geology and botany of the Northern Pacific Railroad country, on New Jersey and Connecticut fossil floras, and even of the flora of the Great Falls coal field of Montana.

It was Hayden who did much to shift the center of paleobotanic interest from Newberry to Lesquereux—not so much by reason of his own explorations in northern interior Territories, as extensive as these were,[16] but by his employment of Lesquereux to systematize abundant fossil material obtained by the United States Geological and Geographical Survey of the Territories. When in 1853, under patronage of James Hall, New York state paleontologist, Hayden had gone with F. B. Meek to the Bad Lands of the Dakota region, the memoir issuing from discoveries of fossil invertebrate forms showed "for the first time the order of succession of the different beds of the Cretaceous in the Upper Missouri Region." In 1854 Hayden returned to the region (for the American Fur Company)—this time for the most part by himself and without aid— and went far into Montana and the Yellowstone regions, often going by foot. However, from the standpoint of paleobotany, his important exploration was under General Warren to country bordering the upper Missouri and continued several years. Of the 1855 expedition, Newberry wrote: "In the great mass of interesting materials brought by Dr. Hayden, were a number of angiospermous leaves obtained from a red sandstone lying at the base of the Cretaceous formation at Blackbird Hill, in Nebraska."

In 1858, accompanied by Meek, Hayden again went to Nebraska and Kansas to collect. Apparently during these years he turned almost exclusively to Newberry. He did not participate in Lesquereux's and Newberry's controversies—as to whether certain western materials were Cretaceous or Tertiary—but, although his collections from Yellowstone River tributaries and mountains under General Raynolds went to Newberry for systematization, within a little more than a decade most, if not all, of Hayden's collections were going to Lesquereux.

Under the United States Geological and Geographical Survey, explorers went to survey as well as explore. They went into interior lands, not river margin areas for the most part. Remote lands such as those deep in

---

[16] See Henry Newton and Walter P. Jenney, *Report on the Geology and Resources of the Black Hills of Dakota*, where introductory remarks beginning on page 5 give an excellent account of Hayden's northern interior territorial explorations (Washington: Government Printing Office, 1880). See also account of Hayden in *Dict. Amer. Biog.*, VIII, pp. 438-440, published by Chas. Scribner's Sons.

the Black Hills were unexplored scientifically. Because of their danger
and inaccessibility, it took many years before science began publishing
elaborate observations and enumerations of the fossil floras of these
regions. Although during the 1870's publications increased, such tardi-
ness, moreover, was similarly true of publications of any botany.

In the summer of 1874, Lieutenant Colonel G. A. Custer was ordered
by government authorities to assemble an exploring party at Fort Abra-
ham Lincoln, Dakota Territory (located now near Bismarck, North
Dakota), to reconnoiter a route from there to Bear Butte in the Black
Hills and explore the country south, southeast, and southwest of that
point. Colonel Custer proceeded through Red Water Valley and past
Sun Dance Hills and, entering the hills beyond, passed along Floral
Valley and Castle Creek, camping on French Creek near the lower canyon. In the course of their explorations they ascended Harneys Peak,
located in what is now western South Dakota, and visited numerous
other unexplored points of interest. Probably the most important part
of the exploration was between Inya Kara Mountain (Wyoming) and
Harneys Peak—"the first expedition that had ever penetrated the fastnesses of the Black Hills." Accompanying Colonel Custer was Professor
A. B. Donaldson who collected botany—a hastily gathered collection
and as a consequence meager—in all, amounting to about eighty species
of plants. The plants were forwarded to Professor N. H. Winchell of
Minnesota, who in turn sent them to John Merle Coulter, by this time
professor of natural sciences of Hanover College, his alma mater. Coulter prepared a report on the plants on December 3 for the government's
Report, prepared by Bvt. Lieut. William Ludlow, and published the
following year.

In November 1875 Coulter founded his *Botanical Bulletin*, conceived
as a botanical journal for the central and western lands of North America
and not competitive with the *Bulletin of the Torrey Botanical Club*
founded some five years earlier. In its first issue was presented the "List
of Plants Collected in the Black Hills During the Summer of 1874" and
the collection became a part of the Hanover College herbarium. The
editor believed that "it may be of interest to know what botanical work
was done upon an expedition otherwise rather famous" and, as he did
not regard himself as more than an "amateur" botanist, species of uncertain determination were sent by Coulter to Porter of Lafayette College,
for several years a student of western botany. However, Coulter's determinations were sound. In his report he announced, "I arrange them in
the order of Gray." The trees and shrubs were listed by Winchell.

Explorations were still pursuing new paths and new trails. But now it was not so usual to read of expeditions returning with long lists of new species of plants. Botanists less often visited unbroken trails or explored fastnesses. Finds of new genera were less frequent. Revisions of existing botanical concepts were creating new genera and new species in the great herbaria. Still, there were many new regions in Canada and Mexico.

In March 1875 Sereno Watson received a letter dated March 15, from John Macoun of Belleville, Ontario, Canada, where he held the chair of natural history at Albert University:

... I have been appointed botanist to the Expedition which is intended to explore the Rocky Mountain passes north of Lat[itude] 54°—a region which is positively a terra incognita to the botanical world. I start for San Francisco four weeks from today and shall spend nearly a month in Vancouver before proceeding up the country.

The contemplated trip to Lake Superior will therefore have to be given up this season but I shall go next year if all is well. . . .

I shall send the plants by way of Owen Sound. I may not be able to furnish many but [what] I have you will receive. I shall send a small specimen of C. montana from Cariboo. I expect to collect it in abundance this summer.

John Macoun[17] was, as Ernest Thompson Seton has since said, "the pioneer naturalist of Canada, with official recognition as such. . . ." Born April 17, 1831, in the parish of Maralin, Ireland, he emigrated to Canada in 1850 with members of his family, his father having died when John was six years of age. When yet a small boy an uncle had once taken him to an orchard, shown him a row of filbert trees and pointing to the aments or barren flowers said, "Jock, these that you see here will all fall off and in the autumn it is on these trees we get the nuts that we use at Christmas time." After arriving in America John was splitting rails one morning when he noticed some hazel bushes and he went to examine them. He discovered, he said, "that these were identical with what my uncle had shown me in Ireland. I discovered that he did not seem to have known that on these same bushes there were other little objects that were pink and these I found to be only on the bushes that held the aments. Later, I knew that these were the female flowers and that the nuts were produced by these being pollenized by the male flowers. These were the first studies I made in Botany."

John studied a list of plants prepared in England and based on the Linnaean system of classification and as he discovered plants in Canada

[17] *Autobiography of John Macoun, M.A. Canadian Explorer and Naturalist*, Ottawa Field-Naturalists' Club 1922. For a "Review of Canadian Botany," generally discussed, see an article by David P. Penhallow, *Trans. Royal Soc. of Canada* (2nd ser.), III (1897-1898), Section 4, where Part II is contained, pp. 56 with bibliography. Published as reprint.

he would try to find where it stood in the system. He also studied a book
—*Mistress Lincoln's Botany*—and Louis Agassiz's *Lake Superior* which
contained an account of plants. He was a farmer for six years; partly for
the purpose of studying botany, then decided to become a teacher. After
doing some teaching and botanizing, he went to Toronto to attend the
normal school there and, meeting a friend, "a prize man in botany,"
began going with him on botanical excursions. For years he studied,
always retaining an interest in the science; and although for some time
without the aid of a microscope or glass of any kind, he improved his
knowledge of plants and soon was interested in physical geography and
animals. He collected Carices and sent them to Chester Dewey of
Rochester, New York. He collected mosses and liverworts and sent them
to William S. Sullivant in Columbus, Ohio. He sent Hepaticae to Coe F.
Austin. And gradually he became acquainted with the naturalists of
Canada; particularly, Professor George Lawson of Queen's College,
Kingston, whom he called "the father of Canadian Botany." Dr. John
Bell had collected plants on the Gaspé Peninsula and Macoun was asked
to "decipher" them. George Vasey began making exchanges with him.
He started a correspondence with George Engelmann. But his most difficult specimens he sent to Sir Joseph Hooker of the Kew Gardens in
England. However, since Hooker had practically left the North American field to Asa Gray, Macoun turned to Gray and at first found what
other young botanists often found—a severe but encouraging critic. Of
course, there were others with whom he corresponded; as, for example,
Edward Tuckerman of Massachusetts and Robbins of Vermont. But
these were the principal ones and with them he worked till the field of
botanical exploration on a large scale in Canada was opened to him.

Early in July 1869, he sailed from Collingwood on an exploration of
Lake Superior where he gathered a large collection of plants containing
many rare species. Agassiz had characterized the flora there as "mostly
subarctic," Macoun said, "but I found that the statement only held close
to the lake, while I found the plants a few hundred yards back from the
lake almost identical with those north of Belleville. I saw the cause at
once, the lake water according to Agassiz was 48° F. at midsummer and
120 miles of cold water accounted for the change in flora on its shores."
On this trip lumber to build the first house at Port Arthur accompanied
Macoun. The following year he botanized with one of his students near
North Hastings and the next year, wishing to visit Lake Huron, he
accepted an invitation to visit Royston Park near Owen Sound where he
collected flowering plants and mosses (which he sent to Thomas P.

James), and a number of ferns. In 1872, however, his explorations took on larger proportions.

He met Sandford Fleming, chief engineer of the Pacific Railway being built in accordance with an agreement with British Columbia. Fleming invited Macoun to go to the Pacific coast, serving as botanist to a party then en route. Macoun accepted and the party arrived at Port Arthur, then Prince Arthur's Landing, on July 22. They proceeded from there to Lake Shebondowan where they took a water route, going by barges and canoes. Macoun looked with eager interest to the flora of the plains, sometimes, upon landing, searching the new areas by torchlight. In some places, the lands were found scorched by the summer heat. But after they had passed Fort Frances and Lake of the Woods they came to the prairies and found a rare feast of botany—"two or three distinct floras"— where Macoun counted more than 400 different species in one day's ride. On the morning of July 31, the camp was awakened by hearing the botanist exclaim, "Thirty-two new species already; it is a perfect floral garden." The party looked out on the fields and "saw a sea of green sprinkled with yellow, red, lilac, and white" flowers. For nearly a thousand miles, at intervals, similar beautiful landscapes, varying remarkably in colors, appeared to the almost enchanted party. Their route by way of Fort Garry, Portage La Prairie, and Fort Ellice brought them in August to the South Saskatchewan and at Fort Carlton they crossed the North Saskatchewan and took the northern trail for Edmonton.

From the eastern edge of the prairie at Oak Point to the Saskatchewan, a certain sameness in the flora discouraged Macoun as he found few new varieties. But he was chosen by the chief engineer of the Canadian Pacific road to make a reconnaissance of the Peace River valley and so, proceeding, they reached the Athabaska River on September 7 and the Peace River on October 1—"the long-looked-for goal of our hopes"—a river flowing majestically in a winding silence to the Arctic Ocean. Macoun busied himself collecting at all points when possible. But winter came on and plants became less abundant. Some of the way had been through most difficult swamps and marshes. Moreover, with the coming of the cold came also the mountains and over them they had to go. By the time they reached Stuart River they could cross it on the ice. The Nechaco River was more difficult, being fully three hundred yards wide with a current filled with ice hummocks. But they crossed the summit and reached Quesnel and the Fraser River, and eventually Yale and New Westminster from where they could telegraph Victoria. Macoun went to Victoria where he sailed for San Francisco and there took the Union Pacific Railway east, then only four years old. When home, he began more earnestly to study physical geography, climatology, geology, and meteorology. He wrote:

> While crossing the continent between Winnipeg and the Pacific, I noticed a wonderful sameness in the flora and concluded at once that there must be a sameness in the amount of heat given off in each district and, therefore, the plants of one district give a key to the climate of another that produced the same plants and the result was that I published the statement that it was only the growing months of the season that should be counted. Many other problems came before me and, in thinking them out in after years, I came to certain conclusions that were expressed in future years.

In the spring of 1874, Macoun's report on the 1872 expedition was published in the Canadian Railway report. This brought him to the attention of the Canadian Geological Survey and Dr. Selwyn, its head, exclaimed, "I must have that man with me when I go out next year." Accordingly, Macoun was employed to aid in examining again the Peace River Pass and more of the country adjoining for the Mackenzie government authorities. A railway was being planned through Peace River valley.

Macoun went again by railway through the United States and its Territories to Sacramento, California, and evidently San Francisco. En route, a washout at Laramie, Wyoming, detained him and on another occasion the passengers were compelled to walk for several miles. However, on the last day of the train ride, Macoun said, "we took dinner on the top of the Sierra Nevada with fully ten feet of snow on all sides." He went to Victoria and began immediately examining the flora there, collecting on Cedar Hill, Mount Tolmie, and other localities. He noticed the similarity of the flora to that of California, particularly that around San Francisco. "Two facts regarding the climate of Vancouver Island and indicated by the flora are," he said, "dry summers and abundant rainfall. The former is shown by the annuals being all in bud and flower by the first week in May and the latter, by the luxuriant growth of succulent vegetation in the low grounds. The general character of the flora, therefore, proves that the climate is warmer than that of England and that the rainfall is periodic. . . ." From Victoria to Peace River Pass, along the Peace River itself, and for almost 1,000 miles beyond, Macoun followed his instructions to study the flora, climate, and agriculture.

Macoun went by steamer to New Westminster. Observations concerning climate especially interested him although on reaching Harrison River he noticed that the white thorn was in flower and at Yale examining the mountains in the vicinity, rediscovered[18] *Saxifraga ranunculifolia*

18 Described by William Jackson Hooker, father of Joseph Dalton Hooker.

found by David Douglas in the course of his very early explorations. He spent a week on the Thompson River at Spence's Bridge and Cache Creek "and collected many species of rare and interesting plants which were not observed in the low country." A curious resemblance to the flora of Nevada and Utah interested him, especially in view of discoveries of *Astragalus Beckwithii* and *Crepis occidentalis Nutt, var. nevadense,* the former found at Salt Lake, Utah, and Ruby Valley, Nevada, and the latter in Nevada. Eventually he arrived at Quesnel and found many of the common eastern plants in full flower. "Nearly all the species," he said, ". . . were eastern ones or western plants that reach the wooded country west of Lake Superior." On June 4 he crossed the Fraser River and went at once into the wilderness. The land between the Nechaco and Stuart rivers, which he had seen before only in the winter, proved "of the very best quality." Near Stuart Lake and within a few miles of Fort St. James, he beheld that:

Many beautiful flowers that I had not seen since I left the lower Fraser Valley were in full bloom and, on the rocks at the base of the cliff, they made such a charming picture that I sat down in my loneliness—but not alone—and drank in the surpassing beauty of the scene; hunger and weariness were forgotten and I resumed my march with the light, joyous step of the morning, feeling that in the realm of Nature, God's hand was ever open to strew one's paths with beauties and fill one's heart with praise. While others cursed the road and the flies, I, in my simplicity, saw nothing but Nature decked out in the springtime loveliness and, instead of grumbling at the difficulties of the way, I rejoiced in the activity of the animal and vegetable kingdoms. For nearly a month, I had kept travelling with spring, but now, with one bound, we had passed its portals and stood on the verge of summer.

Onward they went to Fort McLeod, the forks of the Finlay and Parsnip rivers, "Hell's Gate," and "Mount Selwyn," which they climbed. Macoun commented:

Where the heaviest drifts of snow had lain, and where much of it still remained, one or two anemones and *Ranunculus hyperboreus* were blooming and in fine condition. To show the progress of the spring, four yards from the snow the petals had fallen and between that and the snow the plant was in all stages of growth, from its springing out of the soil to the faded flower. A number of drabas and arenarias absolutely plastered the ground with multitudes of flowers. Five hundred feet below the summit, M[oun]t Selwyn stands first, in my imagination, as the highest type of nature's flower garden. None of the plants except the peduncularias, rose above the general level, which was about two inches or possibly less, and all was a flat surface of expanded purple, yellow, white and pinkish flowers. . . .

The vegetation of the Peace River valley proved luxuriant. They came to the Rocky Mountain Canyon. At Hudsons Hope, Macoun wrote:

Wild peas and vetches grow to an amazing height in the poplar woods, and form almost impenetrable thickets in places. Vetches, roses, willow-herb and grasses of the genera Poa, Triticum and Bromus fill the woods and cover the burnt ground, and surprise Canadians by their rankness and almost tropical luxuriance.

They floated down the river on a raft to St. John's where Dr. Selwyn went on an exploration of Pine River without Macoun who spent the time till August 4 drying and packing his plants. Macoun prepared for his next journey seven hundred miles down the Peace River to Fort Chipewyan in a canoe in the company of one man, something up to that time which had never been done. The journey was perilous and daring but on his arrival at the Fort, Macoun, sick, tired, and starving came to full consciousness of the immensity of the Great Northwest. He was 1,300 miles from the Arctic Sea and 1,200 miles from Winnipeg.

Sometimes with a large party and sometimes alone with a guide Macoun returned east by way of the Athabaska River, Buffalo Lake, Clearwater River and Lake, Isle-a-la-Crosse, and Green Lake, and then went across the country to Fort Carlton and on to Winnipeg where he arrived on November 3—a trip made by long, weary tramps on foot and long voyages in a canoe. From Winnipeg to Fargo, North Dakota, he went by stage and then took the train to St. Paul, Minnesota, the Northern Pacific Railroad by that time affording transportation. He returned to find himself a famous man—a public character—even more so than on his earlier returns from previous journeys. He was a famous lecturer and teacher. He was in a position to recommend a route for the proposed railroad west from Winnipeg. Some favored a line past Lac la Biche and north of Little Slave Lake and through either the Peace River lands to Pine Pass or the Peace River Pass itself. Others wanted the line to go by way of Yellow-Head Pass and into the country westward. Macoun's report impressed many but no immediate settlement was effectuated. In 1876 Macoun was asked to write a report on the country between Port Arthur and the Pacific. With its publication soon afterward Canadians began to realize the value of their western lands and migrations increased even to the prairies. Macoun said, "no settler had passed from Manitoba on to what was called the 'Second Prairie Steppe,'" up to 1875.

Within a little more than a decade, a governmentally sponsored movement to breed hardy fruits for the Canadian and northwestern United States plains (also northern New England)—a movement in large part originated by practical breeders and by Charles Gibb, a Canadian, and Joseph Lancaster Budd of Iowa Agricultural College—would be under way.

Macoun's interest as a naturalist, however, was more botanical than horticultural. Nor was he a civil engineer. On August 12, 1876, he wrote Sereno Watson:

During the years 1872 & 1875 I collected plants all the way from Lake Superior to the Pacific, crossing the Rockies about Latitude 56° by the Peace River and reaching the coast at the mouth of the Fraser.

I have been engaged for some time working up those collections and would like very much to have the privilege of comparing my specimens with those in Dr. Gray's Herbarium.

Should the liberty be given I shall go down to Harvard toward the end of this month. . . .

On November 21 he again wrote, saying he regretted he could not furnish the specimens which Watson had requested. The greater part of his Lake Superior specimens went to David A. P. Watt of Montreal who had financed his trip there. However, Macoun said, "I purpose going to Lake Superior on a collecting tour next year if I can raise funds enough. I would guarantee 1,000 specimens of the following species for $100"; and he named eight species, two in Aspidium, two in Woodsia, one in Cystopteris, and three in Botrychium.

However, Macoun evidently did not go to Lake Superior the next summer as planned. He took his son James on his first trip—to Toronto and Niagara. And in 1878 he was chosen to lead one of ten parties to the prairies and the country north of Jasper. At first he refused the appointment, holding out for a permanent position. He wrote Watson on January 26, 1879, from Belleville:

I have this day sent on a parcel of plants addressed to Dr. Gray. They are another installment of our Canadian Flora. I purpose going on until he has got a full set of our whole botanical productions. Next season I shall add largely to what I have already sent and before winter is at an end I expect to send the grasses and sedges and possibly the mosses.

The 60 species of the old set whose names you sent a few weeks since were all collected on that peninsula which extends between Georgian Bay and Lake Huron. They were all collected the last week in July and first weeks in August 1871. . . .

You will find a parcel addressed to Prof. James. It contains mosses. . . .

That year Macoun went west to Fort Ellice on the Assiniboine River from where he went to Long Lake and, crossing the Saskatchewan River, made for Battleford. There he decided he could go to Calgary and Old Bow Fort and so following for the most part Red Deer Valley, "a beautiful stream of clear water," he went by Crowfoot Coulee where he discovered his first exposure of coal and took fuel to burn. They had many interesting experiences; fishing for trout, examining old Indian battlegrounds and learning that the day before their arrival at a telegraph station Dr. George Dawson and Reverend Mr. Gordon had sent a long telegram concerning their investigations in Peace River valley that year. But the most interesting place was the region around Long Lake which Macoun regarded as "The Flower Garden of the North West," writing in his journal the first week of July:

Flowers are a most conspicuous feature of the prairie. Hedysarum and various Astragali vieing with the lily and vetch in loveliness and luxuriance. Often, whole acres would be red and purple with beautiful flowers and the air laden with the perfume of roses. Sometimes, lilies (*Lilium philadelphicum*) are so abundant that they cover an acre of ground, bright red. At others, they are mixed with other liliaceous plants such as *Zygadenus glaucus*, and form a ring around the thickets which we passed. Another time, we come upon a pool of fine, pure water and in it grows *Carex aristata*, which the horses love so well. Around it, where the water is nearly gone, are *Carex marcida* and *lanuginosa*; outside of these a ring of white anemones and, growing where it is slightly drier, another flower, *Potentilla gracilis*, and, as the ground becomes still drier, *Pentstemon confertus* would appear and, lastly the lilies would surround the whole.

Reaching Calgary Macoun went up to Morley and Old Bow Fort at the entrance of the Rocky Mountains. They proceeded up the Bow River as far as Point of Rocks and then returned to Morley and went on toward Edmonton. Winnipeg was, however, Macoun's destination and going by way of Battleford he reached there after a number of adventures. As he proceeded on the railroad from Winnipeg to Fargo he remembered the stage ride he had taken between those points not long since. It was not much more time till he reached Belleville. And from there, on March 28, 1880, he wrote Watson:

I have been here nearly a month and now take the opportunity of thanking you for your kindness in naming my plants and giving me the information in which I stood in need. Thanks for the corrections on the Catalogue. I am engaged on a complete revision of it and purpose using your Botanical Index and all the later works and Revisions for the nomenclature. . . .

Dr. Vasey examined my grasses and found many interesting things. Dr. James and Prof. Tuckerman my Lichens so that I may say with safety that now the greater part of my collections are properly named.

Should I get time this spring I shall send all the carices and grasses found in Canada.

Macoun's great *Catalogue of Canadian Plants* published over a period occupying most of the balance of the century was begun. Macoun had substantially given up his teaching and was devoting himself to this and exploratory work. He was a professor emeritus at Albert College now.[19]

[19] Most of the material of this chapter relating to Macoun has, with the exception of unpublished letters, come from his autobiography.

CHAPTER IV

North Carolina and Florida. The Hooker-Gray Expedition to the West

IN July 1874 George Engelmann wrote Gray asking whether Watson had given up completing his *Catalogue of Western Plants.* "I am glad that your N[orth] Am[erican] Flora Ideas take shape and are likely to give us the long desired work," said Engelmann, "capital to commence where you left off 35 years ago! Why not at the other end?" he inquired, referring to the fact that Gray was beginning with Volume II of the *Synoptical Flora of North America* rather than Volume I.

Almost immediately after completion of the Pacific Railroad botany and portions of the United States Pacific Exploring Expedition botany, Gray had turned to the immense tasks of revising and combining the old and new materials. John Torrey and he had fought three decades for time to complete their famous joint work, *The Flora of North America.* All that had been found possible had been the commencement of revisions of existing genera and species, determining the new and revising the old. Engelmann had taken a number of the plant families. And Watson was collaborating with Gray, the two having begun the preparation of their worthy series, *Contributions to American Botany.* Gray had written on the North American species of Astragalus and Oxytropis (1864), the Eriogoneae with John Torrey (1870), the order Diapensiaceae, the North American Polemoniaceae (1870), Labiatae (1872), Compositae (1873-1874), the North American Thistles, Borraginaceae, the North American species of Physalis (1874), and many other genera and species. In not all instances were the revisions complete—sometimes they were notes only—sometimes reconstructions—sometimes descriptions principally of new characters. Gray was working with Brewer and Watson on the *Flora of California,* doing as his especial part Gamopetalae and, where possible, combining this task with his work on the *Synoptical Flora of North America.* He kept alive to the work of George Bentham, Charles Darwin, and Sir Joseph Hooker in England, to the European continental productions, to all world-wide scientific progress relating to botany. Gray had shouldered a tremendous task. With Torrey and Sullivant gone, he had lost his two oldest friends and in Torrey his most important associate. His had been the inheritance of the fame of "Torrey and Gray." With it came tremendous responsibility and much work. With him, however, there was Sereno Watson, who had already revised the extratropical North American species of the genera Lupinus, Poten-tilla, and Oenothera (1873), the section Avicularia of the genus Poly-gonum (1873), and the North American Chenopodiaceae (1874). The early arrangements of botanical materials had depended on collections of early explorers such as David Douglas, Thomas Nuttall, and a few others. Now the much larger and better arranged herbaria of Gray, Torrey, Eaton, and the Philadelphia Academy of Natural Science, not to mention the herbarium of the United States Department of Agriculture, were available, the results of half a century of North American and world scientific exploration.

On February 8, 1875, Engelmann wrote Gray, "You have done a good work in clearing up Cnicus and Physalis, and the confused Borragina-ceae." Engelmann encouraged Gray although he disagreed with him as to the places of certain species and believed that many times Gray and Watson established too many species. "Yes, do take up Scrophulariaceae a worthy subject: a foeman worthy of your steel but after conquering Pentstemon you will find little difficulty with the smaller genera but until you get quite well, take an easier sure playful task. I will help you in Gentians, and may do Erythraea. . . . Rothrock has sent good speci-mens of an Erythraea," Englemann said. About two weeks later, he added: "That is good news which you give me—you going south and returning by St. Louis! Keep me informed about your movements. You should try to see Ravenel and especially Mellichamp,[1] at Bluffton, be-tween Charleston and Savannah." Engelmann admired Mellichamp. On November 23, 1873, he told Gray, "My friend Mellichamp is on another point. It is the oaks now, which he hunts up (or down) with the same zeal and acumen as he did the Yuccas and Pines. We have established an interesting fact that there is in S[outh] C[arolina] a 'Running oak.' . . ." And to Parry the following February, he said, "Isn't Dr. Hays a correspondent worth having? A western Dr. Mellichamp. He promises to work up the oaks of his region." Engelmann sought to have a genus named for him, Mellichampia. Gray also regarded him well, referred to him as "a good observer," in the course of correspondence concerning *Sarracenia variolaris*, "the best of Sarracenias," and of the pitcher plant family. "Have those carnivorous plants been chemically examined?" asked Engelmann of Gray who in 1845 had written on "The Chemistry of Vegetation." "Do they contain other ingredients than common plants? Newer nitrogen compounds?" Engelmann asked. The subject by 1874 was of keen interest.

The Carolina regions of the "carnivorous plants"—Dionaea, Drosera,

---

[1] Joseph Hinson Mellichamp (1829-1903).

most of the species of Sarracenia, of the "bladder-bearing Utricularias," and the largest species of Pinguicula—and the pine barrens of New Jersey were classic botanical grounds to Gray. Of the pine barrens, he said in his address to the *British Association for the Advancement of Science*, meeting at Montreal in 1884 (August):

To have an idea of this peculiar phytogeographical district, you may suppose a long wedge of the Carolina coast to be thrust up northward quite to New York harbor, bringing into a comparatively cool climate many of the interesting low-country plants of the south, which, at this season, you would not care to seek in their sultry proper homes. Years ago, when Pursh and Leconte and Torrey used to visit it, and in my own younger days, it was wholly primitive and unspoiled. Now, when the shore is lined with huge summer hotels, the Pitch Pines carried off for firewood, the bogs converted into Cranberry-grounds, and much of the light sandy or gravelly soil planted with wine-yards or converted into Melon and Sweet-potato patches, I fear it may have lost some of its botanical attractions.[2]

However, when in 1875, because of Gray's health, Dr. and Mrs. Gray made a trip south to Apalachicola, Florida ("a now almost deserted, but once flourishing town, on the Gulf of Mexico," Gray wrote) and went to the place of growth of the Taxoid conifer, the genus Torreya, Gray named another classic botanic ground in North America. "Apalachicola was heavenly," Gray wrote Canby. And to R. W. Church, he said, "The botanizing was delicious, very many nice things which I had never seen growing before. . . . I had special botanical objects leading me to west Florida, an out-of-the-world region, where we had everything to ourselves." Gray described the genus Torreya in his address, "Sequoia and its History," as:

. . . a noble, Yew-like tree, and very local, being, so far as known, nearly confined to a few miles along the shores of a single river. It seems as if it had somehow been crowded down out of the Alleghanies into its present southern quarters; for in cultivation it evinces a northern hardiness. . . .

The genealogy of the Torreya is still wholly obscure; yet it is not unlikely that the Yew-like trees, named Taxites, which flourished with the Sequoias in the tertiary arctic forests, are the remote ancestors of the three species of Torreya, now severally in Florida, in California, and in Japan.[3]

En route, Dr. and Mrs. Gray visited in Washington, Augusta, and Savannah. Their intention was immediately to go to Apalachicola but while on their way to Chattahoochee, they were compelled to stay over-night at Live Oak. Going on toward Tallahassee and Quincy, they learned that high swamp water had so overflowed a trestle they could not reach Chattahoochee at all by train. All of the next night they slept

---

[2] "Characteristics of the North American Flora," *Scientific Papers of Asa Gray*, selected by C. S. Sargent (New York and Boston: Houghton, Mifflin and Company), II, p. 275.

[3] *Op. cit.*, pp. 149, 161.

on the train (Gray making his toilet next morning from water of the locomotive tank) and then proceeded by steamboat. However, Gray learned before the steamboat arrived that he might go to a locality of Torreya! Guided by a young man, he went to a ridge where, included in a growth of pines and deciduous trees, he found "a thrifty young *Torreya*" and later several of larger size; and with it, as he expected, the curious little herb, *Croomia pauciflora*, just as the discoverer, H. B. Croom, had found it. From one Torreya he took a branch large enough to make an official baton for the presidency of the Torrey Botanical Club. In all, he spent a delightful ten days in the regions, beginning with the voyage first up the Flint River about forty miles and then "down the brimming [Apalachicola] river, bordered with almost unbroken green of every tint, from the dark background of Long-leaved Pines to the tender new verdure of the Liquidambar and other deciduous trees in their freshest development." These were "interspersed with the deep and lustrous hue of *Magnolia grandiflora*, and, when the banks were low, [were] dominated by weird, naked trunks of Southern Cypress (Taxodium), their branches hung with long tufts and streamers of the gray and sombre Southern Moss (Tillandsia) below, while above they were just putting forth their delicate foliage. Along the lower part of the river occasional Palmettos gave a still more tropical aspect."[4]

Gray renewed his acquaintance with Dr. A. H. Chapman, author of the *Flora of the Southern States*. Guided by him he gathered "the stately *Sarracenia Drummondii* in its native habitat" and they must have discussed the matter of a supplement to Chapman's *Flora*. For the next year, on June 8, Chapman wrote Gray: "I don't feel able to get out a new edition with supplement, and still I dislike to take final leave of it with its numerous blunders and errors." The early work had been done under the most difficult circumstances and, though with some known errors, had won the conceded admiration of both Torrey and Gray. Chapman went that fall to the Florida Keys and gathered "a few poor specimens of plants." He sent them to Gray saying that he had selected an uninteresting time of the year for trees and shrubs in flower or fruit, but added, "I mean, somehow, to make a prolonged visit to that region and thoroughly explore it—Everglades and all. I hope to do this, through Vasey, by an appropriation of Congress, if possible, and failing in this I am inclined to go on my own hook."

About this same time, Abram P. Garber of Pennsylvania made one of his botanical rambles in East Florida, going to Palatka and regions of the

St. Johns River and 125 miles south to Lake Monroe, Mellonville, and Enterprise (now Benson Springs). Nothing indicates he saw Gray but, describing his journey, Garber wrote a most vivid description of his exploration for Coulter's *Botanical Gazette* (formerly *Botanical Bulletin*).[5] Garber was south and east of where Gray went. He wanted to reach the headwaters of the Kissimmee and Indian River regions. But dry weather prevented and he returned to St. Augustine, soon to explore again but this time in middle Florida from the St. John's country to Baldwin and Gainesville where he studied the spring flora and found a new Lobelia from Manatee concerning which Chapman wrote Gray August 2, 1876.

Gray, after his visit with Chapman, so planned his return voyage up the Apalachicola River that after sunrise he and Mrs. Gray reached the bluff of Aspalaga, where the Torreya was first found. Many Torreya trees had been cut away for steamboat fuel; nevertheless, while the boat made a sidetrip, Gray had a day with the region and returned to the boat at nightfall with thirty or forty seedling Torreyas. The Grays went on to Stone Mountain, Georgia, where he discovered a Sedum and a Diamorpha (both later sent to Paris), and *Arenaria brevifolia* of Nuttall, and to Lookout Mountain, Tennessee, where he gathered roots of *Silene rotundifolia*. They proceeded thence to Washington where Gray went to the Smithsonian Institution of which he was a regent. In the course of another week they were in Cambridge. Engelmann wrote Gray, "So you had a fine time, [have] seen Torreya and the new Pine and old Chapman! But you could go and *see* that pine and be satisfied with the bits Chapman gave you, the poor bits!! I am afraid you never felt the authority of a collector! You ought to have had me along—how I would have got bark, wood, young branches & no doubt flowering just then, and old cones, which must have been in abundance on the tree: nothing, nothing but a few old leaves and a cone. . . ." Severity was not always all on Gray's side.

The year 1875 produced Watson's *Contribution to American Botany V*, "Revision of the Genus Ceanothus and Descriptions of New Plants with a Synopsis of the Western Species of Silene"; Gray's "Conspectus of the North American Hydrophyllaceae"; Farlow's "List of the Marine Algae of the United States, with Notes of New and Imperfectly Known Species"; Redfield's "Geographical Distribution of the Ferns of North America"; and other articles of importance. The regents of the University of Minnesota, determined upon having a thorough and systematic

[5] "Botanical Rambles in East Florida," II, Number 3, p. 70; Number 4, p. 82; "Botanical Rambles in Middle Florida," II, Number 6, pp. 102-103.

examination of the flora of that state, placed N. H. Winchell in charge, and he in turn issued a circular letter informing the state's botanists how to proceed with the work systematically. And geological surveys in both Indiana and Missouri reported progress.

On June 24 of the next year, Engelmann wrote Gray, "Yes, we hope to come east in Aug[ust] & Sept[ember]—but do not know exactly how, yet. You go to the M[oun]t[ain]s of N[orth] Carol[ina] with Mrs. G[ray]—how would it do to join you? What is your plan, time etc.?" But Engelmann was afraid of the heat and crowd for Mrs. Engelmann and so he asked Gray to be on the lookout for *Abies Fraseri* for him. However, by July with the heat in St. Louis, Engelmann wrote, "I suppose we could be at the Hot Springs on French Broad River in 2 or between 2 and 3 days, via Nashville, Chattanooga & Knoxville and be with you a week or more. Our intention is to go from there to Philadelphia." August arrived and Engelmann wrote saying that he would be much disappointed "if the plan to go to the Black Mountains should fall through but more sorry if" Gray would be prevented from going. "I hope it is your stomach and not your heart which is at fault," he told Gray.

Great were the preparations once it was finally decided that everyone would be there. Canby was more or less placed in charge by Engelmann and Gray, and he wrote Redfield:

I am delighted to hear that you are going with us on our Mountain journey.

I think very well indeed of Dr. Gray's suggestion. I have never been up the "Valley of Virginia" but Harpers Ferry is as you know very picturesque and is besides a good botanical station. Nor have I been to the White Sulphur Springs— yet it would be pleasant to see it, even if botanical pursuits were not consulted. There are two excellent botanical localities. . . . The one, Salt pond M[oun]t[ain]— one of the higher elevations of that district with a good road over and near the top so that the splendid view therefrom is easily accessible, and with a curiously formed lake, near which grows . . . rare plants,—and New River White Sulphur Springs, about 8 miles from the M[oun]t[ain] where grows . . . rare plants. Without consulting a better map than I have I cannot get at the distance from the Springs across the country to the R[ail] R[oad] but that can easily be found out hereafter. There I suppose our route would be down E[ast] Tenn[essee] R[ail] R[oad] to Wytheville and from there as determined upon.

Engelmann wanted to see Kentucky and Tennessee landscapes and so he agreed to join the party at Warm Spring on the French Broad River. The party went first to New River Springs, then to the French Broad Hot Springs, and round by a rough trip to Asheville, joining Dr. and Mrs. Engelmann there. They continued through the mountains to Caesars Head where they took a railroad through South Carolina and

Georgia to Jonesboro. From that point they went on a camping and exploring excursion of Roan Mountain.

So enthusiastic was Engelmann about the trip he wrote Parry on September 4 from Atlanta, Georgia:

> You see I am here "mid Sherman," no, with Canby bound for Stone Mountain! We have had a hasty roundabout hunt, no rest, no ease, no repose!
>
> At first we hunted up Gray and party (Mrs. G[ray], Canby & Redfield—Hooker had excused himself because just entering second marriage!) here and there until we overtook them in Buncombe [Asheville] but instead of going up the big mountains it was resolved to take them with *Abies Fraseri* on the return trip and go to "Cesars Head" an outlier reaching into the Palmetto State. A few pleasant, but also, very busy days botanizing—of [which] I will only mention *Pinus pungens.* Splendid vegetation. . . .
>
> While the party were enjoying themselves or working yesterday, Canby & myself came here to do Stone Mountain, will all then go to the N[orth] W[est] slope of the M[oun]t[ain]s and ascend the Roan. . . .
>
> I write also to let you know not to be uneasy about me, that I should be tempted to visit Dr. Mellichamp as I am in the Carolinas. . . .

Engelmann enjoyed himself with the firs, the oaks, and the quillworts. He and Mrs. Engelmann went to Washington where Vasey showed him "a good deal in the coniferous line, Newberry's types etc. etc—and many new things from California and elsewhere." They spent a while at the Philadelphia Centennial and after short visits with Canby at Wilmington and in New Jersey with some "disciples of Linnaeus" they went to Cambridge where Engelmann conferred with Sargent on forest trees.

Gray returned, longing "to revisit those [North Carolina] mountains when the Rhododendrons and Kalmias are in bloom. . . ." The high Alleghenies in Virginia, Carolina, and Tennessee had interested him long, as there thirty years and more ago he had roamed and botanized. But by winter he was "deep in routine work" at Cambridge, "and with a printer not far behind me," he said, "I can think of little else." That year Gray added new *Contributions,* "Characters of Canbya (n. gen.) and Arctomecon," "On the Character of a New Genus of Papavaraceae, Canbya; Also of Certain Other New Californian Species of Plants"; and that year Gray's *Darwiniana: Essays and Reviews Pertaining to Darwinism*[6] was published.

Watson presented three parts of a new *Contribution* of plants collected by Edward Palmer on Guadalupe Island, Lower California, and other California collections with certain revisions of genera.

Engelmann published in *The American Journal of Science and Arts* an article entitled, "Morphology of the Carpellary Scales of Coniferae."

[6] *Op. cit.* (New York: D. Appleton and Co., 1876).

His report on the botany of the Simpson expedition of 1859, including Cactaceae (a report long held and based on collections of his brother Henry Engelmann, made while accompanying Colonel J. H. Simpson's exploration for a direct wagon route across the Great Basin of Utah from Camp Floyd to Genoa in Carson Valley), was also published that year at Washington. North American botany was not suffering for lack of works from able men.

Although Edward Tuckerman's brilliance as a lichenologist was somewhat dimming, he was carrying on his work at Amherst, Massachusetts, planning a *Synopsis of the North American Lichens*. It was to be a work along lines similar to Lesquereux and James's *Manual of Mosses*. For a period of a decade and a half, Tuckerman had been publishing in the *Proceedings of the American Academy of Arts and Sciences* his valuable "Observations on North American and Other Lichens," following his *Genera Lichenum*, an arrangement of North American lichens, published at Amherst in 1872. His first work of importance had been his *Enumeration of North American Lichenes*,[7] with a preliminary view of the structure and general history of the plants and of the Friesian system (of which he was a disciple) in which was composed an essay on the natural systems of Oken, Fries, and Endlicher. He had added to the enumeration, publishing lichens of several of the more important North American exploring expeditions, of California, of Oregon, of the Rocky Mountains, of Hawaii, and of Annanactook Harbor, Cumberland Sound. In 1875, with Charles C. Frost,[8] was published, *A Catalogue of Plants Growing without Cultivation within Thirty Miles of Amherst, Massachusetts*. Tuckerman's *Lichenes Americae Septentrionalis Exsiccati*, in six fasciculi, or three volumes, and his *Lichenes Caroli Wrightii Cubae curante E. Tuckerman* were fortunately to be possessed by the most important herbaria of North America. He was a great student of history, one of the earliest scientific explorers of the White and Green Mountains of New Hampshire and Vermont, and, after Cutler and Bigelow, one of New England's great botanists—in lichens, one of the world's most eminent students. His "Synopsis of the Lichens of New England, the other Northern States, and British America" endures in the literature of this section of North America, and occupied in publication the most part of the first volume of the *Proceedings of the American Academy*.

Gray said of him, ". . . his botanical model was Elias Fries . . . [and]

---

[7] Published in 1845 at Cambridge.

[8] An eminent New England botanist of Brattleboro, Vermont, especially well known in cryptogamic botany.

he took broad views of genera and species. So he was quite unlike that numerous race of specialists who, in place of characterizing species, describe specimens, and to whom 'genus' means the lowest recognizable group of species . . . it was most natural that, at his time of life, he did not take kindly to the Algo-fungal notion of Lichens, and that he was convinced of its falsity by questionable evidence."[9] Tuckerman's letters to Gray show this. On December 16, 1873, Tuckerman said:

The "Synopsis" is . . . still where it was but I am trying to get a hand-book put together by a friend (name not to be told till he is sure he can do it) with the help of my herbarium &c. Such a book is needed, but the difficulties are vastly increased by Dr. Nylander's constant contraction of new species on (to my view) wholly insufficient grounds. It looks as if the Arrangement of the Lichens w[oul]d become impossible to anybody but himself—and in fact that the study of Lichens would end in collecting and sending the specimens to *Him*. . . .

Again the following May he wrote:

I shall try hard to carry on my work on the *Synopsis* as rapidly as possible. That reminds me that it will not be the first book of the sort supplied to students by me —the earlier synopsis (1848) having proved of no little use to our few lichenists, and as offering the first English version of Fries's admirable diagnoses, deserving of remembrance. Mistakes comparatively very few!

I am thankful to see that Dr. Farlow is also lichenising—but hope fervently he will preserve a little judgment as to species-limits; and this is now perhaps difficult to do, in Europe. But Dr. Mueller seems to be one of the sober sort.

Tuckerman wrote Gray the following November telling him of his and Frost's *Catalogue* and adding, as to Lichens, that he had "quoted Sachs's method merely as the newest—& do not at all agree to his relegating the Lichens to the fungi. . . ." He disliked intensely ordering Lichens, Algae, and Fungi as of the same rank, and he said so more than once. Always with much analysis and argument. But Gray made him happy in his approval of his *Catalogue*—Gray later saying of it: "In matter and form, as well as in typography (in which Professor Tuckerman had exquisite taste), this catalogue is one of the very best."[10] Tuckerman, like every botanist of the period, held Gray in great esteem. When Gray agreed to attend a festival at Amherst, Tuckerman told him:

Botany has long been, & is cultivated here with no little interest and care, & you have long been the teacher of teachers & taught alike. It is much however to have among us the living Botanist whose works are our daily & hourly guide.

[9] Biographical sketch of Edward Tuckerman written by Gray for the *Proceedings of the American Academy of Arts and Sciences* (new ser.; 1886), pp. xiii, 539; also *Scientific Papers of Asa Gray*, II, pp. 495-496.

[10] Gray's biographical sketch of Tuckerman, *ibid.*, p. 493. For an excellent discussion of Tuckerman's and Nylander's work, see Bruce Fink's article, "Two Centuries of North American Lichenology," *Proc. Ia. Acad. Sci.*, XI (1904), pp. 11 ff.

Tuckerman, as a consequence, could not resist exploding to Gray concerning the new systematic methods in lichenology:

. . . I assure you that for everything I may do in determination of lichens (though I have had 30 years of experience since my determinations passed pretty well the ordeal of Fries, &c) will be exposed to contradiction from Dr. Nylander. To him everything that comes along new appears to look like a new species—that is the first, while to me it is the last supposition. I distinctly declare that the larger part of these new species have not even sufficient (assumed) *characters* to stand on. [T]here is not even *prima facie* evidence of their distinctness. And yet if I name a lichen by the name of the larger species which would once take it in, & neglect the Nylanderian new name, I should be condemned. I am wholly sick of it, & getting to feel less & less respect for what the Germans call Systematik—child's play indeed it is in such hands as Nylanders—& the German anatomists very fairly laughed at it & so I suspect the French; . . .

Tuckerman gave Nylander credit for knowledge but claimed he lived "in a glass house." Tuckerman preferred "a humble student of nature in this field" and not an "arrogant autocrat of lichenology; who cannot bear any difference from himself." Differences as to chemical reaction or slight measurable variances in spores did not constitute bases on which to constitute *new species*, Tuckerman said, and this was what Nylander was doing. "They are not species: or 'species' are hardly worthy of our serious study," he said. Basically, the trouble was, Tuckerman thought, a theological one. On April 26, 1877, he told Gray:

German botany has assumed a savage tone of controversy of late which I have sometimes thought might be due to the running of protestant theology, & with it of religion itself, into the ground there, & the evil influence of such writers as Vogt and Haeckel & their disciples—turning the land of idealism into the opposite—and their old soaring in the clouds to wallowing in the mud; but Nylander's tone has always been the same.

Nevertheless, there were more pleasant letters from Tuckerman. Concerning himself only with Volume I of the *Botany of California* of which Gray was a coauthor, Tuckerman wrote Gray on July 10, 1876:

Immediately on receipt of the noble volume of the Calif[ornia] flora, I determined to ascertain the authors of the species, with the following results, which I regard as truly exhibiting the relative rank in this respect of the authors named, though the figures may not always be exact. I have included also species not yet found in California, but named by you as possibly to be expected, and have not reckoned species common to unassociable sections of our country or the foreign countries except the South American. And in accordance with my view of the true author of a species, I always reckon the original describer as entitled to the credit of his plant. I find them as follows:

1. Gray,   ab[ou]t 458 sp[ecies].
2. Nuttall,   ab[ou]t 212 sp[ecies].
3. Bentham,   "   155   "

4. Watson, ab[ou]t   117 sp[ecies].
5. Torrey,   "   93   "   .
6. Torr[ey] & Gray, ab[ou]t 78 sp[ecies].
7. Douglas, ab[ou]t 67 sp[ecies].
8. Hooker,   "   63 sp[ecies].
8. Hooker & Arn[ott], ab[ou]t 63 sp[ecies].
9. DeCandolle,   ab[ou]t   56 sp[ecies].
10. Engelmann,   "   44   "   .
11. Pursh,   "   40   "   .
12. Lindley,   "   29   "   .
13. Fischer   "   21   "   .
    [& Meyer]
    Kellogg,   "   17   "   .
14. Chamisso,   "   16   "   .
15. Eaton,   "   10   "   .
15. Durand,   "   10   "   .
15. H. B. K.   "   10   "   .

with some thirty others who have described less than 10.

How beautiful the volume is. It makes one wish to go to California, & herborize and I am not sure that I should not do it, had I nothing to keep me at home and busy were my power twice what it is. . . .

Sorry for my poor Tuckermannia. But it cannot be helped I am well aware.

Many years before, Thomas Nuttall had dedicated to Tuckerman what Gray described as "one of the handsomer of the Californian Compositae." Revisions now made necessary its change to a subgenus.[11]

Gray had made pilgrimages to two rich and classic floral regions of North America—North Carolina and Florida. Before Torrey's death, Torrey had gotten to California and Colorado. And so had Gray. Yet, new developments had taken place. On May 24, 1877, Gray wrote G. Frederick Wright, "Hooker is coming over, and we are going in summer to the Rocky Mountains together, according to an old promise of mine. To do it I ought to complete the printing of the part of my 'Flora' which I am upon, else I shall suffer in various ways, and there is great danger that I fail."

Tuckerman was wrong only as to time. His California botanizing recommendation given indirectly to Gray in 1876 was taken seriously the next year and Gray went to California with Sir Joseph Dalton Hooker at the invitation of Hayden and the United States Geological and Geographical Survey of the Territories. Between Tuckerman and Gray there lasted a friendship until Tuckerman's death. Unfortunately the same may not be said of Tuckerman's and Torrey's friendship which suffered some misunderstanding although one of Torrey's last actions was to instruct his curator to send Tuckerman plants.

[11] See Gray's biographical sketch of Tuckerman, *op. cit.*, p. 497.

Sir Joseph Dalton Hooker was president of the Royal Society of London and, of course, still director of the Gardens of Kew. Excepting possibly George Bentham, he was the world's most renowned botanist, and much interested in geographical plant distribution on the North American continent.

No one probably will ever be able to gainsay the fact that the combined effect of Darwin's studies in plant and animal evolutionary development, and Sir Charles Lyell's epoch-making work of the first half of the century *Principles of Geology*, had had most to do with bringing about what Gray styled in 1882 "the new mode of thought which now prevails," the purport of which Gray recognized years earlier. Of course there were other masters of great eminence who contributed. Two to which reference may be made were Wallace of England and James Dwight Dana in America, Gray's great personal friend with whom he associated and corresponded many years, both of whom had contributed notably to the new knowledge, a dynamic and biological interpretation of evolutionary development. There was, moreover, the profound history—largely laid in Europe—of morphological, anatomical, and physiological studies of plant cells, tissues, cytoplasm, "the physical basis of life," et cetera, studies correlating plant organs and functions and enlarging and widening through microscopic analysis the great developmental concepts for more than systematic usages. The laboratory was no longer confined to the herbarium sheet and a static interpretation of plants as units. The enlarged searches for affinities and relationships concerned in organic development were under way. A biological interpretation inevitably became related more and more to a renewed study of living plants in the field. As a consequence, the name of Sir Joseph Dalton Hooker had been also written indelibly into the history of developmental concepts, for, as F. O. Bower has pointed out, Hooker was not only one of the first protagonists of the Darwinian theory but also in great mid-century studies of the Antarctic and Arctic and other world region floras he had practiced as early as 1840 in Antarctic regions "Ecology on the grand scale," investigating floras by actual exploration and subsequent careful laboratory examinations on the results of which he predicated great essays and addresses as a "philosophical biologist." In these great phyto-geographical studies, constituting as they did one origin of the later more scientifically developed branch of botanical study, ecology, Gray was definitely associated with Hooker. For their studies lent much support to the belief that "species are derivative, and mutable."

Gray had already visited England several times and Sir Joseph and

Gray both wished to study together their long pondered connection of eastern United States plants and those of eastern Asia and Japan, and the differences of division between the Arctic floras of America and Greenland. Gray had worked on the former study and Hooker on the latter. Both thought the explanation was to be found in the glacial periods when there was believed to have been an early Arctic land connection between the continents on which Asiatic plants had migrated into North America in the east or west or both. The immediate problem, however, was the effects of glaciation in the western mountain chains, accounting for, if they could, the existence of the few eastern Asiatic types of plants among Mexican and more southern types. Obviously there had been a plant migration from the Mexican highlands north into the western United States mountain chains. But why there were only a few "pockets" of eastern Asiatic plants in the West whereas in eastern United States such plants were found in comparative abundance was unknown. Had the high lands of the West been submerged since the glacial periods? What were the evidences since then concerning suitable climate conditions for plant growth? What had been the effects of the Great Salt Lake that it was believed covered during the glacial period the whole saline region of the West? What accounted for the intervening prairies? Lesquereux had given an answer to the last question but he himself regarded his answer as more of an idea than a verifiable theory.[12]

Leonard Huxley, author of *Life and Letters of Sir Joseph Dalton Hooker*,[13] stated the problem thus:

Considering that the high mountains would have kept the glacial cap long after it had retired from the other levels of North America, the plants of East Asiatic type could have got no foothold there save in [certain specially] favoured areas, and by the time that the general change of climate had melted this belated ice-cap, it would also have affected the now treeless prairie district, exterminating these plants and leaving the survivors isolated in the more congenial forest district of the Eastern States, with no possibility of re-invading the Rocky Mountain area, which was thus left open to the plants advancing from the Mexican highlands until they met, not temperate, but Boreal forms. . . .

There was much evidence of glaciation in California. "Glaciers in California!" wrote Gray to DeCandolle in 1873, "Why, there is a fair remnant of one now, on the north side of Shasta,—and more in the

---

[12] *American Journal of Science and Arts*, XXXIX, second series (May 1865), pp. 317 ff., where Lesquereux published a translation of a letter on the subject to Professor Desor.

[13] London, Albemarle St. W.: John Murray, 1918. Volume II, Chapter XXXVIII, contains an entire account of Hooker's American journey. Quotation at p. 206.

southern part of the Sierra; and as to glacial marks, the geologists note them abundantly."

Hooker arrived in Boston on the night of July 8. After visiting Gray and Sargent for a few days, the party started for Cincinnati and St. Louis, in the latter of which cities Lambourne, Leidy, Hayden, and Stevenson joined them. Engelmann wished to accompany them but he was in a hopeless "muddle," botanically. Gray wanted his treatment of Cuscuta for the *Synoptical Flora of North America*. Rothrock was pressing him for completion of the Wheeler survey botany. Engelmann was, as Hooker said, "still hot on Pines, Oaks, Yuccas, and Euphorbias." Engelmann had written Parry the May before:

I believe I shall have to claim the benefit of the bankrupt act! The difficulty and misfortune is that I entered all these obligations voluntarily—and now? Well I hope the pressure will become so violent, that I shall burst and spout out all these beautiful things and more too, for when once unchained, Abies and Juniperus and Arceuthobium may run off like Cholera Injections!

Engelmann had several correspondents of his own—Butler[14] of the Indian Territory (Oklahoma), a zealous young fellow who evidently lived for a while near St. Louis, and a few others of lesser note today. And, of course, several of greater prominence. Palmer and Parry kept sending him plants although Engelmann told the latter that if he had no other correspondent but him he might hunger and starve. Palmer as a collector was improving. At first, Engelmann said, his plants "scared" him but no more. His, like Gray's, loyalty to Palmer remained firm. And Engelmann's correspondence with Mellichamp, Chapman, Greene, Walter, and Sargent was continuing. His gentian paper was in the printer's hands and an abundance of publications would issue from his pen the following year—1878. On June 13, 1877, Watson wrote Engelmann:

Dr. Gray has told me of his endeavors to persuade you to join his party with Dr. Hooker to Colorado. I hope that you will go & envy you the pleasure of the trip. But *what* will become of your contrib[utions] to [the] *Bot[any] of California* which will be wanted before you can possibly get back? Can you not put it together before you go?

So Engelmann decided that he could not leave his duties and work promised long ago, and he wrote Gray on June 21: "The opportunity to spend a few weeks with you, and in a fine botanical country, in the mountains, may not come back—very certainly never will! And the possibility of having Hooker along! It makes me half crazy to think of it. . . ." Watson urged Engelmann once again to go, saying the oaks

14 George Dexter Butler (1850-1910)[?].

and conifers could wait but, he admitted, Loranthaceae and Euphor-biaceae might cause trouble. And the second volume of the *Botany of California* was important to all North American botanists. "I do not like to publish poor work when better can be had," said Watson, "but it is too much to ask or expect that you should do it at so much cost to yourself." Nevertheless, Engelmann did not leave St. Louis to go to the West with Hooker's and Gray's party.

Nor did Parry, who was in Boston at the time of Hooker's arrival, go with Hooker and Gray, although he was urged to do so. On July 8, Parry wrote Engelmann humorously from Boston, and advanced an-other temptation: "... I too am 'too old' & 'stiff' to join that *young party* to Colorado *via St. Louis.* I expect to stay right here, study some, play more; and then when I get ready work leisurely home to Davenport [Iowa] via *Wisconsin* (August)? I am resolving a deeper plan, viz., to go down into *old Mexico* next winter and botanize the slopes of the Sierra Madre!! Seats not all engaged [C]ome along & get young! We look for Dr. Hooker tomorrow. . . ." Gray promised Parry that the Hooker party would stop off in Davenport on their return from the West in September.

Parry, after his Wasatch Range trip of 1875 in Utah, had spent some time in San Francisco meeting Bolander, Palmer, Kellogg, and others there, and worked at the herbarium of the California Academy of Sci-ences, which he found "in a sad muddle." Late that October Parry had written Engelmann and told him of his plans to move to San Bernardino and Engelmann asked him to study certain Abies for him. Engelmann asked Gray to send all there was of "his pets"—Cuscutae, Arceuthobium, Coniferae, Opuntia, Euphorbiaceae—from Palmer's Guadalupe Island, Southern California, and Cantillas Mountains collection. Agaves were then his first need, said Engelmann, and he planned to work up Cupres-sus. Parry had settled in San Bernardino where the new Southern Pacific Railroad was rapidly being pushed on to the Colorado. He had begun doing some trading with the Indians and planned an ambitious spring trip into Arizona returning via San Gorgonio and the high California mountains, avoiding where possible, Rothrock's districts. But Palmer, who, during November and December of 1875 went into southwestern Utah, came to San Bernardino and he and Parry had begun making desert trips collecting. It was planned that Palmer should do some early spring collecting on the lower Colorado, on Bill Williams and Providence Mountains, and then go to Lower California on the Gulf side. (Palmer's collections of both the years 1876 and 1877 in Arizona and Utah are more fully discussed in Rogers McVaugh's study of the subject, *op. cit.,* pages 773-775.) On May 29, 1876, Parry had written Gray:

I have just finished going over Palmer's plants from Arizona & S[outh] E[astern] Cal[iforni]a and selecting such as might be desirable for you to examine at once. [T]hey number up to 156 sp[ecies] and are sent by this mail in two parcels. . . .

Botany a little slack just now except in the M[oun]t[ain]s. We had a nice trip over to the Mojave, up Devils Cañon, just starting now for summit of [the] San Bernardino [Mountain]. . . . I hear from Green[e] at foot of *Mount Shasta,* a good location. . . . The enclosed *Papaveraceous* plant from Palmer seems very *unique.* I cannot think of any genus to fit it so propose our good friend Canby, Canbya, or Canbyella! [I]t seems somewhat allied to Arctomecon as far as *Capsule & seeds* go. . . . We have a *new*! Oxytheca from Mojave & a new Phelipaea. . . .

Palmer, however, while climbing Mount San Bernardino, fell and injured himself. As soon as his injuries permitted, he took a team and went over to the Mojave Desert to the cactus field but, the heat proving too intense, he returned and made for San Luis Obispo at Parry's suggestion, going later to San Francisco and in December 1876 to St. George, Utah, again for archeological investigations. Parry evidently remained around San Bernardino till July and then went north to the Sierras and eventually with Palmer's and his collections to Davenport where he permitted himself to be elected secretary to escape the presidency of the Academy of Sciences, wishing, as he said, to avoid a *"Hayes & Tilden* muddle." Palmer visited him in Davenport for a while before proceeding to St. George, the Colorado River and Salt Lake regions, and Red Creek or Paragonah, Utah, for the Peabody Museum. Except for a brief trip to southern Utah, evidently in the company of Palmer, all that winter Parry had worked with botany, determining his 1875 Utah collections along with others and then decided to go east; among other places, to Philadelphia and Cambridge. There he met Hooker and, possibly to incite Engelmann to go on the Western Territories trip with Hooker and Gray, wrote him jesting concerning their advanced ages (about which Engelmann was more sensitive than Parry, being older). At any rate Parry had wanted Engelmann to return with him to the West the following winter but when he decided on the Mexico trip he urged Engelmann to accompany him there. Failing in this, Parry turned to Palmer again. The first volume of the *Botany of California* was now in the hands of California botanists and had them all, as Parry described, in "Botanical Clover." As a consequence, California was happily prepared to receive Hooker and Gray and their party on their arrival there.

Hooker's and Gray's examinations extended over large areas of Colorado, Wyoming, Utah, Nevada, and California. In the Rocky and Sierra Nevada mountains their investigations were in alpine flora and tree vegetation. They left Engelmann in St. Louis about July 18 and after traveling west two nights and two days on the Santa Fe Railroad along the Arkansas River to Pueblo, Colorado, the Leidys went north while the others went on to Canon City. Engelmann received a postal card from Gray at Kansas City and a letter from Canon City reading:

We had yesterday a good day (with Brandegee) at the Arkansas Cañon; it is grand, surely.

To-day Hooker and the Stracheys[15] drive across and down Wet Mountain Valley to La Veta (two long days), while we, Mrs. Gray, Dr. Hayden, and I, return by railroad to Pueblo, and thence to La Veta, by sunset to-day. Tomorrow up to a camp on La Veta Pass of Sangre de Christo Mountains, which Captain Stevenson is preparing.

Our English friends begin already to feel in a hurry, and for a wonder I am the hold-back member of the party. . . .

Engelmann replied:

We have had cool and pleasant weather, even cooler than it was during your short stay, and see no necessity for leaving for health's or comfort's sake—but thus!

You say nothing of the oaks of the Cañon rocks—I should like to hear your impressions. . . .

I sent two letters for Mrs. Gray to Colorado Springs, where you will find them with this. . . .

Mrs. E[ngelmann] will be glad to hear from Mrs. Gray, and hopes that she will have a quiet day here and there, to rest and breath[e].

I ought to be with you!

The party established a camp at 9,000 feet altitude near La Veta Pass at the edge of the great pine forest. On July 26 they moved on to Fort Garland—Gray gathering material all the while to write Engelmann—which he did from Salt Lake City August 8:

Glad you have had nice weather; but you have no air like that of Colorado and Utah. . . .

Well, much as we miss and want you, yet we should have hurried you too much. We want to go over a good deal of ground cursorily, rather than a little thoroughly and leisurely.

I do not write you about the oaks at Cañon City, because we had nothing new to say. We agree with you in the complete running together of the oaks down to *undulata*. . . .

From Cañon City we—Mrs. Gray, Hayden, and I—went in one day south to La Veta by rail, and the next day, toward evening, up to La Veta Pass, 10,300 feet,

[15] Major General (Sir) Richard Strachey, R.E., and his wife. Strachey was a Himalayan traveler and an old friend of Hooker who accompanied him from England.

and over and 300 feet or so lower, where we camped, nice tents having been provided by Fort Lyon en route, and other furnishings from Fort Garland. . . .

Botanizing up there and in Sangre de Christo Pass good, but only moderate; nothing new, and no great variety. We enjoyed camp life very well; but after three days broke up, and went over to Fort Garland, and thence, while the ladies and General Strachey went off to a Mexican village, we had a two days' trip up the Sierra Blanca.[16] Alpine plants the same as on Grays Peak, but scanty, owing to more southern latitude and greater dryness. A longer time and a searching of the interior of this very rough range might, and doubtless would, furnish much we did not see.

Returning from Fort Garland to the railroad, we went back to Colorado Springs and drove up to Manitou. Next day, we went up Ute Pass—nothing—and looked about. Next day, to Garden of the Gods, to General Palmer's to early dinner, and thence to railroad and to Denver. Next day, Denver. Next by railroad through Clear Creek Cañon and to Georgetown, or within a mile, and thence up to Kelso's Cabin, now a well-kept house, to sleep. Next day, Grays Peak, and I crossed over to the top of Torreys [Peak]. Next day, after morning botanizing, came down to Georgetown and visited Empire City and the Pecks. Next day, Sunday, a restful morning, and then by rail back to Denver in the afternoon and evening. Monday, off at half past seven to Cheyenne, and after dinner took railroad to Ogden, and came up here last evening. Today, a broken day, sight-seeing, etc. Tomorrow, we, or some of us, are going south to American Fork Cañon; up that and over the pass into Cottonwood Cañon; down that, and back here, in time to go on that afternoon to Ogden and thence west to Reno, thence Virginia City, Carson, etc., and the Groves, Yosemite, etc. We shall see, and I will let you know.

Mrs. Gray is out with the party, to see things, and Brigham Young. *I will not.* She would be sending love to Mrs. Engelmann and you, if here. She is very well, and enjoying this travel hugely. I am strong, and ever yours. . . .

From Spring Lake, Utah, on August 5 Edward Palmer told Engelmann by letter, "I should like to meet Dr. Gray and Hooker when they visit Salt Lake." But eight days later he wrote again, relating that when he arrived in Salt Lake City he found Gray and Hooker had been there and gone. "He[a]rd Gray & Hooker had quite an adventure [T]hey had a twelve mile walk over a new road and did not reach their journeys end until midnight [T]hey was quite worn out," said Palmer.

Gray wrote Engelmann again from Yosemite, California, on August 21:

Did I write to you from Utah? We left direct route at Reno, went to Carson City, with detour to Virginia City,—queer place; first got hold of *Pinus monophylla,* but there no fruit.

Hired conveyance to take us from Carson right across the Sierra Nevada via Silver Mountain to Calaveras Big Trees,—a good way for studying the tree vegetation, and other, only all other is mainly destroyed by drought and sheep, and the ground is powdered dust. . . . Losing the [big-cone *Pinus ponderosa*] as we de-

---

[16] Said to be the highest of the Rockies, 14,300 feet.

scended to Calaveras, we come on it again in the Sierra here, when we get up to seven thousand to eight thousand feet. Here it passes for *P. Jaffreyi* or *Jeffreyi.* Is it so? Is it distinct? On bare side of Silver Mountain we found *P. monophylla* with cones, both maturing and this year's. . . .

And from Rancho Chico, the famous ranch of General and Mrs. Bidwell, Gray wrote, "We are keeping lively; are on the way to Shasta. . . ."

In California Gray and Hooker met John Muir. On account of his familiarity with the Mount Shasta region, they persuaded Muir to accompany them. One September evening, encamped on the flanks of the mountain in a forest of silver firs, a log fire was built and storytelling began.[17] Gray told of his explorations in the Alleghenies; Hooker of his in the Himalayas; and they talked of trees, arguing relationships of various species, and Sir Joseph admitted to Muir that "in grandeur, variety, and beauty, no forest on the globe rivalled the great coniferous forests of [Muir's] much loved Sierra." The next day Muir took Hooker on a short exploration westward across an upper tributary of the Sacramento River. There on a bank of a small stream Hooker found the beautiful small evergreen trailer *Linnaea borealis*, something which Gray had surmised the night before would be found in the region. Muir, with his poetically scientific mind, believed that Gray with almost uncanny intuition had "felt its presence the night before on the mountain ten miles away."

Engelmann answered Gray's letters enthusiastically. He had had much to do with California botany by this time—Cactaceae, Yuccas, and other plants of the deserts and mountains there. He had traced in most thorough fashion the tree distributions and accumulated enough knowledge so that many questions were in his mind. Consequently, his letters in answer to Gray were full of instructions concerning investigation methods in pines, firs, and oaks—from Colorado and Wyoming west to California and Oregon. ". . . But the white oaks!!" He asked "[w]hether *Douglassii* does not run into *undulata* on one side (of the mountains) or into *Garryana* on the other etc. etc." So persistent must have been his inquiries that special attention was given oaks. On General Bidwell's ranch in Butte County, California, a California white oak was found 100 feet high, 7 feet in diameter of trunk and 150 feet in spread of dome, and given the name, "The Sir Joseph Hooker Oak."[18]

Engelmann hoped that Gray and Hooker had seen Bolander and

[17] William Frederic Badé, *The Life and Letters of John Muir* (Boston and New York: Houghton Mifflin Company, 1924), II, pp. 80-84.
[18] Julia Ellen Rogers, *The Tree Book* (New York: Doubleday, Page & Co., 1906), p. 194.

"shaked him up—also Hilgard, who is a good fellow, though he may have done Grant some harm at Vicksburg." With characteristic generosity he also wished for the two great world botanists acquaintance with John Gill Lemmon. "Hope you are with Lemmon and he has the pleasure of showing the beauties of his mountains and forests," Engelmann wrote. And added, "wish I was with you."

Somewhat further knowledge of these California botanists is required at this point. Eugene Woldemar Hilgard was Engelmann's favorite cousin and the affection held by the one was mutually sustained by the other. Like Engelmann, Hilgard was of European birth, having migrated to this country to a farm near Belleville, Illinois (not far from St. Louis), while still a child. When Engelmann came to the United States, he was much older than was young Hilgard—the former was twenty-three years of age—but in each had been quickened early a zeal for natural history studies, especially botany. Engelmann and Hilgard had both returned to Europe to study in great universities there and in 1853 Hilgard had received the degree Ph.D. *summa cum laude* from the University of Heidelberg, later reissued as a "golden degree" in recognition of half a century's excellent work in science. Prior to going to Europe and during a brief stay in Washington, Hilgard had attended lectures on chemistry, acquiring such a proficiency in the subject that he had been made a lecture assistant. A splendid example of a great character in early experimental American science was this son of a chief justice of a court of appeals in Rhenish Bavaria. Although ill health pursued him many years of his life, scientific investigations in the out-of-doors combined with a remarkable creative vision influenced by European scholarship brought him not only vigor of body but also enduring fame by virtue of lastingly influential contributions made in America to several branches of science, notably in American geology, botany, and agriculture. In fact, as an agricultural chemist and as pioneer in American and world soils investigation, Hilgard has no American superior.[19]

For a while after his return to America following European study, Hilgard served as chemist in the laboratory of the Smithsonian Institution. His work came to the attention of Frederick A. Barnard and others, and he received a position first as an assistant of the geological survey, and later, as state geologist or mineralogist of Mississippi where, in years seriously interrupted by Civil War between the states, he acquired considerable reputation for his geological studies in Mississippi and Louisi-

[19] Attention is called to Eugene A. Smith's "Memorial of Eugene Woldemar Hilgard," *Bull. Geol. Soc. of Amer.*, XXVIII (March 31, 1917), pp. 40-67. The University of California has also published a small memorial volume of addresses delivered on Hilgard's life and work by Professor E. J. Wickson and others. An excellent National Academy of Science biographical memoir of Hilgard has also been published.

ana, particularly in explorations involving the Mississippi embayment regions, the lower Mississippi delta, and other special geologic formations. Hilgard became a professor of chemistry, and of experimental and agricultural chemistry in the state university, and director of the state survey. On his way south he had visited Dr. David Dale Owen and his assistants and it is known that, following their methods and practices, like Safford in Tennessee and probably Tuomey in Alabama, he furnished Lesquereux with valuable paleobotanic specimens. Hilgard added to their procedure. Although not the first to make a soil survey or chemical analyses of the soil, he was among the first, if not the first, "to interpret the results of analyses in their relation to plant life and productiveness. He was also the first to maintain the physical properties of a soil are equal in importance to the chemical in determining the cultural value." Mineral discoveries alone, he realized, would not bring the state survey into popular favor. Demonstrating a practical relation of soil to agricultural growth would, however, and so he early began recording surface features along with collecting plants that aided in characterizing soils. He had written in 1860:

> ... I departed pointedly from the then prevailing opinions, by which soil analysis was held to be practically useless. My explorations of the State have shown me such intimate connection between the natural vegetation and the varying chemical nature of the underlying strata that have contributed to soil formation as to greatly encourage the belief that definite results could be eliminated from the discussion of a considerable number of analyses of soils carefully observed and classified with respect both to their origin and their natural vegetation and a comparison of these data with results of cultivation, and that thus it would become possible after all to do that Liebig originally expected could be done, namely, to predict measurably the behavior of soils in cultivation from their chemical composition. . . .

Hilgard saw while at the University of Mississippi recognition of "the right of soil analysis to be considered as an essential and often decisive factor in the *a priori* estimation of the cultural value of virgin soils . . . alongside of the limitations imposed by physical and climatic conditions and by previous intervention of culture." However, it was not in Mississippi, but in California, where his great reputation was acquired. Mississippi, favored as it was with a large growth of native timber and a variety of soils "from the poor sandy long-leaf pine lands of the coast region, the richer loams and black clays of the interior, to the calcareous lands of the bluff region and the remarkably rich alluvial lands of the river," had laid foundations.

In the early 1870's Hilgard went to the University of Michigan as professor of geology and natural history. Nevertheless, inability to pursue soil study in enlarged fashion at this institution and an invitation following a series of lectures delivered at the University of California at Berkeley to become a member of the faculty at the then comparatively small California university induced the still youthful authority on soil investigation to accept President Gilman's offer of a place in the scientifically undeveloped and comparatively unexplored West. Hilgard could have taught any one of several subjects. But, the University of California having established on the university grounds the first agricultural experiment station of the United States, a department of the university and, therefore, not at first state-aided by direct legislative appropriations, placed Hilgard, who had been selected largely for his knowledge of soil study, in the position of professor of agriculture and director of the station. By only a few months, according to Professor Wickson, did Hilgard's station work precede that of Professor Atwater of the Connecticut experiment station, the institution to which is usually ascribed priority of founding, although it should be said that the Connecticut institution was the first station directly granted a money appropriation by the state. Just as the course of the Connecticut station was uneven, hampered by lack of facilities and finances, and not of great significance except as a forerunner of the important movement toward federally aided agricultural experimentation, the California station had to spend years gaining public confidence, especially the support of "practical farmers." But by 1877, the year of the Hooker-Gray expedition to the West on which they paid a visit to Hilgard at Berkeley, the station had had two years of existence, had received a specific legislative grant (the same year the Connecticut station was moved to New Haven to become a land-grant institution with the historically renowned American agriculturist Samuel William Johnson as director), and Hilgard had announced his adherence to laboratory and field methods of instruction, saying:

... a knowledge of facts and principles and not the achievement of manual dexterity, must be the leading object of a truly useful course of instruction in agriculture.

Object teaching should be made the pre-eminent method of instruction in natural and more especially in technical science. Manual exercise should be made the adjunct of the instruction in principles.

For this which led to his great advocacy of improved methods of agricultural instruction, Hilgard's memory is imperishable. A few years later his "monumental" study on cotton culture for the census was a force engrafting, Wickson said, "original research on the instructional work established through the educational land-grant law of Morrill, by the enactment of the Hatch law for experiment stations in all States; and when those institutions were being developed in the latter eighties, Hilgard and the research establishment which he had created in California were the accepted prototypes of men, means, and methods." From the California station emanated his further great studies in soils[20] —alkali, humid, arid, sandy, and numerous other combinations and kinds. Practical problems of crop productivity in relation to climate and other environic factors were dealt with scientifically. And from this work sprang into being an improved horticulture and agriculture for California and other western regions.

Another important California botanist with whom Hooker and Gray visited was John Gill Lemmon. Lemmon was by then a close friend of Parry and a correspondent with Gray and Engelmann. He was born in Lima, Michigan, educated in common schools, and went to the University of Michigan. During the Civil War, he was imprisoned in the Andersonville Prison and, when released, he came, much embittered and in ill-health from the experience, to the Sierra Valley, California.[21] The presence of interesting plants near his cabin revived an enthusiasm for botany and, gathering some of them, he summoned courage and went to Henry N. Bolander who wrote Gray February 2, 1873:

> By mail I send you with this letter a small parcel, containing plants collected by J. G. Lemmon, a teacher, in Sierra Valley. His specimens are poor; but still they may interest you. In [the] future he may do better; he is quite an enthusiast, and a good mountaineer; he may be able to find many new plants yet in those mountain recesses.
>
> He stays a good part of his time with Dr. Webber the owner of Webber Lake. [T]he old gentleman has no children; he adopted five, and had them educated, some in Europe, and now he is forsaken by all of them, and leads a retired life.
>
> In connection with this noble character, I would most humbly ask you to dedicate a species to each of these Gentlemen, if there are any new ones.

Gray dedicated not only species but a genus to Lemmon and with the passing of another year eastern botanists such as Canby and others were ordering sets of Lemmon's collections. Engelmann did not hear of him until 1875 when he asked Parry to send him a species of Abies which Lemmon had found. On October 11, Engelmann wrote Parry, "I am corresponding with Lemmon, from whom I hope much about the Abies

---

[20] See, for example, Howard S. Reed's comments on work of Hilgard in *A Short History of the Plant Sciences*, Chronica Botanica Co., 1942, p. 247, where he said, ". . . Hilgard, a man trained in botany and chemistry, gave (1892) an illuminating discussion on the relations of soils to climate. For breadth of view and wealth of information, Hilgard's work cannot be surpassed, yet it is painful to note how little influence it had at the time on investigations in plant nutrition." An excellent article on the "Rise of the American Experiment Stations" is contained in L. H. Bailey's *Cyclopedia of American Agriculture* (Macmillan Co., 1910), IV, pp. 423-425. Authority for some of the facts of the foregoing paragraphs is taken from this work.

[21] *The Dictionary of American Biography*, XI, p. 162.

question. Will you see him?" The following November Parry planned to visit Lemmon in Sierra County after going to Yosemite, probably to visit Muir as they were acquainted and had spent more than a week together around Lake Tahoe after Gray's and Torrey's early visits to Muir in 1872.[22]

Parry and Lemmon conceived plans to explore the Sierras from San Bernardino to the Columbia River! Lemmon arrived at San Bernardino early in 1876 and Parry arranged a desert trip in March to "work off some of his steam." He was in need of finances and Parry sought by a letter to Gray to secure an advancement of $100 or $200 and a considerable sum from Engelmann. Parry and Palmer sought to induce Lemmon to go to Colorado and collect. But he refused and, since Parry had to remain near San Bernardino, Lemmon left for Sierra Valley going by way of Santa Barbara. ". . . He is *active*," said Parry, "but excessively *nervous* & *fidgety*, does not like to stick to steady work, and likes to make a display of what he does besides being short of funds. He is a thoroughly *good fellow* and I would like to see him do well." For a year he was unwell. Parry took his Sierra plants to Davenport and on February 16, 1877, wrote Engelmann:

> Yes, still in the muddle of sorting, and quite discouraged at getting on so *slow*—, the S[outhern] Cal[iforni]a get *untouched*! I have Lemmon's Sierra plants in sets ticketed; ready for distribution, the full set number[s] 272 sp[ecies] and sums down to 180. There are but 8 sets he wants to realize on. . . . He is now buried in 8 f[ee]t snow alone at Webber Lake. . . .

But when Hooker and Gray came to California, Lemmon went to meet them and was made ecstatic by Gray's praise of him both as collector and teacher or "Professor" as he called himself.[23]

Edward Lee Greene did not meet Hooker and Gray, although he wished to. From Georgetown, Colorado, he had written Gray on February 27, 1876, telling him that he expected to go to California the following April and his address would be "Yreka, Siskiyou Co[unty] away up between M[oun]t Shasta & Klamath River!! I can hardly sleep of nights since I have secured my appointment to that field of Missionary labor, so delighted am I," said Greene. By July he was there, had done some collecting of which he sent Gray a parcel, and said he was anxious to climb M[oun]t Shasta. On Goose Nest he had found *Spiraea millefolium* but it was not in flower and the season was about over for the year. He too was interested in California oaks having sent Engelmann

22 See Badé's *Life and Letters of John Muir*, I, pp. 337, 343; II, p. 243.
23 Keeping himself "on the *qui vive* for the great botanists," Lemmon took three day excursions to localities such as Downieville and wrote on "The Great Basin" and "The Big Trees" for Coulter's *Botanical Gazette*, III, Numbers 3, 10, and 11, pp. 24, 87, 91.

"good things . . . one especially," as early as the autumn of 1874. And in August 1876 Engelmann by letter had disclosed to him "the mysteries of Abies, and informed him that he [Greene] was on the dividing line of *A. grandis, concolor, subalpina* (Clear Creek *grandis*) with *nobilis* thrown in—and that he must find out all about them." Greene had written Gray that Engelmann had asked "so many questions about oaks and conifers of this region that it will take me all the rest of the autumn to become able to answer the half of them." But he told Gray of his ascent of the Scott Mountains not far from Mount Shasta, where he had obtained "quite a number of the plants that I should have looked for on Mt. Shasta." Greene could have informed Hooker and Gray of much Wyoming, Colorado, Utah, and California botany; and, in turn, learned much from them. But early in the year of their exploration, Greene had informed Gray that he "was leaving Yreka and going across Arizona. I am now in the south western corner of New Mexico and likely to stay here through the spring and summer. Please therefore address me at Silver City, Grant County, New Mexico. . . . I have picked up quite a number of plants in passing through Arizona and shall try to send you a package within a few days. I do not know how promising a field this may be where I now am, but I shall surely find it interesting." He had found it interesting and in May had written Gray: "Do *you know* that the plains of this S[outh] W[estern] N[ew] Mexico are as yellow with Eschscholtzia at this season of the year as were ever any of the meadows of California, . . ."

Engelmann and he were "now having a hard wrestle about oaks," and Greene was placing before him and Gray much of the tree vegetation knowledge he had acquired in Colorado. So interesting did he find New Mexico that, similarly as he had done in respect to Wyoming, he wrote an article for the *American Naturalist*,[24] entitled, "Rambles of a Botanist in New Mexico." But that summer he received a "pastoral call" to Pueblo and so, hearing of Hooker's and Gray's proposed trip there, immediately wrote Gray telling how very much he wanted to meet them there. Neither there nor in California, however, did he meet them. On November 7, from Criswell, Colorado, he wrote Gray, ". . . I send you a specimen which grew in a spring on my farm in Bergen Park. . . . I reached Fort Garland two days after you were gone, and was too nearly worn out to follow you up though not a little disappointed at not seeing you and Dr. Hooker."

Gray and Hooker returned to Cambridge and from there Gray wrote Engelmann September 24:

[24] XII (April 1878), pp. 172-176; 208-211.

We are just back via Niagara; Hooker and I via New York, and the former having the Sunday with Eaton at New Haven. All well and happy to get home after a prosperous and, as you may imagine, laborious journey of ten and a half weeks. The trip to Shasta involved long stagecoach journeys, but they were most interesting. Returning to Sacramento we went on to Truckee, where Lemmon joined us by appointment. We gave one day to Mount Stanford and one to Tahoe, then took the overland train as it came on at midnight, and thence had no stationary bed till we reached Niagara. And we live to tell the story!

I want to tell you what we are led to think about Firs and Spruces. I will give in this my own opinions. . . . Hooker comes to the same conclusions or nearly. . . . Your reply will come to hand before Sir Joseph sails. . . .

Before Hooker and Gray separated, there was much work to be done at the herbarium. On October 19, Hooker wrote Charles Darwin:

I have indeed had a splendid journey; and thanks to A. Gray a most profitable one—nothing could or can ever reach his unwearied exertions to make me master of all I saw throughout the breadth and not a little of the length of the U[nited] States. The Geographical Distribution of the Flora is wonderfully interesting, and its outlines are not yet drawn. We have material for a most interesting Essay. I have brought home upwards of 1000 species of dried specimens for comparison of the Rocky and Sierra Nevada and Coast Range Floras, an investigation of which should give the key to the American Flora migrations. . . .

While still in western North America Hooker had come to the conclusion that the floras of the eastern United States and that of the West are of two continents. Returning to England, he planned an address before the Royal Institution and wrote Gray:

Well done your hypothesis! It is splendid. It fits in splendidly to a Friday evening Lecture . . . entitled "On the Distribution of Plants in N[orth] America." . . . I have made Meridional Distribution my principal theme, and had intended to treat of Pliocene Flora, &c., and the effect of the Alps as compared with the American M[oun]t[ain]s, in the latter directing the course of migration, and in the former favoring the extinction of N[orth] Pliocene forms; but I had not come to the formulating of the subject as you have done. . . .

. . . I intend to show, first how your researches on the Japan Flora and mine on the Arctic each come in, and are foundations upon which we meet in theory (one of us in England, the other in America), and how we coalesce as to results in our present labors after travelling together. . . .

Talking of the E[ast] and W[est] Floras of N[orth] America, I am surprised to find so many Asiatic types in W[est] America that are not in East; and the Western American representatives of Asia seem to belong to a different type from the Eastern representatives. Can both (the East and the West Asiatic types) have branched off from one Asiatic migration into N[orth] America? or were there two migrations at very different periods, one into East, the other into West? if so which first? . . .

Sir Joseph built his address evidently along lines of his article published in *Nature* on October 25, an extract from which was published in Coulter's *Botanical Gazette*:[25]

The net result of our joint investigation and of Dr. Gray's previous intimate knowledge of the elements of the American flora is, that the vegetation of the middle latitudes of the continent resolves itself into three principal meridional floras, incomparably more diverse than those presented by any similar meridians in the old world, being, in fact, as far as the trees, shrubs, and many genera of herbaceous plants are concerned, absolutely distinct. These are the two humid and the dry intermediate regions. Each of these again is sub-divisible into three, as follows:

(A) The Atlantic slope plus Mississippi region, sub-divisible in (1) an Atlantic, (2) a Mississippi valley; and (3) an interposed mountain region with a temperate and sub-alpine flora.

(B) The Pacific slope, sub-divisible into (1) a very humid forest-clad coast range; (2) the great hot drier California Valley, formed by the San Joaquin River flowing to the north, and the Sacramento River flowing to the south, both into the Bay of San Francisco; and (3) the Sierra Nevada flora, temperate, sub-alpine, and alpine.

(C) The Rocky Mountain region (in its widest sense, extending from the Mississippi beyond the forest region to the Sierra Nevada), sub-divisible into (1) a prairie flora (2) a desert or saline flora; a Rocky Mountain proper flora, temperate, sub-alpine, and alpine.

Hooker anticipated a great botanical question of the future. Gray went abroad a few years later and spent two months at Kew, going also with the Hookers on a journey on the European continent. The two the while prepared together their work, "The Vegetation of the Rocky Mountain Region and a Comparison with That of Other Parts of the World," published as a *Bulletin of the United States Geological and Geographical Survey of the Territories*,[26] Hooker, preparing another address, the great presidential address delivered at York that year, on "The Geographical Distribution of Organic Beings," wrote Darwin:

I am doubtful about going into the Flora of past ages, beyond the Tertiary. I quite believe in the sudden development of the mass of Phanerogams being due to the introduction of flower-feeding insects, though we must not forget that insects occur in the coal and may have been flower-feeding too. . . .

It appears to me that the great Botanical question to settle is, whether the main endemic Southern temperate types originated there and spread Northwards, or whether they originated in the North and have only just reached the South, and have increased and multiplied there (to be turned out in time by the Northern perhaps). The balance of evidence seems to favor the latter view, and if Palaeontologists are to be believed in crediting our tertiaries (even polar ones?) with Proteaceae, it would tend to confirm this view, as do the Cycadeae, now about extinct in the N[orth] Hemisphere and swarming in the South.

25 III, Number 2 (February 1878), p. 14.
26 VI (1882), Number 1. February 11, 1881 (Article I), pp. 1-77.

Buffon's and Saporta's views of life originating at a pole, because a pole must have first cooled low enough to admit of it, is perhaps more ingenious than true—but is there any reason opposed to it? If conceded, the question arises, did life originate at both Poles or one only? or if at both was it simultaneously? . . .

Saporta was a correspondent of Gray and Leo Lesquereux. Europe was responding to the work of Asa Gray—and Lesquereux. Engelmann continued with his researches, publishing in the journal of the *Proceedings of the St. Louis Academy of Sciences* on November 19, 1877, his "Geographical Distribution of the North American Flora." The fact that this was published at this time shows how great was Engelmann's interest in Gray and Hooker's work. And how much he sacrificed by doing his "duties!"

In April 1879 the *Botanical Gazette*, commenting on "The Distribution of the North American Flora" by Hooker, had tersely said: "Hence to state it all in one sentence, our Eastern flora has come from the North and our Western flora from the South."[27]

Great as were botanists' interests in American plant distribution, this, nevertheless, was not the only subject of growing interest. In 1879, Gray noticed in *The American Journal of Science and Arts* a study by Eduard Strasburger on "Polyembryony, True and False, and Its Relation to Parthenogenesis," an investigation of the embryo sac of angiosperms. Strasburger was at Bonn, Germany, establishing a great center for plant cytological research. In 1875 had been published his *Zellbildung und Zelltheilung*, after which came his important studies in nuclear organization and mitosis. Following the great genius Wilhelm Hofmeister, Strasburger took the center of European botanical interest enlarging work in subjects such as alternation of generations and the embryo sac of gymnosperms and angiosperms. He it was who was largely responsible for development of cytological technique, the course of which was to influence much plant morphological study in America in less than a decade as American students began going to his laboratory. Julius von Sachs a few years earlier had given to the science his famous text, *Lehrbuch der Botanik*, presenting an adequate treatment of the whole vegetable kingdom. This work, and others of the German masters, soon brought a number of English and in time many American botanists to study on the continent. With their return to further studies, botany was to begin a new era of investigation, developing many new branches of research.

However, already a new type of work had begun in America—the study of plant life histories. One year before—in 1878—Gray had re-

[27] *Botanical Gazette*, IV, Number 4, pp. 147-148.

viewed Darwin's *The Different Forms of Flowers of the Same Species*, a work dedicated to Gray. Coulter, after reading it, commented:

It was refreshing to see all through the book the notice that was taken of American botanists, for it is a sign that they are not all completely absorbed in Systematic Botany, which, in a country comparatively new, very justly has a controlling interest, but are beginning to study life histories.

Study of plant life histories was to be an important step in developing a scientific approach toward another mode of inquiry in the plant world —forestry. Forestry, almost unknown in America in the early 1870's, was first conceived as an economy of resources, a conservative lumbering essentially practical, a means of preserving and restoring forest growth. But a small leadership, aiming to make forestry more, was arising: urging reservations of state owned land under management, and to make the entire work, public and private, scientific, including investigation along silvicultural and other lines of scientific study of forest productions. In this no one leader was to assume greater prestige than Bernhard Eduard Fernow. Since the author of this book plans writing a biography of Fernow, consideration here will be confined to participation of American botanists in the movement which, within a few decades and a generation of American foresters, was to sweep the North American continent from Mexico to Alaska. Indeed, its orbit became the world. Illustrated by work of Beal and Spalding in Michigan, Bessey in Nebraska, Charles Mohr in Alabama, and Rothrock in Pennsylvania, and others, botanical science laid a basis for the forester's identification of tree species, and, moreover, aided the still greater work of "agitating" need for spread of knowledge of forestry practice and principles. In 1882 the first American Forestry Congresses were held in Cincinnati and Montreal, making the American movement at once international. Of the committeemen of the American Association for the Advancement of Science which memorialized the federal government to create a division of forestry, Gray, Newberry, Brewer, and Hilgard were botanists. While they were not the real leaders of the forestry movement in America, study of soils, waters, light, and climate in relation to plant life, including the forest cover, gradually became fundamental and was transitional in botany. During the two decades here considered, real American forestry got little more than a start but directions taken laid sound, scientifically conceived foundations for the future.

CHAPTER V

Other Southern and New Explorations
in Mexico

STIMULATED by the developing philosophical phases in botany, arising from sound systematic investigations, North American botanists continued with renewed zeal the exploration of their continent.

In 1878 Sereno Watson's great *Bibliographical Index to North American Botany*, Part I, *Polypetalae*, was made available as a Smithsonian *Contribution*.[1] The year before he had presented another *Contribution to American Botany*, a "Descriptions of New Species of Plants, with Revisions of Lychnis, Eriogonum, and Chorizanthe"; and in 1878 the *Contribution*, "The Poplars of North America."

Gray, busy as he was, presented in 1877 a paper entitled, "Characters of Some Little-Known or New Genera of Plants," including Lemmonia and other interesting genera; and that year appeared his morphological paper, "Mode of Germination in the Genus Megarhiza." About this time was published a part of Goodale and Isaac Sprague's *Wild Flowers of America*, a work conceded "magnificent" by all botanists. In 1878 came another *Contribution* by Gray, including in its scope *Elatines Americanae*, discovered by "the sharp-sighted and enthusiastic Mr. Lemmon"; two new genera, one named for Charles Wright and another for Hezekiah Gates, an Alabama collector—both genera of Acanthaceae; new Astragali; and some miscellanea. Included among Gray's determinations were many plants from Parry, Palmer, Lemmon, and William Cusick of Oregon; and *Erigeron miser* collected at Donner Lake, California, by Greene.

As for Engelmann, sacrifice and indeed some suffering yielded their compensation. For all he gave up by not going with Hooker and Gray, he was repaid by one of his most productive years of publishing activity. The *Transactions of the Academy of Sciences of St. Louis* presented his papers: "About the Oaks of the United States," "The American Junipers of the Section Sabino," "Notes on Agave," "Notes on the Genus Yucca," "Oak and Grape Fungi," "Synopsis of the American Firs." He knew he had much work to finish. *Botanical Gazette* presented his articles, "The Species of *Isoetes* of the Indian Territory" and "*Baptisia sulphurea* n.sp." And this year some of his reports for Rothrock's publication of the

[1] *Miscellaneous Collections*, XV, Number 258 (Washington), p. 476.

Wheeler expedition botany were completed. True, some of these articles were continuations of other articles already published. But they were all, for the most part, intended for publication in Gray's great *Synoptical Flora of North America* and, in part, for Watson's second volume of the *Botany of California*.

In 1878 a *Bulletin of the United States Geological and Geographical Survey of the Territories*[2] published a "Catalogue of phaenogamous and vascular cryptogamous plants collected during the summers of 1873 and 1874 in Dakota and Montana along the 49th parallel by Dr. Elliott Coues, U.S.A.: with which [were] incorporated those collected in the same region at the same time by Mr. George M. Dawson, by Prof[essor] J. W. Chickering." The Dells of Wisconsin, the New River region in southwestern Virginia, the New and Guyandot rivers of West Virginia, portions of Missouri and Kansas, the Great Basin of Utah, the states around the Gulf of Mexico, Willoughby Lake in northern Vermont, and many other localities had been visited by sporadic explorations during the few preceding years. Leo Lesquereux had issued additions to his Arkansas flora, including botany and paleobotany, originally published in 1860 as part of the geological survey of that state. Additions were made similarly to the published Iowa flora, treating both higher and lower orders of plants, and including Bessey's excellent treatise on the blights or Erysiphei. Charles H. Peck was investigating the Fungi of New York state with much thoroughness. A. E. Johnson had issued a mycological flora[3] of Minnesota, listing 559 species, all new to the state and two new to science. Oliver R. Willis had published a *Catalogus Plantarum in Nova Caesarea Repertarum* which earned praise as a most worthy work on New Jersey plants. Daniel Cady Eaton was preparing to publish his great work, *The Ferns of North America*. Lichens were being studied in several regions, particularly Illinois. Austin was busy with hepaticology and W. G. Farlow's, C. L. Anderson's, and D. C. Eaton's *Algae Exsiccatae Americae Borealis*, Fasciculus I, dealt ably with Algae. It contained fifty authentically named North American seaweeds, many of which were rare species from California and Key West, Florida.

The important explorations of this year, however, were in the southern United States and Mexico.

On August 17, 1878, Dr. A. Gattinger of Nashville, Tennessee, wrote Gray:

[2] IV (December 11, 1878), Article XXXIV, pp. 801-830.

[3] Published in the *Bulletin of the Minnesota Academy of Natural Sciences*, an essay of 100 pages, determining 559 species, all new to the state. See *American Naturalist*, XII (1876), p. 466.

Four weeks ago I started on a tour in the mountains of East Tennessee where I visited the Big Frog Mountain situated on the State line between Georgia & Tennessee and not far from the North Carolina line. The vegetation made up from very diverse species was, in the upper part of the mountain exceedingly dense and luxurious rendering the ascent very tiresome. 6-7 feet was the average h[e]ight of the tangles. *Lilium superbum* abounded reaching a h[e]ight of 7 feet. I have counted 25 flowers on one shaft but could not see any with more. The summit is timbered with Chestnuts in full bloom at the time, with Black oak, Carmine Ash, &c and any thing like an alpine character of vegetation was not to be noticed at least on the summit (5000') and western Slope from which I ascended. That this region is very wild you may presume from the fact that during the night we biv[o]ua[c]ked on the mountain a deer came to us to gaze at our camp fire and in the morning before we reached the summit we saw a bear and two cubs not more than 10 steps to our right. The northeastern side of [the] mountains forms inaccessible walls & in the depths of these gorges winds the Ocoee river and along it the copper mine road through the mountains. . . .

Gattinger continued going through the Cumberland Mountains and later explored in the Cedarbarrens along the Nashville and Chattanooga Railroad. Two years later he was to return to the latter place to gather "the golden flowered Leavenworthia,"[4] and each season he sought to collect. But Gattinger's explorations for the most part were financed by himself with small encouragement as compensation. When in 1901 he characterized his years in Tennessee, he termed them "a school of endurance." To him fell the task of compiling a state flora,[5] and he very creditably performed it.

Of course, botanical exploration was going on in practically all of the states. Still, the work in many regions was that of amateurs with little or no professional skill and with little or no knowledge of scientific systematization. In mountainous regions in states with remote areas such as Kentucky and even Virginia there were many regions *terra incognita* to the struggling science striving to keep pace with those branches of science more remunerative and of more quickly recognizable utilitarian value.

Florida was yet a difficult land for exploring. On April 21, 1877, Garber had written Gray from Miami sending some new plants for determination, asking the results of his Tampa exploration, and saying: "The country here is wild and the botany interesting but it is difficult to walk over the land on account of the saw Palmetto and entangling vines." Still from Miami, Garber wrote Vasey in May:

. . . The vegetation is markedly different here from that of Middle North and

4 See Asa Gray, "The Genus Leavenworthia," *Botanical Gazette*, V (1880), Number 3, p. 25.
5 See notice of the Tennessee flora's distribution in the *Botanical Gazette*, XII (1887), Number 4, p. 98; also *The Tennessee Flora*, Nashville, 1887, by August Gattinger.

West Florida, especially so in the woody growths. Then, too, a greater variation in size—Erythrina, which I have met in the latitude of Cedar Keys and Mellonville, was always a shrub four to five feet high—here it is common and generally of the same size, but also not uncommonly assumes the tree form and attains a height of twenty to thirty feet. . . .

I encounter a good many disadvantages in exploring and drying here, but altogether my success was good and I am very well satisfied with the progress. I think I will have some new to our flora and possibly to science . . . it is not unlikely that I will meet many of the same plants at Ft. Myers and [Peace?] Creek where I now propose to go . . . probably a month will enable me [to] collect all there, and thence to Manatee and Tampa. . . .

The following April Garber mailed Gray a small parcel of plants but said, "My collection of '78 is small—about 50 species and not many specimens." In May, however, he sent Gray a set of his south Florida plants and told him he had found Chapmannia abundant in the pine woods near Sarasota, and said:

I am now making arrangements for a 6 mo[nth]s cruise around the coast from here [Manatee] to [the] north end of Biscayne Bay and back in the fall, going up the rivers and lakes of the mainland as far as possible. I observed last year on the Miami, Little, and Hillsboro rivers after striking purely fresh water, there appeared considerable change in the vegetation, so that in going up and down streams I will have the advantage of meeting plant[s] of fresh water habit[at]—the brackish water plants and the saline or coast plants.

While exploring, Garber often went by boat. Having spent considerable time around Cedar Keys, Florida, and in the "jungles" along the Gulf Coast, he had gone from there by schooner to Tampa; and from Tampa he proceeded by land into the distant highlands beyond. Florida, of course, had other explorers. Much of the same ground over which Garber had gone had also been explored by W. W. Calkins who in August 1877 communicated his "Notes on Winter Flora of Florida."[6]

Chapman, who had recently published an "Enumeration of Some Plants Chiefly from the Semi-tropical Regions of Florida,"[7] continued exploring as much as his advanced age and health permitted. In 1877 he had visited his "old haunts of Chattahooche & Quincy," of which latter place he had published about 1845 a plant list,[8] and found that in the more than thirty years wonderful changes had taken place. Fields were where woods had been. Near Quincy pine woods had been transformed "into a noble growth of Oaks. Along the fence where you got your Torreyas," he told Gray, "I found for the first time *Gonolobus Baldwinianum* and the flowers of the paniculate umbels were white,

6 Coulter's *Botanical Gazette*, II (August 1877), Number 10, p. 128.
7 *Ibid.*, III (January 1878), Number 1, p. 2.
8 *West. Jour. Med. & Surg.*, III, Number 6, and reprint, 1845.

pure white." When in May 1878 Chapman received the first part of Volume II, containing *Gamopetalae,* of Gray's *Synoptical Flora of North America,* he wrote Gray:

... [I] have been looking over your book—a masterpiece of research, and generally judicious (not always) It is capital & call it a Synopsis or what not the characters are long enough for accurate determination which is all that is required in such a work. The antique type makes the best Flora and I am glad to see you continue its use. . . .

... Thank you for the book—hope you may live to see the end of it—although like the excellent Elliott you may have to consecrate the result of your labor to the dead. . . .

Could anyone succeed to Chapman's place in southern United States botany, that man was A. H. Curtiss. When in 1873 he had published his *Catalogue of the Phaenogamous and Vascular Cryptogamous Plants of Canada and the Northeastern Portion of the United States,* Curtiss had indicated the geographical range of each plant in what was then believed "the three most dissimilar districts, viz.: (1) Canada; (2) Illinois; (3) Virginia."[9] The regions near Liberty, Bedford County, Virginia, were then one of the major fields of his explorations which also had included in 1867 a trip to the celebrated Peaks of Otter about which he had written an account to Gray in August of that year. Curtiss, however, after these early explorations, turned his attention to the South and he is best known for his Florida explorations. On June 10, 1878, he wrote Gray. The letter was described as from "Near Jacksonville, Fla.":

... I expect to spend the last half of this month up the St. Johns [River], travelling in a small boat with a man to assist—hope he will not serve me like Bartram's boatman, get tired of travelling with a "puck puggy" after a few days, give a whoop & vanish in a convenient forest.

What a noble river is this broad, placid, blue St. Johns—now encrimsoned by the setting sun! Here she is old & near her last resting place—the ocean; ere I write you again I hope to become acquainted with her younger self.

Two years passed and Curtiss's explorations became even more extensive than Garber's. On August 12, 1880 Curtiss wrote Gray a lengthy letter narrating his important explorations between Key West and Tampa where the luxuriant tropical forests of southern Florida had astonished him. Said Curtiss:

... I will soon overhaul my collections made between Key West & Tampa—& send you samples—you will find them of much interest. I solved the great botanical mystery of South Fl[orid]a by penetrating to the "Royal Palm Hammock" as it is called by those who explored the Big Cypress during the last Indian war. I could

[9] *The American Journal of Science and Arts* (3rd ser.), VI, p. 230, for Gray's comments and review of the publication.

find but one person who knew the way to it & he lost the way twice. When at last I stood at the foot of those trees I beheld the most wonderful tree—except as to size —in America, & at the same time the most wonderful Orchid growing on them— both companions in the tropics no doubt.

Curtiss found the coconut growing wild, many new species of plants, and much else of botanical interest though not without suffering some hardship. While on land, at points like Miami and the Everglades which Curtiss entered twice, he found the forests needful of careful examination—his party suffered exceedingly from fever and insects. While on water, they endured squalls, shoals, seasickness, and the glare of the sun; for their route took them by boat by way of Bay Biscayne, Cape Sable, Punta Rassa, Ft. Myers, the Caloosa River so-called by Curtiss, Gasparilla Pass, to Tampa Bay. But Curtiss exulted in his journey. He told Gray:

. . . I never heard of any one working around the Everglades & Reefs in summer except parchment-skinned sponges and botanists. A botanist can keep well & hearty for 40 days subsisting entirely on esthetic pleasure—give him a handful of Cocoa Plums occasionally & a bit of al[l]igator meat or jerked rattlesnake & he will work the year round.

The year 1880 saw other Florida explorations. That year John Donnell Smith botanized the Peace Creek region of South Florida, paying attention to Florida Wolffiella. During January and February, Dr. J. J. Brown and W. W. Calkins began at Sand Point opposite Cape Canaveral and, going a distance of 150 miles to Jupiter Inlet, collected 106 species in flower, later comparing notes with Curtiss at his beautiful home, Talleyrand Place, on the St. Johns River near Jacksonville. At Jupiter Inlet Calkins found *Epidendrum cochleatum*, L., an orchid, the first time the species had been found in the United States, Watson said. Another interesting discovery of about this time was the floating fern, *Ceratopteris thalictroides*, found by Curtiss in southern Florida.

Curtiss visited all the most interesting Florida regions. He visited Chapman at Apalachicola and rambled through the Torreya region Gray had explored. He planned a trip to the Florida Reefs and interior of the state. Marine Algae which he collected at Key West were published by Farlow and a list of Florida ferns and localities was regarded so valuable by Eaton the latter thought it should be published. Curtiss's valuable fascicles of Florida collections, his continued thorough explorations there and extending to islands of the West Indies, his distribution of subtropical and tropical plants to taxonomic institutions throughout the world rank him as one of America's most important botanical explorers. An untimely death took Garber in 1881, robbing the state of one of its most enthusiastic explorers. So rich a floral region as Florida was bound to have other able naturalists, good botanists, and collectors. Mary C. Reynolds may be cited as an example.

Indeed, the South, from early beginnings of United States history, had always produced able botanists who, laboring against poor communication facilities and lack of access to larger herbaria, produced over the years notable botanic works.[10] Charles Mohr of Alabama, explorer and systematist, presented to science a work on the flora of that state, regarded at the time of publication as one of the great taxonomic productions of American botanical history. A product of a lifetime's work, it had been started during the last decades of the century by visits to such regions as "the southern spurs of the Cumberland mountains bordering on the Tennessee" River, to other places during successive seasons—the Cahaba River headwaters was planned as another—and finally in 1901, after Mohr's death, his great account of *Plant Life of Alabama* was published as a United States National Herbarium *Contribution*. Forty years of exploration had been utilized in preparation of the work.

Curtiss was not the pioneer Floridian collector. A number had preceded him and most of their work had been systematized by Torrey. Alvan Wentworth Chapman still stood as the great survivor of the first important group. His famous *Flora of the Southern United States*, which Gray had aided him in compiling, still was the leading taxonomic authority. On April 18, 1881, Chapman wrote Watson:

> ... A new edition of my *Flora* has long been needed and I have at last "woke up" and prepared all additional matter so far as I know in the form, for the present, of a Supplement. Now there are in the Herb[ariu]m of which you have charge a goodly number [of] species belonging to my beat that must come in, and the question is how can this be done. I am old (oldest of the lot I believe) and don't want to go on if I can help it. Are your hands so full that you cannot help one here. If I knew how many species you have in your Herb[ariu]m, not included in the List I send with this, I could better estimate what amount of labor is demanded to include them. . . . I know of three or four orchids down in the peninsula which I have no means of determining, & Garber found two Tillandsia. . . . I wrote to Eaton to ask him to do up a dozen Ferns that have turned up & I have sent a copy of the List to Canby & shall send another to Engelmann, for I want to bring up the work complete to date as far as it is possible. . . .

Two years later the supplement, containing flowering plants and ferns of Tennessee, North and South Carolina, Georgia, Alabama, and Flor-

---

[10] See F. Lamson Scribner's article on this subject, *Bulletin of the Torrey Botanical Club*, XX (August 10, 1893), pp. 315 ff.

ida, made its appearance. Another edition would appear in 1897 from Cambridge.

Chapman died in 1899. Curtiss, however, lived until 1907 and was able to do much in preparation for more thorough Florida explorations which began after the start of the twentieth century. Dr. John K. Small commenced valuable studies around the Florida Keys, Biscayne Bay, the Everglades, and other near by regions. The establishment of a subtropical laboratory, first at Eustis and later Miami, aided. In his early exploring, Curtiss satisfied himself concerning the ranges of the cypress, pine, and cedar trees for Charles Sprague Sargent (who was interested in forestry at Cambridge) and as far as possible geographical ranges of plants for Gray, and reported them. Thus was laid the foundation for the later more thorough investigation by Sargent of the arboreal vegetation of southern Florida, especially that of the Florida Keys.

Sargent was at this time engaged as an agent of the United States census, bringing together data in regard to American forests and forest resources. His work was to culminate not only in the very valuable government publication of the census on the subject but also in the remarkable Jesup collection of wood specimens of American trees and the important forestry herbaria of the Natural History Museum of New York and the Arnold Arboretum founded at Cambridge less than a decade previous. In the early 1880's Sargent, as a part of the census investigation, and later as a member of the Northern Pacific Transcontinental Survey in Montana, would extend his forestry searches to cover the United States on a nationwide scale. The great glaciers of northern Montana would in the course of the latter be discovered and Sargent would have much to do with the eventual establishment many years later of Glacier National Park.

Of however much consequence botanical exploration in southern United States in the year 1880, and previous thereto, was, investigations into Mexico and regions outside of the United States in the western hemisphere were even more important. Exploration by North American explorers to regions outside the United States and its Territories had begun almost contemporaneously with inauguration of world-wide exploration by the United States government: the United States Exploring Expedition under Captain Charles Wilkes; Commodore Perry's expedition to the Japan and China seas in 1852-1854; the North Pacific Exploring Expedition under Captain John Rodgers; all of the first half and early second half of the century. Botanical exploration had continued, with increasing but not complete thoroughness, to Alaska, Greenland, the Arctic regions on the north and to far southern regions of South

America. For the most part, however, foreign and not North American explorers had gone to South America and islands of both the Pacific and Atlantic oceans. True, explorations had not gone deep into the interiors in most instances and many areas remained totally unexplored, especially in Mexico and South America.

Scientific exploration, moreover, had several times gone to regions such as Panama, Venezuela, Cuba, and other more commonly visited places; also to Nicaragua which was on the route of many explorers across the Isthmus of Panama to Western United States. Only recently American botanical exploration had been extended by Augustus Fendler, an able, notable, and worthy collector, to the Island of Trinidad. Fendler had been living near Canby at Wilmington, Delaware, but in the autumn of 1876 had notified Gray of his intention to leave the United States. He had been a keen scientific observer in many ways, studying such matters of vegetable physiology as the influences of excessive atmospheric humidity on plant health; insect, weather (especially storms), and other special agencies in plant fertilization, and the like; but he had gotten off into studying and published in 1874 a work *The Mechanism of the Universe, and Its Primary Effort-Exerting Powers. The Nature of Forces, and the Constitution of Matter; with Remarks on the Essence and Attributes of the All-Intelligent.*[11] And Gray considered him "a gone goose." Gray wished that he would let "Cosmical Science" alone. But when he heard from Fendler at Trinidad in 1877 and learned that he had collected "not less than 2800 specimens of ferns besides a number of phanerogamic plants" with seventy-one fern species being prepared in fifty sets,[12] Gray's interest was once again revived in this man whom he had once sought to be curator of the Gray Herbarium. He asked Daniel Cady Eaton to name Fendler's ferns and immediately Eaton agreed, saying he wanted also a set for George Wall who had sent him "a very fine set of Ceylon Ferns."

Botanical exploration, however, by explorers from the United States, had not gone deep into Mexico. At the time the boundary line between Mexico and the United States had been determined, and for a few years before the war with Mexico, botanical explorers such as Dr. Wislizenus, Dr. Josiah Gregg, George Thurber, Charles Wright, Parry, and others had explored the northern portions—especially Sonora and Chihuahua —but very few comparatively had gone to the central areas.

---

[11] Published at Wilmington with plates and 188 pages.

[12] Fendler's very interesting letters from Trinidad have been published by Canby in Volume X of the *Botanical Gazette* (June 1885), Number 6, p. 285. "An Autobiography and Some Reminiscences of the Late August Fendler." See also Gray's biographical sketch, *Scientific Papers*, II, p. 465.

On September 2, 1877, at Davenport, Iowa, Palmer wrote Engelmann:

... As Parry & myself are talking of going to *Mexico* this coming winter I am desirous of making a collection that is large and payable Mexico seem[s] to be the place for that.—The collections made in the dry barren sections though of importance are so limited that they will not pay the outlay Many of the places visited this summer were 80 miles apart and nothing to be obtained between the places because it is all desert waste and a team [had to be] hired to [make most] of the journey.[13] So am desirous to try a more fertile country. . . .

Parry and Palmer were waiting for Gray and Hooker to arrive in Davenport to aid in laying the cornerstone for the new Academy of Sciences building there. Palmer's mound explorations had not proved "very productive of specimens but much valuable information was derived therefrom." He waited in Davenport until October 8 and then started for Cambridge "to arrange some business matters. [W]ill then determine my next field of operation. [A]m thinking of Mexico if Parry would go along and money can be obtained," he said. In Cambridge he found that his 1877 collection had "turned out very well in new & rare species so says Dr. Gray," and, as a consequence, he turned with more determination to the Mexican trip. He had wanted to go to Cape Flattery and the Sandwich Islands but this had been abandoned.

Parry meanwhile had been making plans, writing Engelmann October 30:

Palmer is to go along to collect in archaeology & assist in Botany. The Peabody Museum appropriates $1000—to him for that purpose. . . . If we go I should hope to start early in December & go direct to City of Mexico via Vera Cruz, thence work north along the high slopes of Sierra Madre till we reach a good location for a central collecting point, perhaps near *San Luis Potosi*, then work out as the season closes via Texas &c&c I want to confer with you & Wislizenus on the subject and in case it would not inconvenience you would stop over with you for a week perhaps taking Mrs Parry with me that far whence she will go East to winter in Philadelphia & N[ew] Y[ork] So you see the grand scheme as it now stands and can advise me in the matter *soon*.

Parry read Humboldt's work on "New Spain" and a history of Lower California and by December 5 was in St. Louis where he spent "a very pleasant & profitable week" enjoying Shaw's Garden and Dr. Engelmann; and revising his plan so as to go north from San Luis Potosi to Chihuahua. By December 16 he was in New Orleans amusing himself "with rambles on the w[h]arf and the French Market" and planning "some excursions into the suburbs among the live oaks & cypress??" he told Engelmann. He reflected "with constant pleasure on the delightful

---

[13] Referring to Palmer's southern Utah—St. George, Paragonah, Mt. Trumbull (Arizona), Beaver, Colorado River, Spring Lake—journey where he had found *Agave utahensis* and his collections netted eighteen new species this year. See McVaugh's article, *op. cit.*, p. 775.

visit with" Engelmann, "occultations thrown in *gratis*." He wrote, "I hope to be able to make some small returns eventually properly to express the gratitude of Mrs. P[arry] and myself to your good wife & you. Agave Quercus Yucca??!!! Junipers Arceuthobia Pines &c&c I say." The Mexican War scare did not frighten him. "I have heard the wolf cry too often," he said.

On January 2, he wrote Gray from Vera Cruz:

> I arrived here after some delay Dec[ember] 28th. 6 days from New Orleans. [F]ound that Palmer had gone on to City of Mexico. As it was cool & pleasant here I concluded to stay over a few days and have been making short excursions in the vicinity. . . .
>
> I expect to go to City of Mexico by evening train. [W]ill stop over a day at Orizaba. Imagine it is cool on the table land and not much to see in the way of live botany. [W]ill look over the collections & sights of the city and vicinity before moving North to San Luis Potosi. . . .

To Engelmann he told of tramping over the sand hills and finding strange sights and strange people; and of finding one Cuscuta. "This is a remarkably *clean* town," he wrote, "beats Philadelphia all to pieces, is well watered, paved & lighted—a great traf[f]ic by sea with all parts of the world."

The trip from Vera Cruz to Mexico City was very interesting, especially after reaching the slopes of Orizaba. "The ascent to the table land 8000 f[ee]t is a succession of wonderful curves & zigzags," he commented, exulting in the pine groves near the summit and determining to return there to make an ascent of one or two high mountains—"not a very serious job," he said, "as it has been accomplished by ladies." Parry met Palmer in Mexico City which proved to be the place to find out everything about Mexico except botany "which has to be hunted up in its native haunts." There a society of natural history flourished and Parry was invited to attend a meeting where, when he could, he planned to present Engelmann's paper on junipers. A difficulty had immediately beset them. At every stopping place Parry and Palmer had found custom officials whom, they realized, would make collecting difficult, at least, when the collections were transported. Parry went to Chapultapec. He examined several species of Cupressus but the botany at that season was scant and so they decided to leave for San Luis Potosi, deciding their further plans after they learned "the lay of the land" there. On February 17, 1878, Parry wrote:

> . . . We have been here nearly 4 weeks, and are getting to feel quite at home. We have convenient *roomy* quarters and board at reasonable rates in the house of a *Presbyterian missionary*!! from Penn[sylvania]!! Barroeta[14] is very friendly &

[14] Gregorio Barroeta.

useful in introducing us to proper parties to facilitate our researches. We have also *another Botanist* (German) Dr. Shafner who has collected quite extensively and corresponds with Prof[essor] *Schultz*.[15] On the whole the prospect seems good for plenty of hard work, and I mean to take time for it, and do what I can but what *ought* to be done is quite appalling! I find the botanists here as in Mexico[16] quite loose in their determinations like Kellogg in San Francisco inclined to make species without knowing what has been done. [F]ortunately Barroeta has a fair library including Hook[er] & Benth[am] & Endlicher. I have been puz[z]ling over Cactus. . . . I wanted to stay longer in Mexico & clear up several things, but Palmer was fidgety & uneasy to get off. So we left with the inten[t]ion of going back, but I conclude to *stop here* where the field is less known and work up the locality thoroughly staying probably till Oct[ober]. Though the immediate vicinity is not promising the facilities of reaching mountain districts and visiting Haciendas is excellent. Palmer starts this week for Rio Verde on the borders of the tierra Caliente. The journey from Mexico here was by Dilligence. 4 days was interesting though rough and only able to catch tantalizing views of vegetation mostly at rest. Some of the Cactus views were wonderfully strange, great Tuna forests, Organ Cactus and another giant Cereus, now just coming into flower of which more anon. I must say before I forget, that I did not study as I should the Mexican Cypress (Taxodium) I was twice in the Chapultepec groves and took a good look at it, concluding in my own way that it was *identical* with our *bald Cypress, not evergreen* and as brown as any in Louisiana from a *late frost*. I think it is only the *absence of frost* that makes it *sempervirent* in exceptional winter. [S]ome of the *under* protected shoots were green, otherwise dead brown. The trees were too high to secure branches as I should have done. I do not know if I shall meet it again, but Palmer may on his return to Mexico. I noticed & collected 2 sp[ecies] Cupressus, one of which also grows here. I have not yet seen a Juniperus unless on the R[ail]·R[oad] from Vera Cruz, but am on the track of some. . . .

So continued Parry's interesting letter to Engelmann concerning pines, agaves, cacti, and oaks of the region. That March Parry sent him "several scraps of Euphorbia for [him] to growl over." Collecting continued, Parry going to a near by hacienda getting nice things at the foot of the mountains and going high enough to reach the outskirts of "Piñon." He tried several times to establish cordial relationships with Schaffner, the German botanist, but each time was treated shabbily and though he attempted what he termed lessons in politeness on several occasions finally gave up. Never once did the foreign botanist invite Parry to his house or grant him an interview to compare specimens. Parry grew lonely—the land was dry—and though Palmer returned he was soon away again to Mexico City and Zacatecas to study ruins. Parry would have turned over gladly the remainder of the task to Palmer but Palmer, while "persistent and industrious," was "not enterprizing" and Parry was not satisfied with his way of doing the work. "Besides," Parry wrote, "he has enough to do in ethnology & zoology."

[15] C. H. Schultz.   [16] City of Mexico.

Parry believed he was doing important preliminary work, "perhaps laying a stepping-stone" to help others. "The relation of this table-land flora to Central U[nited] S[tates]," he wrote Gray, "is very intimate & suggestive as I hope to show...."

Parry decided, nevertheless, to pull up stakes in July and move northward toward Saltillo and the Texas frontier, stopping at good botanical localities along the way. San Antonio and St. Louis were to be his destinations, hoping to reach the latter in time for the Solar Eclipse. "If I do," he wrote Engelmann, "I shall leave here full direction[s] for Palmer to finish up collection here. . . . I have made a final attack on our friend? Schaffner I addressed him a polite note which I delivered *personally* at his house, stating that I had been requested by my botanical correspondents in the U[nited] S[tates] to make a special study of Agaves. . . . This will I imagine force him to *show his hand* and then I shall be ready to show mine either for *peace or war*!!" Parry became convinced he was "at *least* secretly hostile. . . . I notice in a late publication to a Mexican journal he promises to send to the Nat[iona]l *Museo* a set of over 1000 sp[ecies] of plants collected near San Luis—that will beat me by 500!! at least but I will steal a march on him in the publication as he is apparently doing nothing and Prof[essor] Schultze being dead has no one to have recourse to in Europe. Prospect soon of getting over my lazy fit & getting into the M[oun]t[ain]s among the pines & oaks." Parry told Engelmann, "I am anxious to hear what you make of my *Agave Engelmanni* or shall we say *A. Schaffneri*!?" In July Parry wrote Gray concerning his "nice pot plant [which] grows in *dry rock crevices* [which he] called *Agave Engelmanni*" and said, "My work here will I think prove more interesting in its relation to U[nited] S[tates] Flora than you suppose; equally in the *presence* and *absence* of western interior sp[ecies]. Regards to Mrs G[ray] & Watson & 'Max' [Dr. Gray's black and tan terrier which, although Parry did not know it, had died the September previous]."

Parry, hoping "to pick up some nice things besides getting a connected series of observations," left San Luis for Monterey and Saltillo, taking the overland route which led by way of Brownsville to San Antonio, Texas. He evidently became ill and had to abandon some of his intended explorations but on October 30 he wrote Engelmann:

What has become of you? time the *quarantine* was raised, and you reported on plants sent, fulfilled promises &c&c&c

I am right well again   good ap[p]etite & lively. [H]ave sent Gray a set of Compositae, of which he reports over 10 per cent n[ew] sp[ecies]! So I feel encouraged. My Boxes should now come on soon & then I shall be full of business so as well to clear up what I have on hand Can you name the two Asclepias? sent or shall I send to Gray. I expect something rich in that line when the full collections come in. I have not heard from Palmer who should be leaving San Luis Potosi about this time for *Tampico*. Mrs. Putnam has got up a splurge in moving my cases to the Academy, which is more than half *"Buncombe."* I see no chance for quiet steady work there, and will put up a *shanty* in the country where Mrs. Parry can help me *arrange*—the *show part* to stay at the Academy. . . .

That district between Monterey & Saltillo should be thoroughly explored, the high M[oun]t[ain]s & valleys adjoining would *"pan out"* splendid. Where is there a *young* active collector to undertake it? easily reached now via San Antonio. I would say Palmer, but I find he is like the rest of us good at shirking hard work!

We have had a snow storm & cold weather Do you know where Greene is? [H]e stayed here 3 or 4 days last August when we were all away. . . .

Engelmann heard from Palmer on December 4 by letter written from San Luis Potosi November 10. He was sending Parry three large boxes of plants. Among them were pines, oaks, and agaves, and "a plant of a large tree Yucca 20 to 30 feet high," also hickory nuts. Palmer said:

. . . have done all I can do in this locality [A]m tired out. [E]xpect to leave hear in a day or two for the mountains between this place and Tampeco [W]ill not be able to finish before the middle or latter part of December [T]hen shall take steamer for New Orleans via San Luis M[issour]i & Deavenport—if my funds holds out and nothing prevents. . . .

By January of the next year (1879) Parry's and Palmer's plants were in Davenport and Parry's shanty was piled high "with suggestive bundles no doubt some 'instructive' ones." Parry wrote to Engelmann: "Greatly to my surprize & delight Palmers live plants come through unfrosted, having slipped in between the sharp *zero* snaps and there is turning up a goodly number of choice things among more indifferent ones, including Agaves, Cacti, &c&c. I send in mail parcel of a few specimens of Juglans. . . . I hope between the two collections we can complete most of the Cacti & Agave &c&c The plants as far as I have been over them are in Palmers best style for which I hope he may receive due credit. . . . But where in the world is Palmer? . . . he is good at noting dates, localities &c&c."

Palmer turned up in Cambridge after a silence of four months. Soon he wanted to be off again and, as Parry suggested northern Mexico, he went to Texas, a land where Julien Reverchon[17] was exploring and beginning to build an herbarium of local specimens to be the best collection of plants of that state yet in existence—20,000 specimens of more than 2,600 species of Texas flora. Reverchon was a correspondent of Gray, Engelmann, and others. Perhaps that correspondence inspired

[17] See an excellent account of Reverchon in Samuel Wood Geiser, *Naturalists of the Frontier* (Southern Methodist University, 1937), pp. 275-288, especially pages 284-285.

Palmer's choice of localities. At any rate, he went to western Texas beginning in San Antonio late in the summer of 1879 after the prevalent epidemics of yellow fever and smallpox had subsided. Covering a wide scope of country, he went to New Braunfels, Lindheimer's region, Austin, Indianola, Longview, and other places in search of Indian mounds and plants. However, his plans included going to Matamoros, Mexico, and from there through the northern Mexican states, especially Nuevo Leon and Coahuila, to El Paso where he expected to cover northern and north central Texas—the Rio Grande and the country around Fort Concho, much of the areas being unexplored. Palmer eventually was to go where Parry wanted an explorer sent. On March 20, 1880, he arrived at Saltillo, Mexico, having already spent some time around Monterey and the intervening country. He found another new Agave near Monterey but an impending Mexican revolution kept his explorations near the frontier. In a box he sent he told Engelmann he would find "an eye opener (a fine large plant of *Agave Victoria*)." From Saltillo he planned to "go to the section about Par[r]as, Lagona [Laguna], and Balsa Mapseé the unknown sections to spend some time returning by way of Mount Clover to spend some days & then Fort Clark Texas &c. . . ." A portion of Palmer's route may have been over the one made famous by Dr. Wislizenus a number of years before when he entered Chihuahua and became a Mexican War prisoner. In spite of the fact that the Peabody Museum dropped him from their payroll because of lack of funds, Palmer completed his journey directing most of his attention to botany and covering the region from San Antonio to Laredo and Eagle Pass and areas in Coahuila, Nuevo Leon, and San Luis Potosi in Mexico.

In 1879, the same year Gray published "Characters of Some New Genera and Species of Plants, Chiefly of California and Oregon," he made known the botany of Parry's and Palmer's collection in "Characters of New Species of Plants from Mexico, Collected by Dr. E. Palmer and Dr. C. C. Parry," both published in Volume XIV of the *Proceedings of the American Academy of Arts and Sciences*. In Volume XV appeared Gray's special *Contribution* on the new species of Compositae[18] in Parry's and Palmer's San Luis Potosi collection of 1878, including two new genera, one of which was dedicated to Barroeta, the Mexican botanist who aided Parry, Barroetea, along with a second part on new North American genera and species, most of which were from

---

[18] On February 23, 1879, Parry wrote Engelmann, ". . . I am busy sorting up, just through with Compositae numbering up to 555. [T]he Compositae about 242 sp[ecies] will easily make up 1000 sp[ecimen]."

California and Oregon explorations. In 1880 followed another *Contribution*, dedicating among many others, a genus to Greene, Greenella, a genus to Reverchon, Reverchonia, and describing two species of Asclepias omitted from a former published conspectus and found by Greene in New Mexico where he was by then stationed at Silver City. The year before Gray had dedicated a new Saxifragaceous genus to William Suksdorf, an Oregon explorer, Suksdorfia, and a new genus of the lobelias to Thomas Howell, Howellia, also of Oregon.

After Sereno Watson presented on May 14, 1879, his most able *Revision of the North American Liliaceae*,[19] Watson presented May 5, 1882 as another of his *Contributions to American Botany* a "List of Plants from Southwestern Texas and Northern Mexico, Collected Chiefly by Dr. E. Palmer in 1879-80." Determinations of collections of other Mexican botanists were included, notably those made by Dr. J. G. Schaffner in the State of San Luis Potosi which had been also in part determined at Kew. How this must have interested Parry! But he must have applauded Watson's use of the recent catalogue issued on the known Mexican flora by W. B. Hemsley as part of the great work *Biologia Centrali-Americana* by Godman and Salvin. Watson published "Polypetalae" in Volume XVII of the *Proceedings of the American Academy of Arts and Sciences* and "Gamopetalae to Acotyledones" in Volume XVIII.

Daniel Cady Eaton did the Ferns and other vascular Acrogens, Thomas P. James the mosses, and W. G. Farlow the remaining lower cryptogamous plants of all the Palmer plants, including those of Parry's and Palmer's. Engelmann did the Cactaceae. But a great bereavement had entered his life—Mrs. Engelmann, beloved since his youth, died. Engelmann could do little in botany for almost a year. He had taken her for her health to Lake Superior.[20] But she had steadily failed.

With each of the Palmer plant publications and the revision of Liliaceae, Watson accompanied as part of the *Contribution* "Descriptions of New Species of Plants" ranging over North America, one set being specially allocated to the Western Territories and including species collected by Greene, S. B. and W. F. Parish of San Bernardino, California, Thomas and Joseph Howell of Oregon, Dr. Havard of the United States Army stationed in Texas, Lemmon, A. S. Packard Jr., and many others. Especially interesting were collections arriving from near

[19] Ably reviewed by Dr. Gray in *The American Journal of Science and Arts* (3rd ser.), xviii, 313 pp.; also Gray's *Scientific Papers*, I, pp. 278-282.

[20] Engelmann published in 1880, "Vegetation along the Lakes," *Trans. Acad. Sci. St. L.*, IV, p. xx. Prepared during Mrs. Engelmann's illness.

Jacksonville and the St. Johns River country, Florida, from A. H. Curtiss. And C. G. Pringle of Vermont was by then sending collected plants to Watson for determination.

Not all North American exploration, however, was in the Southwest and West. On April 25, 1879, Gray wrote Canby concerning another North Carolina exploration:

... About scheme: it is rather my notion to go via Statesville to Newton, explore down one fork of Catawba, till we find Darbya, or find Curtis's locality,[21] and back by the other; two days. But perhaps, to save time, you would prefer to keep on the railroad from Statesville to Lincolnton (where, by the way, *Magnolia macrophylla* grows), pick up Darbya, and then come up to us at Statesville or Marion. Then we will see locality of Shortia.

Then, my notion is to get some good searches along the flanks of the mountains, from Swananoa Gap to Linville Falls (find Shortia for ourselves, etc.), and even up to Deep Gap, which you see is pretty north. Then make Cowles tote us to Bakersville, and then end on Roan Mountain. . . .

Less than a month later, he wrote Engelmann:

... We go on a trip south to the mountains of Carolina with Canby, Redfield, and this time Sargent.

It was to have been done whenever Shortia blossomed. But that stole a march on us by flowering in April. So now we time it for the Rhododendrons, and will see Shortia out of blossom, and we hope to find new stations. Then I want to look up Darbya, of which only the male is known. Curtis seems to have got it, without flowers, near Lincolnton. Then we are to explore the east side of the Blue Ridge, from the base of Black Mountain to Grandfather, and then cross to the Roan, on which is now the Cloudland Hotel.

Oh dear! now that the time draws near, I wish I could stay at home and finish Parry and Palmer's Mexican Compositae, which abound with new or interesting species! . . .

I send you by mail a copy of my new "Text-Book."[22] You see I relegate to other hands the anatomy, physiology, and cryptogamia,—glad to be rid of them. I send, too, one of the few copies of the Shortia paper. . . .

The western North Carolina mountains were to Gray especially interesting. There the Atlantic forest, "especially its deciduous-leaved portion, is still to be seen to greatest advantage," he said, "nearly in pristine condition, and composed of a greater variety of genera and species than in any other temperate region, excepting Japan. And in their shade are the greatest variety and abundance of shrubs, and a good share of the most peculiar herbaceous genera. This is the special home of our Rhododendrons, Azaleas, and Kalmias; at least here they flourish in greatest number and in most luxuriant growth. . . . On these mountain-tops

[21] Moses Ashley Curtis, not A. H. Curtis.
[22] Sixth edition. The fifth edition was styled *Introduction to Structural and Systematic Botany;* the sixth, *Structural Botany or Organography on the Basis of Morphology.*

we meet with a curious anomaly in geographical distribution. With rarest exception, plants which are common to this country and to Europe extend well northward. But on these summits from southern Virginia to Carolina, yet nowhere else, we find—undoubtedly indigenous and undoubtedly identical with the European species—the Lily-of-the-Valley!"[23] Gray and his party visited the Yellow Mountains, Roan Mountain, the Blue Ridge, Negro Mountain, and Iron Mountain, among other places, finding on each plants of interest.[24] He returned by way of Lynchburg and Washington to Cambridge after resting with Mrs. Gray on Roan Mountain, a favorite locality.

Late that year Gray agreed "in an unguarded moment" to deliver two lectures to theological students of Yale College. These, delivered the winter of 1880, were first read to Dr. Oliver Wendell Holmes, a great friend, and became classics of the literature of the subject—*Natural Science and Religion.*[25] Gray had been long a member of an eminent Cambridge group—James Russell Lowell, Emerson, Longfellow, and others. Once calling on Longfellow, it is told, Gray criticized a poetic flower description by Longfellow. The latter replied, "I was writing a poem; not a book of botany." Gray was entering his seventieth year in 1880 and was still working on the *Synoptical Flora*, becoming aware, however, he would have to go to Europe before much more time elapsed. Sereno Watson was completing the second volume of the *Botany of California.* He was awaiting Thurber's completion of some grass work since specialists (for example, Bebb in willows) had been employed. Watson needed a vacation; Goodale did also. Gray, though sympathetic, did not worry. Time, he knew, would solve the matters. What worried him were the asters.

The year 1879 had accomplished much for American botany. Rothrock's reports of the Wheeler survey were finally completed. John Merle Coulter and his associate, Charles R. Barnes, were planning and had begun a flora of Indiana[26] covering the sand hills and grassy plains of Lake Michigan, the wet grassy meadows and swamps, the lakes proper, the tamarack and sphagnous swamps, and the prairie flora. J. C. Arthur had added to the compilation of the Iowa flora being done at Iowa Agricultural College.

[23] "Characteristics of the North American Flora," in Sargent's *Scientific Papers of Asa Gray,* II, pp. 276-277.
[24] See Redfield's "Notes of a botanical excursion into North Carolina," *Bull. Torr. Bot. Club,* VI (1879), pp. 331-339.
[25] Published in New York in 1880.
[26] *Catalogue of the Phaenogamous and Vascular Cryptogamous Plants of Indiana* (Crawfordsville, Indiana, 1881).

George Davenport was preparing a valuable *Catalogue of North American Ferns*. Ferns of Kentucky had been the subject of a work of John Williamson. There was much study of ferns at this time. Subsequent parts of Eaton's *Ferns of North America* were being published. Even Thomas Meehan had published a work entitled *The Native Flowers and Ferns of the United States*. Garber for whom the genus Garberia[27] this year was named was sending ferns from South Florida. John Donnell Smith had gone along the Gulf Coast region from Cedar Keys to Charlotte Harbor, and up the Caloosahatchee River ninety miles to its source near Lake Okeechobee studying a *"pseudo* Fern," *Ophioglossum palmatum* of Linnaeus. Mary C. Reynolds also had contributed knowledge concerning Florida ferns, collecting in the Indian River regions and around St. Augustine. Curtiss had also collected there as well as in southern parts and the Shell Islands of Florida. And Calkins had prepared a list of Florida plants.

Studies from the Red River of the North to Fort Custer in Montana by Dr. P. F. Harvey; studies of the autumnal flora of Fortress Monroe by J. W. Chickering; studies of Tennessee plants by Dr. Gattinger of Nashville; plant studies by G. C. Broadhead in Missouri; continued studies of Fungi by Peck; studies of fresh-water Algae by Francis Wolle; studies by Farlow on sea weeds of Salt Lake and Algae collected at points in Cumberland Sound in 1877; studies of the flora of the Blue Ridge, Virginia, by Howard Shriver; beginning studies of West Virginia flora along the Kanawha and New rivers; studies of foreign plants introduced into the gulf states by Charles Mohr; studies by Marcus Jones in Colorado; these and much else added to the botanical scholarship of that time.

Model works were being produced in those years, works to merit a lasting influence in American botanical systematization, as well as point the way to more exact and enlarged services in research. On April 4, 1881, Burrill, thanking Farlow for his recently published "Gymnosporangia or cedar apples of the United States," commented:

> Your resume of the literature is of much interest and value while the critical examination and comparison of species is, to my mind, the best work of the kind so far produced in America. Not less interesting though less conclusive is the account of your culture experiments. I sincerely hope these may be repeated by yourself and others stimulated by your example.

Farlow's research work had been productive for several years now. Since Charles H. Peck had written him in December 1875, expressing

[27] See Asa Gray, "On the Genus Garberia," *Proceedings of the Philadelphia Academy of Sciences* (1879), pp. 379-380.

appreciation of his study of the potato rot, others had followed Peck in addressing similar expressions of interest. "We need more such papers as the one on potato rot for it touches a point that most people can appreciate," said Peck, "why dry science has for many but little interest." He urged, "Let me say by all means follow up this line of investigation." Pathology had had its American origins in the rugged individualistic work in bacteria, initiated by Burrill, confirmed and extended by others. Nevertheless, the most important work of Farlow, although for the most part mycological in character, had fundamental bearings in plant pathological development in America. His was not the same type of laboratory work and field experimentation as that commenced in plant investigations of the United States Department of Agriculture later under Erwin F. Smith.

For a number of years students interested in plant disease study would survey conditions in the field, particularly in the farm field, garden, and orchard, and, bringing their problems into the laboratory, devise techniques to combat or solve widespread alarming destructive pestilences produced by insect ravages and other more inscrutable sources and causes of diseases. It was a work looming of equal rank and importance with a growing interest in plant breeding, long an art practiced by skillful growers of plants, but now being slowly professionalized, to considerable extent in the few early agricultural experiment stations and numerous college departments of plant study—some with laboratories —in America.

Mycology was essentially descriptive, describing more or less taxonomically diseases and disease-producing organisms when discernible. Pathology went at its problems as a doctor or research student attacks problems in animal and human diseases to learn the cure or method of control. Indeed, additional to fundamental work in plant physiology and anatomy, these two phases of study—plant pathology and plant breeding—would largely differentiate the work of the older generation and that of the new in plant science research, in botany, horticulture, and agriculture. Taxonomy had laid foundations, not only in higher orders of plants but also in the lower plant orders. Scientists knew fairly adequately *what* the plants were.

Furthermore, they were learning *what* were the parasitic growths and conditions; by microscopic analysis the relation of an alarming plant ravage and tiny organisms known as bacteria. To illustrate, not until the middle 1880's would "absolute proof" be established that an alarming disease known as pear blight was attributable to these tiny organisms, from the study of which a tremendous branch of scholarly scientific investigation would spring. Supplementing and amplifying the great discovery in this respect by Thomas Jonathan Burrill, Joseph Charles Arthur, botanist at the New York experiment station, would receive in 1886 at Cornell University the first doctorate in science degree (D.Sc.) conferred in America for a conclusive research on the pathological aspects of a plant disease. The tenor of scientific investigation would change to *how* to deal with immense new branches of research developed by scientists working quietly and unostentatiously in the field and laboratory—the branches having to do with breeding new plant varieties, new races and strains of plants, plants resistant to deleterious circumstances of environment, and the problems, equally important economically and financially, of saving crops from destruction. Many an eager mind filled with a zeal for plant research would be reinvigorated by the presence of these problems. And within a few decades these comparatively new branches of study would expand tremendously the orbit of plant science investigation.

Farlow's work, however, "was primarily concerned with questions of taxonomy and the life histories of the parasitic fungi."[28] Remedial work in the control of each disease was not his chief concern, although he understood its value and province, especially as the Millardetian school emanating from France "stimulated the beginnings of the Federal work in Washington" some years later. Farlow trained students in the fundamental work of mycology and many of his students supplied most significant contributions to the development in America of an exact science in plant pathology.[29]

[28] See L. R. Jones's excellent biographical memoir of Smith, *Nat. Acad. of Sci. Biog. Mem.*, XXI, first memoir, p. 6.

[29] For an illuminating discussion of this subject, especially with reference to the discovery of Bordeaux mixture which "more than any other one thing influenced and shaped the development of the science of plant pathology during the quarter century following its discovery," and its introduction and utilization in America, see Herbert Hice Whetzel, *An Outline of the History of Phytopathology*. Philadelphia and London: W. B. Saunders Co., 1918, pp. 58 ff.

CHAPTER VI

Engelmann and Parry in Oregon and California.
Gray Goes to Europe

S O OVERWHELMING was the death of Mrs. Engelmann to Engelmann, it took him almost a year to recover from the shock. They had been united for half a century and Engelmann without her looked only toward a life in the past. His destiny seemed dark and botany could not revive his zeal for living. Parry heard of the news indirectly and not hearing from Engelmann wrote Dr. Wislizenus who confirmed the fact. Young George Engelmann sought to requicken his father's will to carry on by soliciting Gray to invite him to Cambridge for a visit. At first Engelmann refused but time revived Engelmann's interest in botany and he accepted Gray's invitation to visit after Gray went on his North Carolina Mountain excursion with Canby, Redfield, Sargent, and others.

Parry—the restless man—had written Engelmann as early as January 1879 telling him of the Arizona Railroad's (Southern Pacific Railroad's) pushing eastward from Fort Yuma to Maricopa, a distance of 150 miles over the very difficult stretch of desert country there, with the ostensible object of joining the Atchison, Topeka, and Santa Fe line. Parry urged that before they became too old, he and Engelmann should go over that interesting travel route together. But young George Engelmann was ill and Mrs. Engelmann at that time none too well. So Engelmann had not responded. However, after her death and young Engelmann's recovery, Engelmann retook life slowly and decided to talk to Parry once Parry had written him again telling of the Southern Pacific's progress into regions of the *Cereus giganteus* and the Utah Southern's pushing into Arizona. "The best way to settle the question (*Abies grandis*) is for you to go *this summer*," said Parry, "1st to Salt Lake, then a trip down the S[outhern] Utah R[ail] R[oad] now rapidly progressing toward Arizona—also N[orthern] Utah[1] now well north of Snake River—then to Sierra Valley &c&c. [P]ity we were not young again. . . ." Engelmann decided to take the boat up the Mississippi River to Davenport and then go by way of the railroad or lake route across to Cambridge. After going to Narragansett Pier[2] where he evidently spent a while with his son and daughter-in-law and where he botanized with renewed vigor, he visited the Grays in Cambridge and then returned by way of New York and

[1] Utah Northern Railway.
[2] Narragansett Pier, Rhode Island, evidently, where W. W. Bailey visited in 1880.

Philadelphia to St. Louis. While in Cambridge Engelmann enjoyed going to the art museums—they brought back to his mind the museums of Italy—so, returning "happy and sad" to his home, he was prepared to answer a long letter from Bentham treating of pines and junipers, in which there were certain notions with which he disagreed. By February he was finished with Pinus and going immediately to Isoetes of which he hoped to give a history of the genus in North America. "So you go to Oregon?" he wrote Watson February 16, 1880. "May we meet there! What are your plans, when do you leave etc." Engelmann found, to his discouragement, on returning home, that he could do little medical practice and, although Letterman[3] had brought him "some good things" from Arkansas and much else remained undone, he could do little at botany. Parry wrote him. Gray wrote him. Watson wrote him. Palmer wrote him. But naught, it seemed, mattered. In March Dr. and Mrs. Parry came to visit him and Engelmann wrote Gray that he was more seriously considering the California and Oregon trip—with Parry.

On January 24 Parry had written Engelmann:

Long before this you have tossed my last ("absurd") letter into the waste-basket and are ready for *more* I am now *Master?* of the premises (if not of the situation). . . . I must scold about your faintheartedness. You a young man of 70 . . . talking of laying up on the shelf, *so much to do*, and *you* the only one *capable* of doing it! [T]hrow away your crutches.—and grasp a stout staff with a crook at the end to reach pine boughs—Tell Sargent you are engaged for a *Western trip* and ask him to go along, and climb for us. [T]here's "maturity" for you. . . .

Again on February 10 Parry wrote:

. . . I think of making a sort of a swinging trip round by St. Louis, by River to Cincinnati, stop over to see some relations in South Ohio & then on to Philadelphia N[ew] Y[ork] Boston &c&c I hear nothing lately from Gray or Kew. . . . I am watching what the papers say about R[ail] R[oad] movements in S[outh] W[est]. [T]hey are now pushing South Pacific R[ail] R[oad] to Tucson—to meet Greene, need not stop there for hot weather but keep on to the Rio Grande *for us* next spring They have had copious rains in the desert East of San Bernardino, and should have a good crop of weeds. The Parish broth[er]s seem to be lively and promising. Why have you always ignored *Opuntia Bigelovi* Engel. on the desert. [I]t covers miles & miles east of Sierra Nevada [A]re you waiting to see or feel it?

And four days later he wrote:

My idea is that *you* should not make a hard trip (physically) but get some nice central points on R[ail] R[oad] routes in Oregon &c and let us *youngsters?* do the *heavy* tramping. I am glad Watson will go. [H]e has well earned an *airing*. I think a scheme might be blocked out to divide the territory and let you work up the material, but you must at least *see* the important features. Well I must go down & see you on my way East. . . .

[3] George Washington Letterman, 1840-1913.

So Parry arrived and after inducing Engelmann to go went east to Ohio where he found it too early to botanize but good forest trees and abundant Indian mounds were there. He went on to Washington where he talked with Vasey at the United States Department of Agriculture, and then to Philadelphia and New York. At New York he found the Torrey Herbarium in a still more miserable plight with LeRoy a picture of despair. So he called on Dr. Newberry to discuss the Oregon and North California forests. "Three Sisters" would be the best locality for studying the whole forest range, Newberry said—better than more known localities such as Mount Hood. Parry met some of his old friends on the Kansas (Union) Pacific Railroad survey[4] and arranged for passes on the Union Pacific which, with promises of passes from Governor Stanford while in California, helped solve the financing of Engelmann's and Parry's trip. Gray, realizing what Engelmann termed the "flower's revenge," was deep in the meshes of Aster-determinations when Parry arrived in Cambridge. Parry saw Sargent and, after further arranging plans for their California and Oregon trip, served as a delegate at the meeting of the American Academy of Arts and Sciences where he met Canby, Redfield, Meehan, and Martindale and "performed full duty in destroying '*quantum suff*' of ice-cream &c&c. . . ." It was a warm day. He listened to Gray's speech. And then made additional arrangements with Sargent. Engelmann and Parry would meet Sargent in Salt Lake City or San Francisco, thus enabling Sargent to go to Colorado for a while. Watson was going alone to Montana and Idaho and then along the eastern slope of the Cascade Mountains in Washington and Oregon. Parry talked with Watson about Ephedra, pointed out some errors, but found Watson tired and sensitive and so abandoned discussion. He wrote Engelmann that, after visiting a conifer nurseryman in Waukegan, Illinois, he would meet Engelmann at Omaha and Sargent at Salt Lake. Engelmann could take a sail on Salt Lake and they could go to near by canyons to see *Abies concolor*, perhaps get a pass on the Utah Northern to see the upper forests. "I imagine we will fall in with Lemmon at Oakland and get an account of his Arizona trip," he added, "I note with some anxiety the Indian difficulties about Silver City N[ew] M[exico] & they will at least limit Greene's botanical, if not his pastoral, work. But we will leave Arizona difficulties to be settled between now & next winter & spring when we may have a *personal* interest in them."

Greene for a time had been lecturing in Pennsylvania and other places.

[4] In 1867 Parry had gone through Kansas to Colorado and New Mexico, following the Smoky Hill River and across the Divide to the Arkansas, thence along the 35° parallel from the Rio Grande to the Pacific, in 1868.

But he had returned to Colorado, getting off the train near the South Platte River and going on foot to Greeley and Golden City, botanizing in both western Nebraska and Colorado. He had sent Gray an unusual set of Santa Rosa del Cobre plants. And then proceeded to Yreka. But on January 12 Parry had told Engelmann, "Greene writes from Pueblo, to start soon for Arizona via New Mexico. [W]ill loiter along so as to reach San Francisco by May. [W]ould like to meet us somewhere in North Cal[iforni]a or Oregon." In April Greene made his first expedition to the Mogollon Mountains. Impressed as he was with the Californian character of the spring vegetation on the east side of the Gila River in New Mexico, he settled in Silver City where among the hills and mountains near by he botanized finding new and interesting species which he sent on to Gray. But Gray named an Asclepias of his *A. Wrightii*! "I think you go to an unnecessary extreme in courtesy to your *discipulus*," said Greene to Gray. "By saying in your paper that I had drawn attention to the species as distinct from *A. longicorum*, you would have done me ample justice: or if you wished to do more you could have named it *Greenei*. That was what Dr. Engelmann proposed to do with the little one after it had been printed in the *Gazette*." Differences as to matters of description caused Greene to disagree with Gray more than once, so much that it was not long before Greene was a distinct source of irritation. Fortunately for Engelmann and Parry they had not learned of this. But when Greene went to the San Francisco Mountains and did not join them, Parry caught on and he said to Engelmann, "I hope he will *cool off* on the species making."

Lemmon had talked of locating at Yosemite and entering into the business of making amateur sets of plants for tourists. Parry had written him, told him to build a shanty there, but go to Oregon where Greene had left off, to collect. Lemmon, however, had gone to San Bernardino and, getting passes on the railroad there, had begun to explore the mountain peaks near by, and the vicinity. He had met a Miss Plummer[5] and gone on several enjoyable botanizing trips with her. In addition he went to the Mount Shasta regions "so near to Oregon that [he had] several plants not described in *Bot[any of] Cal[ifornia]* & I hope," he wrote Gray, ". . . some that are new." He had "scaled Grey-back, the highest peak in S[outhern] Cal[ifornia] but found few rare things." Lemmon, nevertheless, proved to be an able collector, collecting bulbs, roots, seeds, and dried specimens of rare California plants. His explorations to Shasta and regions such as Maricopa in Arizona netted many important finds

[5] Miss Sara A. Plummer, for whom Lemmon strongly prevailed on Gray to name a genus and Gray did.

To his friend Asa Gray
as a souvenir of September and October 1856
New York Aug 1868
George Engelmann

George Engelmann

for North American botany and established him as one of a respected group of San Bernardino botanists of which the Parish brothers[6] and W. G. Wright were second only to Parry. His business remained, however, in the Sierra Valley where he regarded himself as "Amateur Botanist, Lecturer, Microscopist and Collector in Natural History." When, in 1879, Clarence King was placed in charge of the United States Geological Survey, Lemmon sought a berth with the work in California. He, however, was informed there was no place for a botanist. And so he continued with botany till later, after he had married Miss Plummer, he was made botanist of the State Board of Forestry of California. Beginning in 1879 he became a substantial contributor to the botanical literature of that state.[7]

Parry and Engelmann did not meet in Omaha but in Council Bluffs. On June 11, Engelmann wrote Gray:

> Well we are almost ready to start, and I feel more cheerful; the whole thing oppressed me somewhat at first—leave home and the children, perhaps never to return!—Yes I could perhaps not have gone without such a trusty friend as Parry, and I do not know what with my age and infirmities might become of me among entire strangers; for Sargent can not be counted upon; he must attend to his duties. . . . I have not been able to finish that Isoetes paper completely, but it may go to press as it is now (though DeCandolle protests against posthumous papers!) if I should not be able to complete it. . . . Since I wrote you last I had a nice letter from Hooker about Conifers etc. and calling my pine paper a "Capital piece of work" which is some consolation when others praise the *pictures*.

Lesquereux had also written saying he had read "and admired that excellent and beautiful Revision of the Pines. . . . You know yourself," said Lesquereux, "how this memoir on a very difficult subject is valuable to Science and I am sure that every botanist will study it, as I shall do with delight. The plates, the descriptions, the typographical work all is perfect. I congratulate you and thank you heartily for this work which goes par with that of your *Cact[ac]eae*." Receiving a letter such as this must have heartened Engelmann. He, too, was a European by birth. Schimper, Alexander Braun, Engelmann's great European scientific friend from his youth, Joseph Henry, Torrey, Sullivant, and others, had died. It was only natural that he should have thought he too would soon be summoned. With characteristic courage, however, he met Parry who took him, as he said, "under (his) wing (*metaphysically*)." On August 8, Engelmann wrote Gray again:

> A few days at Salt Lake where we found Jones a good fellow who will learn and

---

[6] See Willis Linn Jepson, *Samuel Bonsall Parish*, University of California Publications in Botany, XVI (1932), Number 12, pp. 427-444.

[7] A complete account of Lemmon's life is being written at the present time, soon to be published.

improve, met Sargent and Skinner at Ogden and went to San Francisco, where Botany as you know is very unsatisfactory, thence by steamer to Portland and without stopping through the Sound to Victoria and then up Frazer River, where we made the first regular Mountain ascent, near [F]ort Hope at the angle which the river makes coming from the North and turning west. We were right among the snows and were gratified in finding several of the Alpine Conifers among others *Abies amabilis* which is really distinct from *grandis*. . . .

Back through the Sound to Portland. Up to the Cascades and the Dalles, where Sargent alone ascended an Indian trail to Douglas' original locality where *amabilis* is splendid. Newberry's figure in Pac[ific] R[ail] R[oad] Rep[ort] is in the main correct.

And here we are ready to leave tomorrow for the South—Coast Range, Cascades and then Shasta.

Met Brewer here.

Parry left us a week ago to go with Suksdorf to M[oun]t Adams. . . .

A little more than a week later Parry wrote Gray:

I have been several times on the point of writing to communicate some passing information but something has prevented. Now I have a leisure P.M. on my return from the Upper Columbia that I cannot well put in any better especially as you may soon be on the wing across the Atlantic.[8] It would be rather inspiring here on the track of Douglas & Nuttall if I could find more of their plants but this is now the dry season on the lowlands & little to see. To go back a little I must say that in company with Sargent & Dr Engelmann, we made a rapid trip on the ordinary route of travel from the Columbia River to Puget Sound, thence across the straits of Fuca to Vancouver Island, thence across the Georgian Gulf to Fraser River & up that stream to the head of Steamboat navigation, making here & there side excursions to interesting localities including one climb to the snow line on Fraser river w[h]ere we encountered the usual alpine plants of that district & some rare *conifers*. On our return at Victoria V.I. we called on old *Dr. Tolmie*[9] of Hooker's *Fl[ora] Bor[eali] Am[ericani]*. [F]ound him a nice genial old gentleman with a sprightly family of grown up sons & daughters (half breeds), his wife a full blooded *Indian* (princess)? had died about 6 weeks previous. We saw there some interesting mementoes of Sir W[illia]m Hooker in the shape of presentation volumes of early Botany. Dr. Tolmie was kind enough to lend us a copy of Douglas' personal journal (printed in Sandwich Island Journal) by which we have been able to trace out (on the map the direct routes, and dates of collection &c&c) Dr. T[olmie] has forgot all his botany and did not even know his own genus *Tolmeia* which he wanted us to show him! On our return to Portland I suggested to Mr. Sargent to detail me for some *separate work*, as I did not fancy hurrying over the country. [A]nd a good opportunity offering I joined a small party (with *Mr Suksdorf* a brother of the collector) for M[oun]t Adams one of the high snow peaks north of the Columbia in a line with M[oun]t Hood. It turned out that the Suksdorfs, a large family of Germans, were formerly from *Davenport*, and some of them had known me in connexion with the schools. So I was made quite at home in their circle which lay directly in the line of the route to M[oun]t Adams.—So they got

---

[8] Referring to Gray's trip to Europe in September.

[9] W. F. Tolmie.

up an exploring party for my special benefit and from their lower farm station on the Columbia at White Salmon we went directly north over a good road to a prairie district used by them as a summer dairy—there I fell in with W[illia]m S[uksdorf] the collector, a modest intelligent *farm boy*, who was putting in his spare time (not much of that) in collecting & studying plants. They were then milking 70 cows, twice a day, & making 70 lbs of butter per day. It was with difficulty that he could be spared to join our mountain party but I insisted on his going. So by a quite easy ascent over a good trail we reached the south slope of M[oun]t Adams, and made camp among the snow drifts 20 f[ee]t deep: (7000 f[ee]t el[evation]) On account of the unusual dept[h] of snow last winter, the vegetation was backward but still there was a considerable exposure of alpine flora from 6000 to 8000 (feet) elevation. . . . I endeavored to encourage young Suksdorf all I could to continue his collections: and gave him such suggestions as occurred to me to help him. As they belong to a thrifty German family & in promising circumstances for money making I should hope in time he may do something. What he specially needs now is *books, leisure* & *encouragement*. I received from him specimens of Suksdorfia which is now past flowering.

I have not yet met your other correspondent Howell. I also hoped to have met Reverend Nevius, who is stationed in the interior still interested in botany. I am now on my return to Portland where I expect to hear from the rest of the party & decide what course to take either joining them at Shasta or returning to San Francisco. Dr. E[ngelmann] seems quite lively but does not enjoy the hurry & push of Sargent. I think as soon as *he completes* the most important investigations in the *Conifers* he will take his own "slow & sure" gait. Sargent is anxious to get round to San Francisco by 1st [of] Sept[ember] & thence by some southern route home by Oct[ober] 1st. I hardly expect to join him again but will decide on my return to Portland & when I hear more of their movements

Lemmon was to join them at Shasta from San Francisco. [G. D. ?] Butler [is] a correspondent of Dr Engelmann—[He] was also in that district & wanted to join them. So I think I may well be excused and take my own slow "puttering" course which if it does not accomplish much has its attractions.

Of course we are all anxious to see *Bot[any of] Cal[ifornia]* Vol[ume] II! and to know more of your doings & plans. I have been up the river as far as Walla Walla to see the country [T]oo dry at this season for collecting. This is a very picturesque section of the valley on the dividing line between the burned & dry districts. . . . I heard nothing of Watson in the Blue M[oun]t[ain] district. . . .

Eastward from where Engelmann and Parry were located, Sereno Watson was in the Northwest, particularly Montana, investigating tree areas for the forest department of the United States census of 1880.

Still farther east, a large command of the Powell survey under Henry Newton and Walter P. Jenney had with comparative recency reached the Black Hills of Dakota on the east fork of the Beaver River, begun surveying, and established a permanent camp on French Creek near a stockade built the winter previous by miners come to a "new El Dorado of the West." They had worked northward, surveying and mapping a large area between the forks of the Cheyenne, determining its geology and resources and the rights and interests of the Sioux Indians under existing treaties. On October 14 they had reached Fort Laramie, one of their starting points, and their scientific materials were submitted to authorities in the East. The expedition's *Report*[10] was published in 1880 and Gray was shown to have determined the botany, calling particular attention to *Clematis Alpina* Mill., var. *occidentalis*, subvar. *tenuiloba*, and *Calochortus Nuttallii*, Torr. and Gray, noted as an insect-capturing plant. "Those who have the advantage of seeing this and similar species alive, either in their native haunts or in cultivation, should learn whether [their] bristles manifest any irritability," said Gray.

Strangely enough, neither Parry nor Engelmann refer to this expedition; nor to the expedition of 1880 by John Macoun to the Canadian Northwest—to Winnipeg, Grand Valley, Qu'Appelle, Moose Jaw, Old Wives Lakes, Swift Current, Cypress Hills, Fort Walsh, Humboldt, Fort Ellice, and Portage La Prairie—to the east and north of them. The more unusual it is since Parry always noted the activities of all botanists interested in regions he was exploring. John Macoun was becoming a noted naturalist, too. On January 1, 1881, he was appointed by Sir John A. Macdonald botanist to the Geological and Natural History Survey of Canada, due principally to his three years of explorations as far west as where Parry explored.

The far Northwest of the United States eventually was to produce a number of botanists, among whom Suksdorf and William C. Cusick were prominent. In February 1881 Suksdorf sent Watson several hundred plants, and observed:

> I have not yet seen any yellow flowered specimens of *Senecio lugens* here (White Salmon, Washington), but east of the Klickitat river the flowers are bright yellow. The change takes place a mile or two west of that river, and seems to be very sudden.
>
> Suksdorfia is also on the Klickitat river near its mouth.

Five years later the young botanist in whom Parry had interested Gray was given an opportunity to come to Cambridge. At the solicitation of Suksdorf's sister-in-law, Gray offered Suksdorf employment at the herbarium which at first he refused, saying, "My life so far has been a life in the open air rather than indoor life. I am pretty sure that a sedentary or indoor life would not agree with my health very well. . . . I really believe I am better off in the field." Nevertheless, in October 1886 Suksdorf went to Cambridge where, as he later told Davenport, Gray gave

[10] Department of the Interior. United States Geographical and Geological Survey of Rocky Mountains. *Report on the Geology and Resources of the Black Hills of Dakota* (Washington, 1880), pp. 531-537. "Botany," by Asa Gray. List of Plants Collected.

him "an opportunity to become more of a botanist." Suksdorf remained at Cambridge a considerable time but in respect to his health his apprehensions proved not unfounded and he returned to Washington Territory where he continued to serve the botany of the Northwest usefully with improved knowledge and much zeal.

Within a year after the visit of Engelmann and Parry and Watson to the Northwest, Cusick forwarded a bundle of plants from Union, Oregon, of which, he said, almost a hundred were new to his collection. During the summer of 1881, while he had collected only 300 species and spent some time overcoming sickness, he said he had enjoyed the time spent in the mountains and planned "to go, for two or three months, to the m[oun]t[ai]ns of western Oregon, next year." It was not until the year 1886, however, that Cusick expected to be able to spend the entire season botanizing. He, too, evidently struggled to keep in good health. Similarly, while Suksdorf was effectively serving the botany of the Washington regions, Cusick with equal and meritorious ability was aiding Gray, Watson, Engelmann, and the western botanists in furthering completion of the knowledge of the Northwest botanical regions, with special reference to Oregon.

During the middle of September Parry was in San Francisco near the Academy of Sciences. Engelmann had gone on to Southern California and Parry wrote him at San Bernardino, saying, "I enclose this line to you through Mr. Wright at San Bernardino whom you will find a very accommodating intelligent man who will be glad to show all the attention in his power. I hope if you find your strength not *fully* equal to the trip to Arizona &c that you will *stop short* and take your *own time.* . . . Still if your strength is equal to the trip, the present *reconnaissance?* will be useful & *full* of interest and '*instruction*' besides indicating parts that may be desirable to *revisit!* I have concluded all things considered *not to go* East at present but have Mrs. Parry come on as soon as she is ready & join me here, then we will stop a month or so before going south for the winter. In the meantime I expect to see *Gov[ernor] Stanford* . . . and perhaps [take] some short trips to the Sierra &c&c till you return. Lemmon expects to go to Shasta. . . . [H]e has your plants in good condition & *promises* to attend to them. . . ." Engelmann went on to Tucson and there evidently concluded to go by way of New Mexico home to St. Louis. Parry wrote him anxiously: "If you are to come back here [San Francisco] I shall see you and arrange for a *winter campaign!* . . . I have arranged for a trip to M[oun]t Talampais[11] with Kellogg probably to

[11] Mount Tamalpais.

Monterey, less likely to Mendocino, & meet Mrs. P[arry] at Truckee.
. . . Greene writes enquiringly about you. [E]xpects to meet you next
spring [H]ad been to the summit of San Francisco M[oun]t[ain]s, says
little about trees, more about plants & shrubs. [W]ill stay at Silver City
another season (to Christianize the natives & bag plants). Vasey[12] is
running about. [G]ot most of his section except *Pinus Coulteri*. [I]s
leaving that for Sargent or *'Parish boys.'* . . . I shall hope to hear from you
at Tucson under the *shade?* of *Cereus gigant[eus]*." Engelmann went by
way of Ogden, Utah, to St. Louis and waited eagerly for letters from
Parry which were several. He heard of Watson's going to Cedar Mountain and Monterey, examining Cupressus, of Lemmon's marriage, of
Dr. Behr's inquiries concerning Engelmann, of the contemplated completion of a connection between the Atchison, Topeka, and Santa Fe
Railroad and the Southern Pacific—and most of all, of the western
botanists' earnest solicitation for Engelmann's return the following
spring. A new railroad was to be built north of San Diego! And Greene
had transferred *Whipplea utahensis* to Fendlera! A storm of protest was
expected from Watson who by then had returned East. In January 1881
Engelmann wrote several letters[13] to Gray who had gone to Europe:

How it was possible not to write you in 6 months can only be accounted for by
the terrible strain Sargent's restlessness and energy put us under. [H]e perhaps
told you himself that he rarely took or allowed us more than "5 minutes" for any
thing; but he begins to feel the consequences—says himself that he is "demoral-
ized" about the things left undone, the too great hurry and the unsatisfactory out-
come. But I am afraid, indeed I knew it, you and Hooker would have been worse
travelling companions—the "5 minutes" of Sargent would have been reduced to
2 or 3, I fear.

Well, with all that, we had a glorious time; all the little inconveniences, troubles
and mishaps are forgotten long since and the satisfaction and after enjoyment only
remain, tempered, however, very naturally with the regret that not all that might
have been done, was accomplished!

My health has improved wonderfully. . . .

Your kind letter, written before you left, was received in San Francisco. But the
one which "you hoped to write soon" never came. I have however seen a letter
you wrote to Sargent lately and heard also through Hooker of your doings in
Spain and France.

You have heard of our doings through Sargent. He left about the middle of
October and I remained 5 or 6 week[s] longer, partly in the rich vin[e]yards (not
forgetting the wine presses and wine cellars) north of San Francisco, often with
my cousin Hilgard, and partly in the Southern part of the State, with Parry often
indulging my Cactus fancies, not rarely to the detriment of clothes. . . .

I hope soon now to be able to begin the study of some of the more important of
our collections, especially the Conifers and the Oaks, which will give the more

---

[12] George Vasey's son.   [13] The quoted letter is dated January 31.

trouble the more I see of them; the simplicity of the former arrangements won't hold out—as you probably have also found in Asters and Solidagos. By the way I have collected a lot of them which I hope will be useful to you when you come back.

So you are going to visit Italy with Hooker. I wish I could be with you! And I am not sure that such a thing is impossible—I hesitate between recovering here and working in a killing summer heat, or travelling—and if travelling whether go west again, to those tempting Arizona Mountains or east to Europe! My time of life is getting short and my strength and ability to work will not last very long—therefore it ought to be used to the very best purpose—the ever true story of the Sibylline books! . . .

On February 19, Gray replied from Kew Gardens, England:

A few days, or say a week ago, we were gratified by receiving your pleasant letter of the 31st January. I hasten to reply before we get afloat again, when writing becomes precarious. Just now Mrs. Gray and I have our evenings together in our quiet lodgings, that is, whenever we are not dining out or the like, which is pretty often.

You know of our movements, then, up to our return here. The Spanish trip was very pleasant and successful, and the three weeks afterward in Paris both useful and enjoyable. As for botany, it was all given to Aster and Solidago, at the Jardin des Plantes, and at Cosson's, who has the herbarium of Schultz, (of Bavaria) Bip., which abounds with pickings from many an herbarium.

We got over here early in December, and here I have worked almost every week day till now, excepting one short visit down to Gloucestershire, and a recent trip to Cambridge, where, however, a good piece of three mornings was devoted to Lindley's asters. I know the types now of all the older species of North American aster, Linnaean, Lamarckian, Altonian, Willdenovian,—excepting one of Lamarck's, which I could not trace in the old materials at Paris; and Röper writes me that it is not in herbarium Lamarck. As to Nees's asters, most of them are plenty, as named by him directly or indirectly. But where, on the dispersion of his herbarium, the *Compositae* went to nobody seems to know, though I tried hard to find out. Have you any idea? But he made horrid work with the asters, and the Gardens all along, from the very first, have made confusion worse confounded. No cultivated specimen, of the older or the present time, is *per se* of any authority whatever. I am deeply mortified to tell you that, with some little exception, all my botanical work for autumn and winter has been given to Aster (after five or six months at home), and they are not done yet! Never was there so rascally a genus! I know at length what the types of the old species are. But how to settle limits of species, I think I never shall know. There are no characters to go by in the group of Vulgar Asters, the other groups go very well. I give to them one more day; not so much to make up my mind how to treat a set or two, as how to lay them aside, with some memoranda, to try at again on getting home, before beginning to print. The group now left to puzzle me is of Western Pacific Rocky Mountain species. The specimens you have collected for me last summer, when I get them, may help me; or may reduce me to blank despair![14]

---

[14] A complete account, with copious letters, of Dr. and Mrs. Gray's journey may be found in Chapter VIII of *Letters of Asa Gray, op. cit.*, Volume II, pp. 701-724.

Dr. and Mrs. Gray had sailed for Europe the previous September. After a fortnight in England where they were elaborately entertained by Sir Joseph and Lady Hooker, meeting there Bentham, Oliver, Baker, Masters, young Balfour, and DeCandolle who came over from the continent, especially to see Mrs. and Dr. Gray, the Grays spent the autumn in western France and Spain where at Madrid Gray looked over the herbarium. During the winter they returned to Kew and that spring went on a journey through Italy with Sir Joseph and Lady Hooker. The summer was then spent at Kew working in the herbarium and in October the Grays sailed for America. In the course of their journeys they met many celebrities—Robert Browning, Decaisne, and others. But the high point of Gray's trip was the visit with Charles Darwin at Down, although he must also have been much interested in the underground caves of ferns and the alpine garden of Backhouse to whom North American collectors such as Parry and Palmer had been sending seeds for some time. While at the Hookers, Gray worked up Oxytropis. His principal work, as he said, was the Asters and Solidagos, genera which had been bothering botanists since the early days of Torrey and Gray collaborations. He wrote Engelmann again, on December 13, 1881:[15]

Accumulated collections, of Lemmon, Parish, Cusick (of Crowell, Oregon), etc., especially have taken all my time up to now, after getting my home in order, a deal of trouble. And now I can think of getting at my "Flora" work again.

First of all, I am to make complete as I can my manuscript for Solidago and Aster. Solidago I always find rather hopeful. Aster, as to the *Asteres genuini*, is my utter despair! Still I can work my way through for the Rocky Mountain Pacific species.

I will try them once more, though I see not how to limit species, and to describe specimens is endless and hopeless. So send on your things. But first I am to print, *pari passu* with my final elaboration, an article, "Studies in Solidago and Aster,"— taking the former first, giving an account of what I have made out in the old herbaria, stating investigations which I can only give the condensed result of in the "Flora," etc. Considerable change as to some old species. . . .

To A. DeCandolle he wrote:

We, Mr. Watson and I, are still much occupied with the distribution, and therefore in good part the study, of the recent collections which have accumulated here and are still coming in. Much valuable time do they consume. The most interesting are from Arizona, etc., near the Mexican frontier, among which those we have most to do with are by Lemmon and by Pringle.[16] The former, I know,—and I

---

[15] After Gray had returned from Europe.

[16] Cyrus Guernsey Pringle (1838-1911), one of America's greatest botanic collectors, early a specialist in New England botany and a pioneer in plant hybridization. Before 1880, with others, he visited lake and mountain regions of Vermont. His *Life and Work, op. cit.*, on later pages of this book, has been written by Helen B. Davis.

shall soon know as to the latter,—has sets to dispose of, and I think you would like to have them. . . .

I have no other botanical news for you. Dr. Engelmann, who of late has roamed a good deal, is now at home, and busy with botanical work, of various sorts, Isoetes, Cupressus, etc. It is quite probable that he will cross the ocean next spring, in which case you will probably see him. Professor Sargent is busy with his forest reports in connection with the United States Census of 1880. Mr. Watson in this service made a long journey through our northwest region, while I was in Europe, at too late a season for much ordinary botany; . . .

My colleague, Professor Goodale, giving over to Professor Farlow the university lectures, etc., is now abroad with his whole family, to recruit health and acquire information. You will see him at Geneva. . . .

During 1882 Goodale traveled over much of Europe, visiting Kew, Copenhagen, Scandinavia, and many places on the continent. An undated letter to Gray reads:

I have been greatly assisted by Prof[essor] Sachs in every way. He has given me excellent advice regarding apparatus and has put me in the way of getting the most useful things cheaply and well done. . . . But I am under great obligation to all here. . . .

Goodale was much interested in the research experimentation of Wilhelm Pfeffer. Work of Pringsheim, Wiesner, Frank, and others, captured his interest. He brought home new sets of exercises for the "thoroughly furnished" Harvard laboratory, importing also new valuable apparatus for physiological investigation.

In 1882 final publication of Farlow's very important work, "The Marine Algae of New England and Adjacent Coast," appeared. And on February 8 of that year Gray presented two more numbers of his famous *Contributions to North American Botany*: first, his valuable "Studies in Aster and Solidago in the Older Herbaria," and, second, "Novitiae Arizonicae," representing mainly Arizona and adjacent district collections. Watson presented about this time the Palmer (and Parry) northern Texas and Mexico plants, with descriptions of many new species of plants from the Western Territories, and some from Florida.

Engelmann published in the St. Louis Academy *Transactions* his paper, "The Genus Isoetes in North America," which ranked with his "Revision of the Genus Pinus"[17] and his studies for the *Botany of California*, volume II of which had appeared in 1880. The *Gazette* published his "Notes on Western Coniferae." With works of lesser rank, American botany was going forward magnificently.

[17] Containing also description of *Pinus eliottii* (St. Louis: R. P. Studley & Co., printers and binders. 1880); also published in *Transactions of the Academy of St. Louis*, IV, No. 1.

CHAPTER VII

The Development of Morphology. Gray
and Western Botany

WHILE Gray was in Europe, he must have pondered long over the paragraph in Engelmann's letter of January 31, 1881, as to conifers and oaks, which said, "the simplicity of the former arrangements won't hold out—as you probably have also found in Asters and Solidagos." Indeed, Gray probably discussed with Hooker and other foreign botanists recent progress in systematic and other branches of botany both in Europe and America. On August 22, 1878, Sir Joseph had written Gray:

> Assuredly you should try for an English market for your Introduction to Morphology and classification. It is much wanted—but all the world is mad after Physiology and Histology, and Morphology *pure* and classification are despised on the Continent, and Britain is fast following suit.

In North America Gray had pioneered in "Morphology *pure*," as Hooker then described a type of morphology chiefly an aid to taxonomy. Among his earliest publications, *The Botanical Text-Book*, dated 1842, a revised edition of his first work of importance *Elements of Botany,* had been devoted to an introduction to structural and physiological botany and to systematics. Many editions of his famous text had followed and in 1857 he had styled the volume *Introduction to Structural and Systematic Botany, and Vegetable Physiology.* The following year appeared his *Botany for Young People and Common Schools. How Plants Grow* was a "simple introduction to structural botany." Frederick Brendel, in his "Historical Sketch of the Science of Botany in North America from 1635 to 1840,"[1] commented: "[U]ntil Prof[essor] A. Gray's popular book, 'How Plants Grow,' appeared in 1858, not a single work of any importance was published in this country, either on anatomy or on the physiology of plants. ..."

In 1879 Gray presented to the science the sixth edition of his *Botanical Text-Book* conceived on a large scale. This edition was planned to contain four parts: Gray to write the first part on morphology, taxonomy, and phytography; Goodale to write the second part on vegetable physiology and anatomy; Farlow to do the third on cryptogamic botany; and Gray planned the fourth for himself on morphology and economic use

[1] *American Naturalist*, XIII, pp. 754 ff.; XIV, pp. 25 ff. (involving the period from 1840 to 1858).

of the natural orders of phaenogamous plants. The first and second parts appeared much as planned, although Goodale's work on physiological botany was not published until 1885. Farlow did not perform the task assigned to him; and, while Gray hoped rather than expected to write the fourth part, it was never written. However, *Structural Botany, or Organography on the Basis of Morphology,* to which was added principles of taxonomy and phytography, appeared in 1879 and it must have been as to this part of the edition that Hooker wrote Gray in 1878. Gray's *Text-Book,* his *Manual,* and *First Lessons* were standard works.

North American botany can never forget that, while Gray's major interest was not morphology or physiology but taxonomy, he led in America and kept apace with increasing studies in these subjects made in Europe and America during the last years of his life. Except in taxonomy, prior to 1870, little that was important had been done in American botany. Not till students began going in larger numbers abroad to study in the 1880's did any advance in experimentation develop. Here in morphology at first the vascular plants and mature external structures dominated study. Root, leaf, and stem were organs favoritely selected. The idealistic doctrine of metamorphosis prevailed. Type plants being selected, study narrowed to a few representatives of the entire green kingdom. Gradually, however, with work of Farlow and others, the non-vascular were included. Botanists like Penhallow saw need for studying interior structures. Study went from algae to seed plants. Injurious fungi were studied with some consideration given plant diseases. Laboratory microscopes, including compound microscopes, were employed for more than taxonomic uses. With use of reagents and scalpels and razors for hand sectioning, a morphology of minute structures developed. When the microtome would arrive, progress would go forward by leaps and bounds. But let no one be unmindful that microscopic examinations with specimens at hand to study the immature subjects of morphology, physiology, mycology, a kind of pathology, and the like, had not really begun in America until about 1870.

Some restlessness, change, and progress was to be expected in America when the impact of scientific expansion in Europe began to arrive in the United States and Canada. Even taxonomy soon felt the force of new researches.

Gray's and Engelmann's apprehensions as to uncertainty of existing concepts of *species* soon found echoes. "Species . . . are not facts or things, but judgments, and, of course, fallible judgments," had said Gray, "how fallible the working naturalist knows and feels more than any one else." The inevitable tendency in one direction is fixity of concepts with ironclad limits not truly representative of reality. In the opposite direction, the end is the undoing or a total lack of scientific organization of materials and, ultimately, chaos. Between the two extremes are the media, the organization of plant families, orders, genera, and species, along which established lines the science should adhere as far and as much as the limits of truth allow, if orderly progress is to be maintained. As new materials widen or limit the concepts, revisions are necessary. But in the absence of such, or without conclusive demonstration that past concepts are erroneous and not in conformity to the weight of data, the elder botanists maintained the existing structures should stand. In the minds of many, Darwin's evolutionary theory had wrought havoc with the truth of many concepts formed prior thereto. The horizon of scientific investigation was ever widening and new phenomena shown. Anatomical studies with the microscope and other agencies were revealing a hidden universe of plant study. A real science—the beginnings of an experimental science—in morphology and allied subjects was developing.

Furthermore, all plant science study was soon to feel the impact of a gathering momentum seeking to develop an American scientific horticulture and agriculture—movements to take shape with astounding rapidity in the United States and Canada when agricultural experiment stations, modeled in considerable part after those of continental Europe and Rothamstead, England, would be established in every state of the Union and in a half dozen Canadian regions, and would study, along with progressive departments of American colleges, plant growth and development, in health and disease, their functional and physiological processes as well as their morphological history and taxonomic differences and affinities.

Gray knew this. He kept alive to all new investigation to the day of his death. Always with a mind happy to be convinced, he would hear of the new results. In America he felt his responsibility as leader. When Tuckerman violently disagreed with the German botanist Nylander, Gray did not become openly hostile too. When Edward Lee Greene began his barrage of criticisms, based mostly on claims that Gray overlooked differences in structures and habits of plants, Gray did not maintain that Greene was wrong. If he disagreed, he told Greene so; if he agreed, he made his work conform. However, with it all, he fought for order. On January 24, 1879, Gray told DeCandolle that "... You and Bentham have kept orthodox views of nomenclature at the fore in Europe, and I have seconded them here, so that, except among cryptogamists, heterodoxy makes no headway. . . ."

Even the changes among cryptogamists could not have surprised

Gray. His friend, Sullivant, before his death had warned that in all probability most of the bryological systems of classifications would "be upset after awhile, as the knowledge of species (the present desideratum) increases—thus furnishing the materials on which to found a better system than any yet proposed." Elasticity had to be maintained; but also a need for stability was seen.

Studies in effects of light on plants, on physiological relations of coloring matters, on functions of chlorophyll and other substances such as starch-grains and water-movements, studies in self-fertilization and cross-fertilization in plants, studies of reproductive organs—important adjuncts to development of morphology and physiology—all received appropriate commendation from Gray. In fact, he himself spent time investigating all phenomena possible. While in Europe, Gray read Darwin's *Power of Movement in Plants*, and, although he found it "a veritable research, with the details all recorded; and so . . . dull reading," he ably reviewed the work in *The American Journal of Science and Arts*. The importance of Gray's reviews has been overlooked by many botanists. There was betrayed many times his great interest in European study, especially in morphology. For example, already for the same journal Gray had reviewed Darwin's *The Effects of Cross and Self-Fertilization in the Vegetable Kingdom*, having for thesis proof that cross-fertilization is beneficial to the plant—not light reading, said Gray, but worthy of careful study. Gray experimented with yuccas and other plants. "Glad," wrote Engelmann, "you got the bugs to show your yuccas how to do their duty to posterity." Engelmann, through friendships with Alexander Braun, Schimper, and others, followed European morphological study. His own observations on acorns and germination; maturation of oaks, whether biennial or not; and much else interested Gray. Farlow and Gray rejoiced in 1881 that fertilization interest and "other physiological and etiological questions" increased.[2] Especially valuable and progressive were laboratory investigations during these years in Europe, stimulated in part, Farlow believed, by Darwin's researches. Economic and medicinal uses of plants, as well as plant disease study, and studies of ferment, won Gray's interest always.

With the vast array of literature that responded to scientific investigation at this time, it is small wonder that much talk concerning modifying concepts of species and changing nomenclatural usages developed. Not the "talkee talkee" variety of discussion of which Torrey com-

[2] Attention should be directed to an excellent series of articles published by Professor Farlow in the *Smithsonian Reports* 1881-1885 entitled, "Recent Progress in Botany . . ."; also similar studies for years 1879 and 1880 by Charles E. Bessey in the *American Naturalist*. See Volumes XIV and XV.

plained, either. But sound argument based on sound scientific progress. Most of all it can be understood why argument concerning adequacy of descriptions in plant classification should have developed. And why Gray was drawn into the midst of the discussion. One botanist regarded Gray as an "infallible Pope," ruling the botanic sphere of North America. Certainly he was the leading North American botanist to whom Europe looked and who wielded more influence and control than any other had before his time or has since—in taxonomy. But there were three young botanists coming forward. In the West there was Greene—in central United States there was Coulter—in the East, Nathaniel Lord Britton. In February 1880 the *Botanical Gazette* noted:

A new school of botanists is rapidly gaining ground in this country and we are glad to see it. While the country was new and its flora but little known it was very natural for systematic botany to be in the ascendancy. It is a very attractive thing to most men to discover new species, but when the chance for such discovery becomes much lessened there is a turning to the inexhaustible field of physiological botany. Systematists are necessary, but a great number is not an essential thing and it is even better to have but a few entitled to rank as authorities in systematic work. But in studying the life histories of plants or their anatomical structure we can not have too many careful observers. This, at the present day, seems to be the most promising field and one botanist after another is coming to appreciate it. As microscopes are becoming cheaper and hence more common the workers in the histology of plants are becoming more numerous and it is to such the *Gazette* would now address itself. It will be noticed that the notes published heretofore would largely come under the head of systematic botany, and it is our intention to continue to give large space to this subject, but we would like to take a stand in this new school and call for notes from its workers. Dr. Rothrock's paper on "Staining of Vegetable Tissues" was a start in the right direction and the eagerness with which such papers are now read is shown by the fact that that issue of the *Gazette* was entirely exhausted in filling orders. . . . Let not only the results of study with the microscope be noted, but observations on the habits of plants, such as their fertilization, movements, absorption and evaporation of moisture, and many other subjects which are now attracting so much attention. . . .

June 19, 1880, Rothrock went to Strasburg, Germany, to study plant anatomy under De Bary. So impressed was Rothrock with the difference between American and European methods of teaching botany, he wrote on the subject on his return to America, pointing out that European laboratories stressed anatomical and physiological work over systematic and urged all schools to provide experimental work in adequately equipped botanical laboratories. Bessey appreciated the value of this. In 1877, he had told Beal:

A college which proposes to keep up with the current must provide botanical and zoological laboratories. The college which does not provide such laboratories will fall behind the progressive institutions, at least so far as the biological sciences are concerned. A botanical laboratory is just as necessary for the proper teaching of botany as is a chemical laboratory for chemistry.

Rothrock had initiative and vision. Not only did he aid in establishing pioneer experimental laboratories in America and with them new methods of botanical investigation, especially in medical botany, but later he was one of the great American pioneers in the forestry movement which swept the country.

Against great odds, he aided in initiating the American "reforms." At first, handicapped by poor laboratory facilities, uncertainty of a position, lack of finances, and other difficulties, he kept ever steady to his purpose and vision and, with Gray's encouragement and seeming direction, he brought to consummation the fulfillment of his plans and dreams. On September 2, 1880, while studying under De Bary at Strasburg, he wrote Gray:

I suppose you are now or shortly will be in Kew. . . . Thus far I have staid in Strasburg (except 2 excursions . . .) If I knew when you would be in Geneva I should try to meet you there—that is if it were agreeable to you to have me do so. A trip to Switzerland will be the only travel I shall indulge in as time here is to me quite too precious to waste in mere sightseeing. Much as I should enjoy it. . . .

As for your advice to do some original work while here, I want to, but when you consider that I am here not only to learn the language, but to acquire as large a range of facts as possible in a very short time you will readily understand that the field of original research under the circumstances must be a very narrow one. I want to make one objection, my dear D[octo]r, to your implication most kindly put however in your letter to me before leaving home, that I had done no original work. I think those chapters of mine in the first of my report may be fairly considered original and I see they are so considered in Europe. Then, too, though I claim little for the specific, and nothing for the generic descriptions, I think there still remains a body of facts in the book, on altitudes etc. which will be regarded as having a value. In fact I know where less important work has put men into the National Acad[emy]. If DeBary thinks I will have time for some original work, I will most gladly embrace the chance. I like him very much, first because he is the first teacher in Europe and second because (please pardon the direct personal allusion) he reminds me of my good friend Asa Gray very much in his manner, mode of expression etc.

. . . I am enjoying every hour of my time. Now in vacation I give it all to my German and will even during term time keep up my lessons. So that even if I dont get back here I shall at least be able to read what is written here. . . . The three weeks of term time which remained after I reached here were mainly devoted to Peronospora. . . .

Experimentation with observation soon took over botanists. The younger botanists went to the new laboratories. And the older ones became more observing, at times indulging their fancies in experiments. Older men such as Engelmann had done some serious morphological

study, and, with it, some experimentation. But they had not the zeal for
the new methods, although leaders such as Gray and Engelmann seem
to have been sympathetic and actually in favor of most of the European
work. At least they seldom expressed disapproval. Although urged by
Parry to do so, Engelmann did not return to California. Instead he went
to Colorado again and July 24, 1881, wrote Gray:

> Does it take all of Mount Gray to remind me of you? No, indeed. I have all the
> time so much to talk to you about but time flies so fast . . . [A]t the end of Novem-
> ber the arrangement and study of my collections took up all the time that a light
> recommencing professional occupation left me. My health was excellent, but in
> April, just when I proposed to go west again and study western vegetation in spring
> —or go to Europe—I was undecided yet—a severe attack of rheumatic gout threw
> me down and kept me housed for nearly two months. . . . Then the heat drove me
> out of St. Louis and not hoping much for the seacoast I fled to the Mountains and
> to the highest inhabited part. . . .
>
> I hope to stay a little longer here on [Berthouds Pass]. . . . Today I opened some
> flowers of *Gentiana ovata* and found at 9 A.M. anthers just shedding pollen, stig-
> mas spreading and full of pollen! Self fertilization and no proterandry! While I
> took the latter for granted in these Gentians, the Question arises, whether under
> certain conditions the same species will not behave differently and accommodate
> itself to circumstances! I must watch further and at lower altitudes. . . .

Engelmann went the next week to Hot Sulphur Springs, Middle Park,
Colorado, where he enjoyed further efforts "to recruit and get young
again"; and the following month to Empire City. Although he delighted
in the "gorgeous flowers" of Colorado, he was not able to do much in
botany—"the scientific interest comes afterwards," he said, and did what
he said Gray called "puttering"—wondering the while what were Gray's
"movements." He went home by way of Las Vegas, New Mexico, studying roses considerably as he journeyed. Although on occasions for many
years Engelmann was one of a number of North American botanists
who criticized Gray for establishing too many species, he grew more
liberal as age and experience advanced him. In 1880, he wrote Watson,
Otto Kuntze "is right enough that at the present day the idea of species
is shaky and who that studies Rubus, Rosa, Aster, Solidago or any large
genus does not come to the same conclusion and does not have sensations
of unsatisfied chilliness. . . . He is right also that species of very different
value exist, and that groups are recognized. His algebraic method is not
new however: I recollect that Schimper 53 years ago talked about the
same thing, and would in one line supply thus, all the information of a
long description." Englemann had told Gray some time before:

> . . . simplicity is not always the way in which nature proceeds. . . . And why
> should we not in such a large Class [as Cupressus, Pinus, or Taxus] where the
> struggle for higher development is so plainly expressed find different forms?

[D]oes it not look as if we were attempting to force unanimity or uniformity into a world of plants that indicates every where that struggle[?] . . .

But in December 1881 when Gray was drawn into a maze of small conflicts and jealousies among western botanists especially, Engelmann was quick to come to his assistance, with a letter, saying:

So you have your troubles with aspiring botanists! It is a state of things, which had to come sooner or later, and is certainly unpleasant enough, and will make a good deal of annoying labor. You are *perhaps* correct in ignoring the whole thing and in sticking to your legitimate work. Let them do their best or their worst, as the case may be.

Kellogg worked in that way for many years, Greene is doing better, and now Lemmon and others follow. At the same time that nice play thing, the microscope, induces some to favor the admiring mass with their discoveries of new Fungi. Notoriety is at the bottom of all that. Formerly they were satisfied by collecting new plants and by having their names given to genera and species, now they want to see their names as authors to emulate Kellogg's fame! But let me tell you that you are somewhat to blame for this chase after botanical notoriety, as you, "to encourage aspiring collectors" stuck their name[s] to innumerable new species and [thus there] is no end to *Wrightii, Parryi, Lemmoni*, etc. etc. Let us resolve to use always a descriptive name, where it can be done, give a geographical where another is not applicable, and personal names as a last resort. . . .

It is amazing to see the State Pride cropping out in those western botanists. [T]hey seem to want their flora for themselves! And still there is something in it if they can only do it well.

But it seems to me that you misunderstand Lemmon (& Greene). You seem to think that "[Lemmon] is setting to work to interrupt the plants which naturally come to us," etc.

I do not read anything of the kind in his letter; it is the California Academy men and Greene, it appears, who claim the monopoly. Lemmon only wants to describe and name himself some of his plants. Greene's claim, if Lemmon reports him correctly, is absurd (i.e. to have a right to a plant, if he puts his "n.sp." on it).

Suppose you suggest to Lemmon the propriety of suggesting names for his supposed new things, which he may send with the liberty of your not adopting them, if you do not think them proper—a thing which you and others have often done before. . . .

The several new species found by Greene at Silver City, New Mexico, and its vicinity, in the Mogollon Mountains, and in the San Francisco Mountains,[3] let loose the fire of egotism in him; and he sought to set himself up in the West with his own correspondents. Gray, in opposition to Watson, sustained him in transferring *Whipplea utahensis* to Fendlera. Gray in his published *Contributions* praised him as "an enterprising botanist and most acute observer," and never once did it seem to occur to Greene to fear Gray. In fact, Greene told Gray he always said what he

[3] See "New Species of Plants from New Mexico," *Botanical Gazette*, VI, Number 1, p. 156; VI, Number 3, p. 183; VI, Number 6, p. 217.

meant. And on several occasions when Gray did not follow his suggested names for species or failed to establish species or genera when Greene thought such should be established, Greene's letters to Gray became increasingly more critical, verging many times on sarcasm almost tantamount to insult. Early in 1881 he "bamboozled" Cyrus Guernsey Pringle, Engelmann said, into giving him plants to describe. Pringle, at the instigation of Sargent, had come west in 1880 charged with three commissions: (1) as botanical collector for the American Museum of Natural History; (2) to make general collections under the direction of Asa Gray; and (3) as an agent for the United States Census Department, to explore the forests of that region and to collect data for a final report.[4] And Engelmann who early received some of his plants was immediately impressed with him as a collector. Greene was transferred during the spring of 1881 to a pastoral charge on San Francisco Bay at Berkeley, and soon after his arrival received plants from the Howell brothers of Oregon.[5] Even Parry, who had been ranging from the upper Sacramento Valley to San Diego in California and east to portions of Arizona let Greene have some of his plants. H. H. Rusby, a young teacher from the East who lived not far from George Thurber, had come west on an exploration tour to New Mexico[6] and in the midst of an exciting Indian War had become a collector for Greene in the higher Mogollon Mountains. Lemmon, following the example of others, submitted plants to Greene for determination.

Perhaps Gray and Engelmann saw in all this some threat to their established places of authority in North American botanical determinations. Perhaps Engelmann, sensing inevitable clashes and conflicts of one sort or another, in the summer of 1881 seized as an excuse his condition of health and went to Colorado instead of California to avoid any more involvement in them than his own place and his friendship with Gray and Watson required. The probability, however, is neither. Gray and Engelmann both at first encouraged Greene, especially, to make his own determinations of new species. On December 7, 1881, Parry wrote Engelmann: ". . . Greene is rushing into *print* at a great rate. Gray encourages him to publish n[ew] sp[ecies] and so between preaching & printing he is in a fair way to notoriety? . . ." But differences in opinion arose. Gray and Watson, while they acceded many times to Greene's

---

[4] See Helen Burns Davis, *Life and Work of Cyrus Guernsey Pringle* (Burlington: University of Vermont, 1936), p. 7.

[5] See Edward L. Greene, "The Two Howells, Botanists," *American Midland Naturalist*, III (1913), pp. 30-32.

[6] See H. H. Rusby, "Ferns of New Mexico Sent by Syracuse Botanical Club," *Botanical Gazette*, VI, Number 4, p. 192; VI, Number 6, p. 220.

views, sometimes disagreed. Each time Greene defended his position, sometimes in a none too tactful manner. Till at length Greene was counted in the company of S. B. Buckley, Kellogg, and others with whom Gray took issue sharply and sternly. And Greene became not only hurt but angered. When Lemmon did not submit his plants to Gray after Greene had been over them, and when Engelmann wrote Gray on January 16, 1882, ". . . Greene did take rather strange liberties with [Pringle's] collections, and the way he treated Lemmon is not better," a small storm cloud appeared on the botanic horizon.

On December 18, 1881, Greene wrote Gray:

I am very sorry that your acquaintance with Mr. Lemmon is so slight, and your understanding of him so imperfect that you suffer anything he says to prejudice you against *me* whom you have known so long. The man seems to me to be nervous, and excitable in a very unfortunate way, & to a degree which leaves him sometimes hardly responsible for all that he says and does. I so view the case, and therefore do not accuse him of willfully and maliciously lying. That "Kellogg, Harford, Moore, and Harkness are bitter against him for continuing to send his plants to Cambridge for identification," is *possibly* true: nevertheless *I* do not believe a word of it. I have never heard any of them intimate such a feeling, and I cannot conceive how they as rational men, could so feel. The reasons why a very great many of our supposedly new or doubtful species should be submitted to Cambridge Herbarium people, are manifest enough. As for myself, I wish to answer you that if such a feeling did exist, *I* should not share it.

I have no wish, either, to be reckoned among the botanical authors of this coast. Of the several scores of n[ew] sp[ecies] I have published within a year, the type specimens (in several instances the only ones extant) are in *your* herbol at Cambridge. My work upon them was interesting and instructive to myself, but it was chiefly your own & Mr. Watson's long absence that tempted me to do what I have done in that line. My *Senecio Howelli* is I think the only species which I printed this year without waiting for your "imprimatur." About Mr. Lemmon's recent Arizona collection,[7] the truth is simply this. Immediately after his return he twice or thrice besought me to come to his place to inspect, and identify his plants. When I had two or three days to spare I went, and did what he requested. Most of the species were well known to me at sight. A considerable number of those apparently undescribed had been received by me from Mr. Rusby, some weeks earlier, and Mr. R[usby] was about distributing under my manuscript names. I of course, repeated these names on the tickets of Mr. Lemmon, as I had a right to do. Then in the cases of a few which were apparently wholly new I gave names.

Now if being called upon to identify a collection of several hundreds of plants does not imply the right to do all which I did in this case, then all the correspondents *I ever had* at Cambridge or St. Louis have done wrong. I have sent to them in all, scores of species which I took for new, always supposing they had a right to do with them what they pleased. I had no idea that Mr. Lemmon should withhold from you his plants. I only expected that in his communicating them to you, he would, at least in the case of the Rusbyan n[ew] sp[ecies] communicate the names

---

[7] Lemmon and his wife spent much time at Fort Bowie, Arizona.

under which I had allowed Rusby to be distributing them. For my three days' work on Lemmon's plants I received not so much as the offer of a single specimen for my own herbarium. . . .

Mr. Pringle, knowing my acquaintance with the Shasta flora, asked me to go through the bundles he brought from there last August. I did so, naming, as he wished me to do, the species. I found two or three unmistakably new ones, & named them as such but that did by no means imply that he should henceforth submit his collections to me for identification. I do not suppose that because I know a little something about western botany, I have the least claim upon any collectors. You & Mr. Watson being at home, I do not expect to ask that anybody shall be so foolish as to come to *me*. . . .

The sincerity of Greene's attestations was confirmed by Parry, in a letter to Gray dated January 22, 1882:

I keep up correspondence with Mr. Greene, who feels rather sore over your criticisms &c. As far as the matter with Pringle is concerned, I can only say that I am satisfied that he acted from the best *intentions*, and though it was a liberty that I would not have taken, he thought he was really doing Mr. Pringle a kindness in securing priority for his discoveries: especially when you were out of the country & your return uncertain. I also know that he did not retain for himself any specimens which were scant but as I understood sent all he had to you for verification: Mr Pringle never intimated to me any dissatisfaction and I am sure as far as I was concerned I did not exceed the permission he voluntarily offered to look over his plants and take duplicates. When he comes on here I will have a clear understanding and return anything he may desire.

I am on the whole rather interested in Mr Greene's enthusiasm for descriptive work, as I know him to be sharp-sighted and cautious though sometimes stubborn. I only hope no ill feeling may be engendered that may act injuriously in the cause we all have at heart. . . .

Parry had had a busy year during 1881. Late in 1880 he had moved to Colton, California, and from there had begun a series of collecting tours—to Yuma, Arizona, strolling over the Colorado River bottoms and bluffs and renewing acquaintance with the region of the Astronomical Camp of the Mexican Boundary Expedition of 1849. He had gone to the home of Muir, who—now married and returned from a trip to Alaska where he had studied vegetation making also geological observations— was now awaiting the arrival of an heir; to the Mojave Desert and to Yuma again where he had left young Vasey to go into Arizona; to Santa Monica, California, to try the sea air and examine plants along the coast and sea beaches; to the desert again with Parish and Wright, and, soon afterward, with Parish south to San Luis Rey and up the coast, and, later, to the lower mountains; and to Yosemite. Parry had found both the Parish brothers promising fellows, active and ambitious, and good collectors of rare plants. But chills, which he attributed to sleeping on the wet ground, had struck Parry and he had gone north. Ague also had depleted his energy. Parry had spent most of the summer near the Academy of Sciences at San Francisco, working much of the time among the mouldy, worm-eaten materials that had accumulated there, but finding new species, among them, a Gilia. On June 22,[8] he had written Engelmann:

I ought to tell you more about *Yosemite*, but am not equal to any description. You must still see it before it gets *too fashionable*. [I]t can hardly be spoiled. My ladies concluded to enjoy it leisurely & put in a month there. I got drenched 3 times in the spray of the falls hunting *drip* plants, found plenty of *Bolandra* but I am getting suspicious of these numerous mono specific Sax[ifragaceous] Gen[era], i.e. Sullivantia, Bolandra, Suksdorfia!! some of these personal names will have to *go*.

I collected *Juncus triformis*, growing in mere tufts at the base of Yosemite falls. . . . [T]omorrow I have an appointment for a trip with Greene to Park & Cliff [H]ouse. I shall think *sadly* of the anniversary of our start on that glorious Western trip *June 23d 1880*, when you seemed so well and accomplished so much. Lemmon is away with his wife using up the remnant of an expiring R[ail] R[oad] pass by a trip to Salt Lake. I am more lucky mine including wife extended to Dec[ember] 31st 1881! Parish wants me to join him on a M[oun]t[ain] trip in August. . . . Greene seems rather disgusted at leaving his dear Mogollon M[oun]-t[ain]s in the height of the season. Vasey now at Albuquerque & the Sandia M[oun]t[ain]s. . . . I must go & see Dunn's El Paso Cacti. . . .

Parry sent some materials gathered along the foothills of the San Joaquin Valley, where he had searched for Stanfordia, and having promised to join Pringle at Summit Station in northern California began preparations for that journey. While in Oregon, Pringle had met Suksdorf and explored both the Mount Adams and Mount Shasta regions but principally Pringle's duties had been cutting logs for Sargent. Parry was not sure whether to wait or not for him. Parry spent a few days taking a trip to their old station on the Mojave and then planned with Greene to go for Governor Stanford to Summit Station or Soda Springs. Greene and he took a short trip to San Bernardino stopping off a while at Tehachapi. When, however, it came time for them to depart for Summit Station, Greene, having more interest in tar weeds than oaks and pines, did not go. Parry wrote Engelmann of his trip on October 7:

Yours of Oct[ober] 1st finds me just returned from a trip to the summit. I need not tell you how cold it was, with snow on the ground and ice in our sleeping quarters! Soda Sp[rings] Station is abandoned with nothing to eat and the trains only stop to let off such forlorn travellers as expect to find comfortable quarters at midnight. [F]ortunately we took grub & bedding with us and finding a stove & plenty of wood & water made ourselves *comfortable*? The next day we tramped over our old ground where you did *not* get lost, looked into your Isoetes [?] lake without venturing to "break the ice" and exhausted our energies in cutting down

[8] 1881.

Abies magnifica trees which were loaded with cones in excellent condition, with plump seed. The same afternoon I went up with a man to Summit Station which is reoccupied with telegraph &c. but *no grub*. [H]ere following Dr Kellogg's directions we found a grove of Tsuga.⁹ . . . So leaving two men to collect seed & dig trees for Stanford's Park, I beat a retreat to a warmer climate, taking a day freight train. I had a magnificent view of *snow sheds* for 20 miles, but beyond Blue Cañon we came into the open forest and a fine succession of Conifers. . . .

I . . . was quite disgusted with John Muir's articles in Scribner, adopting all the antiquated names and sketches that would answer for anything else. . . . Greene is rather quieting down. . . . We will do what we can to help Gray in Aster. [T]hey are now in good condition, that small one we collected in Los Angeles is a great branching thing on Parish's ranch San Bernardino. I have written them to make good "*instructive*" specimens. Sam Parish had gone to S[an]ta Monica, would look out for San Gabriel Oak, but I suppose I shall have to "watch" it too.

I suppose we may leave for South on or before Nov[ember] 1st. I may go down in advance, then I shall establish botanical head quarters at Colton, and overhaul my whole collection, take an occasional run down into the desert & Arizona, and return East in the Spring to find you in Europe? . . .

Parish had recently found a new Aspidium and a remarkable new species of Oxytheca. Parry himself was studying Oxytheca and Chorizanthe, concerning both of which genera he published new species, and in respect of Chorizanthe revised the genus, in later *Proceedings of the Davenport Academy of Sciences*. The Parish brothers, as a consequence, were helpful to Parry and living near them at Colton was enjoyable.¹⁰ But at San Diego there were Daniel Cleveland, a lawyer much interested in ferns and the flora of San Diego, and Charles Russell Orcutt, a young man seventeen years of age who was beginning to work at botany. In October Parry met Professor Hilgard of the University of California, visited General and Mrs. Bidwell at Rancho Chico, and then went to Colton from where in December he went with Parish on another desert trip as far as Maricopa. Gray wrote him suggesting a "nice name for [Parry's] new genus of Eriogonae. . . ."

Parry had heard from Muir who had again gone to the polar regions around Wrangel Land and Herald Island, bringing home plants which were sent to Gray. But Muir had not yet returned to his home in Martinez while Parry was in the North, although when with the Bidwells, Gray, Hooker, and Muir had been much in Parry's conversation. When Muir arrived at Martinez, he wrote Gray: "I had a fine icy time & gathered a lot of exceedingly interesting facts concerning the formation of Behring Sea & the Arctic Ocean & the configuration of the shores of Siberia & Alaska. Also concerning the forests that used to grow there,

⁹ The Latin name of Hemlock.
¹⁰ See Samuel Bonsall Parish, "Parry and Southern California Botany," *Plant World*, XII (June 1909), Number 7, pp. 158-162.

etc., which I hope someday to discuss with you"—subjects which interested Parry, but not greatly. Late that year, on December 26, Parry announced to Gray:

My own botanical ambition is narrowing down in the desire to perfect our knowledge of imperfectly known plants from living observations, and encouraging younger men to enter actual[ly] the field of discovery that is now being opened by R[ail] R[oad] extension. . . .

The Atlantic and Pacific Railroad (now the Santa Fe Railroad) by this time reached from the Colorado River to the Rio Grande. At Eagle Springs, the Texas Pacific Railroad had established a junction[11] opening a quicker facility from Texas and near by southern states to the west. And the line from Colton to San Diego was making progress, although slow. There were many reasons inducing Parry to remain in the West. He wrote Engelmann urging him to add Ephedra to his work on Isoetes and the very difficult Cupressus. But, although Pringle left and returned to Vermont for the winter and though Palmer was in the central southern states exploring mounds to establish a relationship between the aboriginal inhabitants of the Mexican tablelands and those of the great region of the Mississippi Valley with no prospect of coming west, Parry remained in the West. Palmer had written Parry from Arkansas Post, Arkansas, and Watson had heard from him at New Port, Tennessee, sending lily roots from North Carolina. Palmer had tried to persuade Baird to let him go to Arizona and New Mexico and, eventually, Chihuahua and Durango in Mexico; but without success. Though Parry fought to get Engelmann to come west again, all his arguments concerning wine, quiet, and climate were to no avail. Parry went to work in Colton on his 1881 collections. With pleasure he heard that the California Academy of Sciences had received a gift of $20,000 from a wealthy railroad man.

East of the Mississippi, botanical activity was likewise alert and considerable. At the Department of Agriculture in Washington, on June 18, 1880, George Vasey had written: "I have been anxious for years to be able to know thoroughly all the N[orth] American grasses, but have greatly needed authentic named specimens especially from the Mexican border & Pacific Coast. We have now a set of the Mexican grasses collected by Bourgeau,[12] but they are mostly unnamed. We have probably most of those collected by C[harles] Wright in N[ew] Mexico, but *none* of those collected on the Mex[ican] Boundary Survey. . . ." A week later a box of grasses arrived from Gray, and Vasey replied: "Your letter is

---

[11] With part of the now Southern Pacific Sunset route.
[12] Eugene Bourgeau (1813-1877).

received, as also the box of Grasses. I have looked them over hastily and find that although many are unnamed, yet by bringing the families together I think 75 per cent can be readily named; some sets are pretty fully named, and will name others. I will bring them together, number them, name such [as] are clear, and distribute them according to your suggestions. . . . Mr. Meehan, Prof. Porter, Dr. Leidy & others were here yesterday on their way to the M[oun]t[ain]s of N[orth] Carolina. Mr. E. Hall wrote me some time ago that he was going to spend a few weeks there." Vasey's son had explored in North Carolina and found a rhododendron in Jackson County submitted to Gray for determination early in 1879. And in the summer of 1881 John Donnell Smith made a trip to the mountains of North and South Carolina. Short and sporadic exploration of mountains in eastern United States persisted, receiving many times only casual and not significant notice.

Vasey was serious about his intentions for the United States National Herbarium. On July 28, 1881, he wrote Watson:

I send you today by mail a small package containing some plants from Idaho collected by Dr. E T Wilcox, who makes miserable specimens, but some of the plants are not altogether familiar to me; also a few collected by my son in New Mexico. I will thank you to look at them and confirm or reject. . . .

. . . [T]he work of my division has been put back from want of help. Six months ago my helper was discharged for want of funds, and a great quantity of plants has accumulated which need to be poisoned & mounted for the Herb[arium]. I cannot do everything with my own hands, and it is not necessary, as there is an appropriation for paying for help and buying specimens. . . .

There is also an appropriation for an Assistant Botanist. His duties are not defined, but it was understood under the old Government, that he was especially to take up Cryptogamic Botany. But Dr. Loving seems to have more utilitarian views, and wants an Economic Botanist or perhaps an Agricultural Botanist. I have mentioned Mr. C. G. Pringle. . . .

My ambition has been to bring this Herbarium up to a first class standard, in fullness and perfection of specimens and if allowed to proceed a few years longer, I think I should reach the standard.

Herbaria in the United States at this time were in none too proud a condition. Although in 1876 John H. Redfield had become conservator of the botanical section of the Academy of Natural Sciences at Philadelphia and was giving years of devoted service making that institution's herbarium efficient and modern, North American botany looked to the Gray Herbarium at Harvard. Said the *Botanical Gazette* (which had become a leading botanical journal) in June 1880:

The Botanic Garden at Cambridge is no longer a local, but a national concern. The eyes and thoughts of the botanists of this country are directed to it as naturally as are those of English, in fact, the world's botanists, to Kew Garden. There we find the largest herbarium, the largest library, the largest collection of living plants indigenous to our country, to be found anywhere on the continent. . . .[13]

Moreover, Gray, ever mindful of the memory of John Torrey, consulted authorities of Columbia College and the Central Park Museum to arrange for the proper care of the Torrey Herbarium which had been slowly deteriorating since the days when Parry and Gray corresponded concerning it soon after Torrey's death. Lectures in botany were given at Columbia by John Strong Newberry. But the characteristic aggressive zeal of Torrey was gone. Gray evidently consulted a young graduate of the School of Mines, an assistant in geology under Newberry, and now serving as botanist and assistant geologist on the geological survey of New Jersey—Nathaniel Lord Britton, author in 1881 of *A Preliminary Catalogue of the Flora of New Jersey*, who wrote Gray on March 22, 1882:

As to the new cases for the Herbarium (at the School of Mines) there is nothing definite decided on yet, so far as we can ascertain.

Dr. Newberry is not informed yet whether the new quarters for the Herbarium in the proposed School of Mines building on 49th St[reet] will be erected this summer, or the Herbarium will have to remain in its present situation for some time longer.

As soon as the trustees decide this question, action will be taken about the new cases, &c. . . .

The museum authorities first agreed to take charge of the Torrey Herbarium but, later, realizing that the materials were made over by Torrey during his lifetime to the college, it was decided the college should continue its custody. Years later, when Nathaniel Lord Britton came into control and the New York Botanic Garden was established, the Torrey Herbarium was transferred under contract to the Garden. Thus another great North American herbarium was saved to posterity.

Botanical exploration had continued unabated in eastern United States as had the building of many individual herbaria. Amherst College purchased J. T. Holton's herbarium of 6,895 species, mostly New Granada specimens, and Ericaceae from the Cape of Good Hope. The Gray Herbarium received a large collection of Indian plants, including, probably, many Australian and New Zealand specimens.[14] Published lists and catalogues of plants, based on collections made by a large number of collectors, were being made known in nearly all of the states. And the compilation by Gerard and Britton of the *Lists of State and Local Floras of the United States* was begun. Among the southern states

---

[13] V, Number 6, p. 62.

[14] Brown University established a professorship in botany with money left by Stephen Olney. Besides receiving Olney's important herbarium and letters, the department soon received Bennett's.

the survey was revealing one or two states to have but one such list while among the northern states several states showed many.

Dr. Farlow, reporting progress in botany during the year 1882, observed more activity in the subjects of vegetable physiology and anatomy and more descriptions of new species of phanerogams and fungi than algae and higher cryptogams. Studies of plant assimilation and plant respiration[15] had continued but studies of the action of light on plants and the relation of chlorophyll to plant economy were somewhat diminishing. Medical studies on bacteria were increasing. Engelmann had studied the relation of bacteria to light and air, concluding bacteria collect in heaps where there is a development of oxygen. Contrivances for cross-fertilization were still receiving attention from men like William Trelease, J. E. Todd, and others. Divisibility in botany between work of the field and of the laboratory was becoming more noticeably clear. Rothrock wrote:

I start with microscopic botany, urging that my pupils see for themselves, draw for themselves and come to their own conclusions. After some months in such mental drill, I shall introduce them to systematic botany. . . . Systematic botany must, if it represents a strictly natural system, be founded on a nice appreciation of the entire organization, the life history of the individual and its relation in past and present time to allied plants. This, then, is the highest, all embracing trend botanical thought can assume.[16]

But few years would go by before Coulter in his remarkable address, "The Future of Systematic Botany," would be saying the same as Rothrock.

With the year 1883, Farlow noticed that, while studies in plant physiology and morphology were abundant, no one of them had made any especially striking discovery. None were as elaborate as in former years. While there were studies in plant electrical currents, in nongreen colors of plants, in the connection of protoplasm of adjoining cells through opening in cell walls, and the like, Farlow said, "The countless papers on bacteria can no longer be considered under the head of botany, for by far the greater part of them have a purely medical bearing."

Taxonomic study still held its important place. Hooker and Bentham's great *Genera Plantarum* was at last completed, marking, Gray said, "an era in systematic botany," covering as it did "the whole field of phaenogamous botany." 1882 had brought forth Tuckerman's *A Synopsis of the North American Lichens*, Part I. After publishing in 1876 his "Cata-

---

[15] Particularly a study on this subject by John Merle Coulter.

[16] See complete article on this subject by Rothrock at about the time announcement was made of establishment of a botanical laboratory at the University of Pennsylvania to study medical aspects of botany, *Botanical Gazette*, VII, 1 (January 1882), pp. 7, 8.

logue of the forest trees of the United States which usually attain a height of 16 feet or more," George Vasey published in 1883 his "Grasses of the United States" and his "New Western Grasses"; and during the next year would follow his famous "Agricultural Grasses of the United States" and "Distribution of North American Forest Trees," all government publications and predecessors of a line of very important studies made by this venerable botanist during his last years.

More important still were the stupendous publications in paleobotany being brought forward.

CHAPTER VIII

Lesquereux and the Development of
North American Paleobotany.
Western and Eastern Coal Floras

O N February 12, 1874, Leo Lesquereux sent Hayden the first of
three major reports styled *Contributions to the Fossil Flora of
the Western Territories—Part I, The Cretaceous Flora.*[1] Problems of geographic floral distribution of earlier geologic epochs and of
their origins and multiplication were all involved in Lesquereux's paleobotanical studies. He went at problems by giving attention to a geologic
formation's age, established by implications of specimens at hand. Always relentlessly verifying new and former determinations, he listed
genera and species according to epoch or period, as, for examples, Eocene, Miocene, and Pliocene epochs of the Tertiary period. At first, in the
East, Lesquereux studied the paleobotany of upper Paleozoic periods. In
the West, studies went for the most part to the Cretaceous and Tertiary
periods. "Science" was to him "a high mountain. To go up to its top or
at least high enough to gain free atmosphere and wide horizon," he said,
"necessitates hard climbing, through bushes, thickets, rocks, etc." He
listed new species, compiled tables of distribution according to ages and
in comparison with species found in Europe, Greenland, Alaska, and
other places. Observations were made as to indicia of climate and special
phenomena. Were there temperate zones in the early geologic periods,
such as we have today? In 1877, when reporting *Part II, The Tertiary
Flora of the Territories,*[2] Lesquereux revealed his objective in the following:

This Flora of the North American Lignitic is like a supplement to that of the
Cretaceous Dakota Group. Both together constitute a historical record not less
interesting to Botany than to Geology; for, beside the evidence afforded on the relation of the groups of the formations, they expose, as in a written book, documents
illustrative of the origin and the successive development of some of the predominant
and more interesting types of the present vegetation of this country.

In 1873 and 1874 he had sent Hayden for his annual reports studies on
"The Lignitic Formation and its Fossil Flora,"[3] which were reports on

---

[1] Washington: Government Printing Office, 1874.

[2] Washington: Government Printing Office, 1878.

[3] Sixth and seventh *Annual Reports of the U.S. Geol. and Geog. Surv. of Terr.*, embracing
exploration reports of the survey. In the 1874 volume, Lesquereux's work is found at p. 365. In
the 1873 volume, at p. 317.

the paleobotany of Tertiary formations generally but, particularly, of the Rocky Mountains. Also in 1874, in *The American Journal of Science and Arts*[4] there was published Lesquereux's article, "On the Age of the Lignitic Formations of the Rocky Mountains," and in a *Bulletin*[5] of the territorial survey his observations "On the General Character and Relation of the Flora of the Dakota Group." In a letter dated March 7, 1875, after listing those who aided in preparing the supplement to Sullivant's *Icones Muscorum*, Lesquereux told Gray:

> I got for mail yesterday the 3ᵈ vol[ume] of the Arctic flora of Heer. . . . Heer owes me nothing while I owe much to himself and especially to his works. Did you not get a Copy of the Cretaceous flora of the Dakota group? The comparison of these two floras of the same age is extremely interesting. Our American cretaceous is however more recent, at least in regard to its vegetable types, most of all representing dicotyledonous forms. . . . Anyhow both these monographs of Heer and of myself are most valuable as the first important contributions to the Cretaceous nearly totally unknown as yet. . . . [I am] deep in the preparation of the Lignitic flora and other matters which will take my time for one year at least.

And on February 16, 1875, he told Lesley:

> This Cretaceous flora is the first distinct ray of light of the vegetation of the Me[s]ozoic times of America. The Tertiary flora is a sequence and both are greatly valued by European authors. For the deductions taken upon the origin and distribution of species have been so till now by mere hypothetical speculations and are all wrong. We have from the Cretaceous, the N[orth] A[merican] types of vegetation becoming more and more distinct and more related to those of our present vegetation. . . .

The "great supposed bridge" between European and American floras across the Atlantic was all changed now, said Lesquereux, and new explanations were in the new works. On January 8, 1876, the Survey *Bulletin* (2nd series, No. 5) released three studies by Lesquereux: "A review of the Fossil Flora of North America"; "On Some New Species of Fossil Plants from the Lignitic Formations"; and "New Species of Fossil Plants from the Cretaceous Formation of the Dakota Group." Also, the eighth *Annual Report*, published during 1876, presented his "On the Tertiary Flora of the North American Lignitic, Considered as Evidence of the Age of the Formation" and "A Review of the Cretaceous Flora of North America."

Both Gray and Lesquereux regarded Heer's researches in fossil botany as "very important in their bearings. They made it certain," Gray wrote,[6] "that our actual temperate floras round the world had a common birthplace at the north, where the continents are in proximity; they essen-

---

[4] VII (3rd ser., June 1874), pp. 1-12; also "On the Formation of the Lignite Beds. . . ."
[5] I (1st ser.), Number 2, pp. 52-62.   [6] See *Sci. Pap. of Asa Gray, op. cit.*, II, p. 449.

tially identified the direct or collateral ancestors of our existing forest-trees which flourished within the arctic zone when it enjoyed a climate resembling our own at present; and they leave the similarities and the dissimilarities of the temperate floras of the Old and the New World to be explained as simple consequences of established facts."

Heer's health was always delicate, so much so he could do little exploring in person. Materials had to be sent him. His condition must have aroused sympathetic understanding from Lesquereux, who was also born in a canton of Switzerland; and had furthermore too known the handicaps of physical disability. Lesquereux, with only six years of institutional affiliation in America, braved, like Heer, his difficulties, lost grief in a world of geologic ages, made plants his companions, and accomplished in America what Heer had accomplished in Europe—became a world authority on his subject. Heer, it is said, served as a professor of botany at the University of Zürich. While Lesquereux was often referred to as a "professor," seldom, if ever, in America did he teach. Certainly he was never a duly constituted professor in an American institution of learning, and certainly he never lectured any more times than necessary.

Heer did some early study in American materials of other than Arctic regions. Before Lesquereux began extensive studies of Western United States and Canadian materials, there had occurred a somewhat complicated, but not highly important, history of fossil plant collecting in Kansas and Nebraska, especially in materials found at the mouth of the Big Sioux River and at Blackbird Hill on the Missouri River in Nebraska. The story of American discoveries of fossil plants in the West was related ably by Lesquereux in *Part I, The Cretaceous Flora* but, for our purposes, it may be said that, excepting a very few articles of minor significance published in America, most of the important early determinations were made in Europe, among which figured prominently a memoir by Heer, "Phyllites Crétéces du Nebraska," elaborating a little less than a dozen and a half species. Even Heer was not the first European systematist to publish concepts of American fossil plant determinations. Lesquereux interestingly has pointed out that at the middle of the nineteenth century scarcely eighteen species from American formations had been made known—these being published by Brongniart in his *Végétales fossiles* from specimens sent him by Benjamin Silliman —and, in contrast to this number, within a quarter of a century, that is, by 1875, more than one thousand species from American measures were described. As already mentioned, American's own eminent geologist, James Dwight Dana, was among the first to publish American fossil plant discoveries, materials then regarded of tertiary origin and gathered by the United States Exploring Expedition, 1838-1842, in Washington Territory. Indeed Gray mentioned certain work Heer did on fossil plants of Vancouver's Island and British Columbia as antedating by a few years what Gray called, "the first of that most important series of memoirs upon the ancient floras of arctic America, Greenland, Spitzbergen, Nova Zembla, arctic and subarctic Asia, etc., which, collected, made up the seven quarto volumes of the 'Flora Fossilis Arctica.' " Heer and Lesquereux corresponded frequently, Lesquereux publishing one letter in support of his view that the British Columbia fossil flora—at least that part from Evans's survey materials—was Tertiary. On Heer's death, Lesquereux carried on Heer's work, adding in 1882 a "Contribution to the Miocene Flora of Alaska":[7]

The plants described by Heer, representing 56 species, are of marked interest by their intimate relation with those of Atane, in Greenland, on one side, and with those of Carbon, in Wyoming and of the Bad Lands of Nevada, on the other. They compose a small group which supplies an intermediate point of comparison for considering the march of the vegetation during the Miocene period from the polar circle to the middle of the North American continent, or from the 35th or 40th to the 80th degree of latitude. The remarkable affinity of the Miocene types in their distribution from Spitzbergen and Greenland to the middle of Europe had already been manifested by the celebrated works of Heer. But the Alaska flora has for this continent the great advantage of exposing in the Miocene period, the predominance of vegetable types which have continued to our time and are still present in [our] vegetation. . . .

Having to do with fossil plants from Alaska and its vicinity, collected by Dr. W. H. Dall of the Coast Survey at Coal Harbor, Unga Island, Shumagin;[8] Chugachik Bay, Cook's Inlet; and Chignik Bay, Aliaska Peninsula,[9] Lesquereux's observations continued:

In the valuable collection, which was intrusted to me for examination, I have found a number of species, already described by Heer, from Alaska, a few others described already from the Miocene of Greenland or of Europe, but yet not known from Alaska, and some new species. These last are described . . . with the enumeration of those described already, but not yet known in the flora of Alaska.

Lesquereux kept alive to progress in botany as well as paleobotany. He read Engelmann's botanical papers, "Notes on Agave,"[10] and "The Oaks of the United States," written for publication in the *Transactions of the St. Louis Academy of Sciences*. On July 3, 1876, he wrote Engelmann:

I read them with the greatest interest, especially the oaks and consider them as

---

[7] *Proc. U.S. Museum*, V, p. 443. Alaska plants are now regarded as Upper Eocene.
[8] South side of Alaska.   [9] Southern Alaska.
[10] Gray had reviewed Engelmann's "Notes on the Genus Yucca" in *The American Journal of Science and Arts*, VI (3rd ser.), pp. 468 ff. See *Sci. Pap. Asa Gray*, I, p. 196.

important documents to Botany. Your distribution of the oaks and your remarks on their characters clear the very difficult subject and will certainly help greatly American botanists in the determination of the species.

Lesquereux had agreed to complete "a Synopsis of the U[nited] S[tates] Mosses" from Sullivant's materials, and on January 18 wrote Gray:

Do you object to describing the whole bryol[ogical] flora of the N[orth] A[merican] continent from Alaska to Mexico? We have now so many species from the Rocky M[oun]t[ain]s that we must forcibly consider Drummond's[11] mosses and taking Alaska in the area (botanical) we have to admit the Canad[ian.]

Sullivant and Lesquereux's supplement to the *Icones Muscorum* had met with immediate approval from Europe's great bryologists, Schimper, Hampe, and Lindberg. In fact, Lindberg considered it "one of the best published [works] in Bryology" and Schimper said it was "an admirable complement of an admirable work." Gray, when Sullivant died, asked Agassiz to free Lesquereux from work being done for him to permit Lesquereux to complete the moss manual.

It was so arranged. And Lesquereux was much encouraged. In the Great Plains and Rocky Mountains of the West where paleobotanic searches continued at the hands of the United States Geographical and Geological Survey, and other agencies, much that would reinterpret North American geological history was being brought to light. Lesquereux's participation was a study of its paleobotany. But the more investigations continued, the more enormous and complex became the problems. Three years later, in 1879, Asa Gray was to say:

The interest which we take in the vegetation of former periods is not so much geological as genealogical; and this interest diminishes with the distance from our own time and environment. We know nothing of the earliest plants—the beginnings of vegetable even more than of animal life are beyond our ken; no great satisfaction seems obtainable from the small acquaintance that has been made with the plants which flourished before the carboniferous period. And the botany of that age, notwithstanding its wealth of Ferns[12] and its adumbrations of next higher types, impresses us as much with the sense of strangeness as of wonderful luxuriance. For even the fern-impressions, familiar as they may look to the unprofessional observer, are outlandish. The more the critical student knows of them the less likeness he finds in them, or in the coal vegetation generally, to any species or genera now living.

While Lesquereux's early North American studies had begun with the

[11] Thomas Drummond, a very important early Scotch explorer in North America. See S. W. Geiser's *Naturalists of the Frontier* (Southern Methodist University, 1937) for an excellent account of him, pp. 73 ff.

[12] A majority of carboniferous ferns have since been shown to be seed plants, plants with wholly enclosed seeds.

Leo Lesquereux

upper coal and recent moss flora and he had published much on the positions of coal seams, on fossil marine plants and flora of the upper carboniferous measures—on the coal flora generally—for a number of years his energies also had been directed to the later geological periods— the Cretaceous and the Tertiary,[13] during the former of which flowering plants (Angiospermae—covered seed plants), both dicotyledones and monocotyledones, emerged; progenitors in many instances of our palms, oaks, maples, elms, tulip tree, fig, grasses, and a great variety of herbaceous plants including grains. Of the Cretaceous flora, he had written Joseph Henry on May 26, 1875:

> The Cretaceous flora of North America has furnished already to the history of the vegetation of the world some documents of the highest interest. It still promises more for the future. As well you know, the first appearance of the Dicotyledonous plants coincides with the first land formation of the Cretaceous. We touch there to the moment which separates the vegetation of the primitive periods from that of the recent ones. How did the Dicotyledonous be formed or developed; what have been the first representations of this great division? It is the great problem which every botanist paleontologist should try to solve at any price. I have been therefore encouraged by the most eminent paleontologists of Europe to try by all means to obtain new documents in addition to those published in the fossil flora of the Dakota group. . . .

As a consequence, Lesquereux welcomed every employment of scientific skill in the assemblage of data concerning these most important past geologic periods. And of these, authentic and accurate data was increasingly available.

But of Mesozoic plants prior to the Cretaceous, American knowledge came slowly. On July 30, 1875, Lesquereux had written: "We have as yet few good specimens of the Devonian and the Subcarboniferous. If any, I have rarely had opportunity to examine them. . . ." As late as January 3, 1881, he added: "For the Jurassic or Trias[sic], we have very little to show. I know the plants by Emmons; some have been described by Prof[essor] W. Rogers. . . ." The Triassic flora Lesquereux recognized as "not distinctly represented in the North American geology." Richmond, Virginia, and North Carolina coal referred to this period indicated by their fossil flora, he said, a relation to "the lowest member of the great Jurassic period—the Triasso-Jurassic. . . ." However, excepting Jurassic and Permian floras and in part the Triassic, he found nearly all other known groups "of geologic floras" well marked in North America. Indeed, publishing in 1882 "On Some Specimens of Permian Fossil

[13] An excellent study of the floras of past geologic periods is that by Frank Hall Knowlton, *Plants of the Past* (Princeton University Press, 1927).

Plants from Colorado," found in South Park near Fair Play, he commented:

> Though the specimens are very small, covered with mixed minute fragments of leaves, scales, flowers, and seeds of Conifers, leaflets of Ferns etc., I was able to recognize in all those which could be determined, the characters of a Permian vegetation. . . . The age of a flora is indicated, not only by the presence of certain types, but by the absence of others. And in this, the group of vegetable remains in Fairplay is remarkably free of any fragments of plants characterizing the Triassic period. . . .[14]

Knowledge of North American paleobotany was then principally confined to more recent geologic periods—Cretaceous, Tertiary, and Quaternary—although Lesquereux with almost herculean strength was assembling data of upper coal formations. This work, to the present, is accounted a labor unsurpassed for greatness of accomplishment. On December 21, 1880, he told Spencer Baird of the Smithsonian Institution:

> As far as I can see there are now described from the United States measures[:]

1. Devonian and Carboniferous	600 species.
2. Permo-Carboniferous. Fontaine and White	70 (")
3. Cretaceous	160 (")
4. Tertiary	550 (")
	1380

> With species to be described in the VIII [volume][15] of the Geological Surveys of the Territories by Hayden the number will be about 1500 to 1600. To this may be added the Triassic plants by Emmons & W[illiam] Rogers, number as yet unknown to me. . . .

Lesquereux's work in floras of upper and middle Paleozoic eras, especially their Carboniferous (usually divided in this country into the Mississippian and Pennsylvanian), Devonian, and Silurian periods, stands as a great example of scientific eminence in American botany. Six hundred and seventy known species may not seem today a considerable number. But when one considers that in 1880 he had labored less than three decades, doing by far the largest number of descriptions and much of the exploration for specimens alone, one realizes the breadth of his accomplishments, done against tremendous odds. He had to determine the origins of large groups, working with scarce and insufficient materials. Materials were fragmentary and in many places obscure. Exposed rock strata had to be found many times. New finds of species incessantly displaced former systematizations, requiring revisions. When he wrote up his descriptions, nearly always because writing English was difficult for

---

[14] *Bulletin of the Museum of Comparative Zoology*, VII (Geol. Ser. I; Cambridge, 1880-1884), Number 8.
[15] See pages 177, 189 of this book.

him, he had to rewrite sentences many times. Despite handicaps and hardships, Lesquereux, nevertheless, had been able to conclude in 1873, "It is certain . . . that in the Middle Devonian we have representatives of three distinct groups of vegetables: the Cellular Cryptogams, in a quantity of marine plants, the Vascular Cryptogams, in Lycopodiaceous plants: Lepidodendron, Sigillaria, etc., and the Phaenogamous Gymnosperms, in the Conifers."

There were few other workers. William M. Fontaine and I. C. White entered fields of eastern states, studying "Permian and Upper Carboniferous" vegetations. Sir William Dawson occupied in Canada a position similar to Lesquereux's in the United States. In 1881, Lesquereux said: ". . . the work of Dawson's will afford all the materials known from the Carboniferous to the Devonian" of the plants of Canada. In 1863 Lesquereux told Lesley: ". . . the Nova Scotian basin is a separated member of our great American coal fields. The flora of both the Canadian and U[nited] S[tates] coal fields is apparently the same. . . . Dawson finds in Canada an abundance of fossil coniferous woods." Some of America's earliest work in Paleozoic botany had been done by Dawson. Combined with the work of Heer in more recent periods, Canadian paleobotany also had made a start and need of joining its findings comparatively with those of the United States measures was foreseen. True, some of the most important plant groups known today were unknown then. Still, with as few workers as there were, one wonders how in less than forty years American paleobotany was established.

The answer is, of course, that paleobotanists did not seek to establish relations between plants of our modern flora and those of the very ancient flora. As a matter of fact, Gray and Lesquereux evidently believed they could not do so, even though they tried. Concepts of phylogeny were practically unknown in America. The advanced morphology of Europe had not reached here. A few progressive American students saw advantages to be gained from study of plant life histories. Strange and wonderfully luxuriant as the ancient floras were found to have been, the critical student nevertheless, it was seen, could not establish their genealogical connection with plants of the present period. Indeed, while Lesquereux risked some "deductions," it was not on this that he concentrated the real forces of his mind. He rightly conceived his task as one of systematizing all available data.

Above crystalline rocks and in stratified deposits of the "Lower Silurian" (Ordovician?), plant fossil remains were discernible. Being marine plants of obscure forms "like crushed bundles of filaments," their contours were often "obliterated in a black carboniferous or bituminous mass." They were, in part, fucoids or seaweeds of vascular tissue, great size, and without woody fibers, and foreshadowed, Lesquereux said, the rise of coal plants—their decomposition resulting, he believed, in "deposits of bitumen, or mineral oil, which man's ingenuity uses now to an advantage not equal indeed, but comparable, to that which he derives from the coal." Furthermore, they gave little, if any, indication of temperature or atmosphere covering the globe during Silurian times.[16]

Granting that a more diverse marine vegetation appeared in the next period, the Devonian, what of land plants that were found in the "Lower Devonian" strata? Were there any land plants of the Silurian period? In 1874 Lesquereux published in *The American Journal of Science and Arts*,[17] the article "On Remains of Land Plants in the Lower Silurian." His reason was the extraordinary discovery of remains of a land plant "referable to Sigillaria," a species, so Lesquereux said in 1880, "from visible or external organization of the stems I consider . . . as related to Lycopodiaceae or Lepidodendron. . . ." The believed Silurian "Sigillaria" discovery of the early 1870's was made on Longstreet Creek near Lebanon, Ohio, "in clay beds positively referable to the Cincinnati group of the Lower Silurian." Up to that time, "the geological formations of the United States [had] not afforded . . . any records of land plants earlier than those of the Lower Devonian. . . ." During the long Silurian period, whence came the first then known vegetable fossil remains, water had covered a part of the earth's surface, Lesquereux believed, either in condensed form as fluid or as vapor. This land plant discovery was of great interest. In 1875 in his "Review of the Fossil Flora of North America," he commented:

> More recently . . . fossil remains of two species of vegetables positively recognized as land plants, have been found in the Silurian formation of the Lower Helderberg of Michigan, and attest the existence of land plants, and, consequently animal life also in the Silurian period, a fact which till now had remained uncertain. The presence of land-plants in strata of a lower formation—that of the Cincinnati group —becomes less improbable by the discovery. . . .

Lesquereux affirmed that in the "Lower Devonian" strata, land plants increased in size and were more in number. But the connecting link between Devonian flora and that of the Silurian was almost negligible in North America. Only a slight thread of connection on Gaspé peninsula in Canada existed. He continued his work systematically and on October 19, 1877, read before the American Philosophical Society a paper on five

[16] *Bulletin*, I, Ser. 2, No. 5, pp. 233-248, U.S. Geol. and Geog. Sur. Terr., *op. cit.*
[17] VII (3rd ser.), pp. 31 ff.

new species of land plants discovered in Silurian rocks of the United States.

Devonian land plants do not differ materially from the Carboniferous, the next period of the Paleozoic era. Yet Lesquereux noticed the Devonian flora, while still largely marine, had more and larger plants than the Silurian and ascended in more complex and more "perfect" structures to the Carboniferous, "known by the great quantity of fossil remains, corresponding, in their proportions to the prodigious exuberance of a vegetation which has furnished the compound materials of the Coal strata. Concerning the character of the plants," he wrote, "the Coal epoch has been named the reign or the period of the Acrogens; the flora from the base of the Millstone Grit, or even from the first traces of the lowest beds of the Subcarboniferous to the Permian, being represented, especially by species of this class, Ferns, Equisetaceae, and Lycopodiaceae. . . ." Ferns, that is, seed ferns, alone were represented by nearly 350 European species, although in America by 1875 hardly half that number, "as yet, [was] recorded from the North American coal measures." In 1878 he studied and published *On the Cordaites and Their Related Generic Division in the Carboniferous Formation of the United States* and in 1879 for the American Philosophical Society "On a Branch of Cordaites Bearing Fruit." Lesquereux's work among Cordaites was distinguished, although his "small memoir on Silurian plants" was most congratulated, its value being at once recognizable. As to Cordaites, it was 1877 before he was able to say: "We have now good specimens for a class of plants formerly known only from fragments of their ribbon like leaves generally mostly undeterminable."

The following year, however, he had related that Charles H. Sternberg, one of his western collectors, had "found first a branch of these plants with leaves attached to the stem (silicified)" and on this a "celebrated analysis" had been made. Since then, except a specimen Lesquereux had found near Pottsville, Pennsylvania, no leaves with stems had been found until discoveries made by F. C. Grand'Eury in Europe, enabling the latter to write "a splendid monography of the Cordaites in his Carboniferous flora." After this I. F. Mansfield, a Pennsylvania coal mine owner interested in coal flora, obtained "a splendid series of branches with leaves, even with flowers and leaves, representing in well defined characters numerous species and a new section of this family unknown to Grand'Eury. . . ." In March 1879, Lesquereux told Lesley:

In the last [box received from Mansfield] I find what has been searched for since botanists began to study the coal plants, one of those large nuts [seeds] generally found scattered, never attached to any support, this time in distinct connection to a branch of Cordaites. . . .

Even fungi were found in the coal. On December 7, 1876, Lesquereux observed that, by extensive researches in the "Subcarboniferous, and the upper carboniferous," not only had many European species believed "absent of the N[orth] A[merican] coal measures" been found, but also discoveries had been made "of such kinds of plants which were formerly considered as non existent at the coal epoch." He said:

For example marine or fucoidal species some of them related to Silurian types and also true Fungus, a Rhizomorpha, perfectly well determined, which settles the question of the existence of mushrooms in the Coal—And now, I have to extend the limits of the Carboniferous to the base of the Catskill and of course can not leave out of the flora the few Devonian land plants which are known before this formation. They belong to the history of the Coal flora either as ancestors or even may be recognized in it by some representatives. And also, I can not fix any geographical limits to that ancient flora. It must be considered everywhere it has been observed and from all the documents obtainable from Nova Scotia to the Western limits of the field in Iowa.

Not surprising, furthermore, were the discoveries of fossil plants bearing characters seeming to "indicate a simple structure of the Algae by juxtaposition of elongated cells joined by their ends, as are now the thread-like filaments of the thermal springs," observed Lesquereux in his "Review of the Fossil Flora of North America," a paper prepared for *Penn Monthly* of Philadelphia but regarded so valuable by Hayden the latter insisted on its being published as a government survey *Bulletin*.

Lesquereux sought "to compare *all the species* related to European ones, whose identity is possible or probable but not yet ascertained and make then what Agassiz wished [him] to do, a close comparison of the American Carboniferous flora with the European types of the same epoch. . . ." He worked slowly and carefully. Eye trouble and infirmities of age bothered him at times. Nevertheless he continued. "As we know the fossil plants merely by fragments," he said, "the characters of the plant can not be understood except when we are able to see them exposed in their many transformations in many specimens. . . . The habitat of the species has to be indicated very carefully in order to precise as far as possible the distribution vertical and horizontal of the plants and I have thus to come and come again to my numerous lists of local floras to fin[d] those habitats. . . ."

In most instances when Lesquereux published he did little more than list, describe, and comment on the plants of various geologic periods. He must have deliberately resolved to make his "deductions" as few as possible, believing the time premature for theoretical observing. He said as much. His *Principles of Paleozoic Botany*, published as a part of the thirteenth annual report of the Indiana department of geology and natural history, did not appear until 1883 and not as a completed work until 1884. He aimed not to make his works "for savants" only, but for "miners, private proprietors interested in coal, teachers," and even amateurs since many times they furnished some valuable collecting. The voluminous works of certain German authors he characterized as "tedious and useless for those who have not made a special study of some parts of botany." Their discussions touched "only three or four paleontologists which corresponding on the subjects are already full of them and do not learn anything by the volume." He knew himself to have examined more specimens of plant remains during thirty years of exploration than any other "living paleontologist." When he spoke, he spoke authoritatively, saying, for example, with assurance: ". . . there is not as yet in the Cambrian any remains of a vegetation analogous to that of the middle or upper Silurian, not even positive evidence of true Fucoidal remains." And this was said as late as 1879 of the earliest period of the Paleozoic era.

However, there reached America during these years word of new developments in European paleobotanic study. On October 14, 1878, he told Lesley:

The English authors, especially Williamson and Binney, mostly study the internal structure of vegetable remains which they have silicified.[18] They can therefore have slides prepared by the lapidary and have microscopical studies upon many things still unknown, things to which we have not any access here. For until now we have not discovered any silicified plant but the trunks of tree ferns of Shade river whose internal structure may be generally seen with naked eyes. These authors are thus upon another track of researches than we are here and the help they can afford us is very little. For this reason they wrote me both, on the Cordaites: that they were most interesting but that they could say nothing more on these remains but what I had said myself. . . .

North American paleobotany during Lesquereux's years of prominence stressed taxonomy—systematics based for the most part on differences among the newly discovered plant groups. But let it not be believed that Lesquereux did not understand the value of seeking to establish affinities. When authorities differed as to whether Cordaites were related more to Cycadaceae or Coniferae, Lesquereux, like others basing his conclusions *on the evidence of leaves*, said: "There is some affinity especially in the nervation of leaves of Cordaites to those of species of the above generic divisions. . . . Thus the relation stands partial and our carbonifer-

---

[18] Calcified.

ous plants (Cordaites) continue to show like many others such a mixed analogy with known types either more recent or of the present that they must be considered as prototypes not of one but of many of the essential vegetable groups which have appeared after. . . . It is certain and you will [have] seen it soon more clearly, that fossil remains of plants, left until now mostly as rub[b]ish in the cabinets, will have to be taken into due consideration for data related to the age of the formation. . . ." In 1880 he was aware of Hooker's "beautiful analysis of the woody structure" of trunks and their fructification. "We do not have in the N[orth] A[merican] coal measures as yet a single specimen representing the bark fossilized and the internal woody matter replaced by clay or other mineral matter," he commented. Lesquereux admired much the microscopical work of the French botanist Renault on "the anatomy of the fossil wood." Still, however, basing his work on "visible or external organization" of stems and leaves, he believed that the anatomists "do not see more clearly the relations of some groups of coal plants from the analysis of the internal structure than we can se[e] from outside characters. On February 12, 1881, he told Lesley: ". . . the conclusions of the anatomists are in concordance with those I have derived and published at diverse times on the absence of Conifers in the Carboniferous; on the relation of the Cordaites and especially on the double character and double mode of vegetation of the Stigmarias." Lesquereux's view was obvious. For his work as a taxonomist, internal structure study was neither necessary nor possible in America. Not once does he seem to have fought the new methods. Rather, he seems to have taken a great interest in them.[19] Before his death, new methods as a part of the "new botany" would be well along. Beginning about 1880, when David Pearce Penhallow would return from Japan—from which country Lesquereux was sent fossil specimens by B. S. Lyman—and when as a student of Gray, Penhallow would go in 1883 to McGill University to succeed Sir William Dawson, North American paleobotany, following the lead of European investigators, would initiate studies in external and internal anatomy

---

[19] In support of this claim, reference may be made to "On a Cours de Botanique Fossile by Prof. M. B. Renault," by Lesquereux. Read before the American Philosophical Society, February 18, 1881. In the course of this, Lesquereux said: ". . . For the Cordaites, Prof. Renault has given very detailed anatomical descriptions and splendid illustrations of all the organs of these plants as complete indeed as if they had been made from living vegetables. The development of the plant is followed from the fertilization of the ovule; for grains of pollen have been discovered, by vertically cutting the embryonic bodies, one already enclosed into the pollinic chamber, two of them still on their way downward in the pollinic tube. . . ." Every source I have been able to discover shows that Lesquereux fell in with progressive world studies in fossil botany. Had facilities been afforded him, had Lesquereux been younger, he doubtless would have taken up—at least, tested—every new method of investigation.

and morphology, relating these in time to schemes of phylogeny, and Lesquereux would become aware that, while most of his taxonomic work would survive as the foundation in America, his "deductions" and determinations of relationships would undergo revision. For plant anatomy would fill a most important place in world botany within a very few decades. Penhallow and others would not only study structure and affinities more than had been emphasized but would initiate study of plant groups in evolutionary sequence based on internal structure; for example, as Penhallow did among conifers. Whole new groups of plants yet to be discovered would be approached from the anatomical side, along with the systematic.

Contrary to popular belief, Lesquereux is not noted simply for his studies in paleozoic botany. His great studies in the Cretaceous and Tertiary periods of the Mesozoic and Cenozoic eras from materials of Hayden's territorial surveys also consumed his interest—the Cretaceous found in the Central Plain and Rocky Mountain areas from Mexico to north of the Canadian boundary and subdivided into bed groups such as the Dakota group, et cetera; the Tertiary also subdivided into groups such as the Laramie group and as to epochs of 75,000,000 years length approximately: Eocene, Oligocene, Miocene, Pliocene, all differing as to conditions of sedimentation, elevations, and climate.

Lesquereux's conclusions are overwhelmingly various and numerous and many are of considerable interest. During eastern and middle western geological surveys, he had announced theories, or adherences to theories, concerning origins of coal, clay, and prairies, relating each to his early Swiss studies in peat formation. In fact, Lesquereux believed peat formation could be related to "all fossil combustible material" origins. As to coal, he tended toward a belief that coals were the result of accumulated sphagnous moss and other plants and woody matter in place, and were not drift. But he was many times perplexed: for instance, once by discovery of a large boulder, "polished, nearly round, quite black, found in the middle of a thick bed of coal, and in true coal" in Perry County, Ohio; once by a find by Mansfield of a pebble of quartz in coal; again his own find at Wilkes-Barre, Pennsylvania, of "a nest of rounded pebbles, immediately superposed by the coal." He told Lesley:

What I find most difficult of explanation in the Coal formation is the distribution of the sand covering immense areas with such an uniformity of their compound, nature and size of the particles, and heaped in such extraordinary thickness. I see nothing in the peat bogs which can explain this except the drift covering the vast plains. . . .

And he named locations in Europe. In 1866 in a report of the geological survey of Illinois,[20] he had said:

The plants growing on the bogs are generally of peculiar species, and contain in their tissue a proportionally great amount of woody matter. The wood, under the influence of continued humidity, does not rot or become changed into humus as when it is exposed to atmospheric action. By a kind of slow decomposition, named aqueous combustion, which chemistry most satisfactorily explains, it becomes, by and by, either peat, lignite, coal, anthracite or any other of our mineral combustibles, even diamonds. . . .

Clay formation he regarded "as a joint phenomenon to that of the coal." And concerning prairie origins, Lesquereux, as early as 1856, embodied a theory which he published in French in a report of the Society of Natural Science of Neuchatel. Practically all of it, "with scarcely any modifications," was also incorporated in the Illinois geological survey report[21] and was mentioned in Lesquereux's work of the Arkansas *Report.*[22] He believed that "the prairies of the Mississippi Valley were formed through the slow recession of sheets of water of varying extent, whereby the existing lakes were gradually transformed into swamps and bogs and ultimately into dry land." The black surface soil of the prairies, Lesquereux thought, "to be due to the growth and decomposition of bog vegetation, confervae, etc."[23]

Moreover, Lesquereux's conclusions concerning materials from the Western Territories were of great importance, perhaps even more startling in their originality and dimensions. Certainly they were numerous, created wide interest, and spoke influentially in controversies then pending. For example, he studied the nature of sedimentary deposits—whether the deposit had been made by carriage in water from place to place, or deposited at the point of growth without change of location. Above the Cretaceous beds of the Dakota Group, he found the Lower Lignitic, shown at Point of Rocks, Wyoming, and other places to have palms, figs, magnolias, oaks, sycamore, persimmon, viburnum, and other plants referable to the Eocene Epoch of the Tertiary era. From almost the beginning of his studies, Lesquereux had been inclined to regard the lignite beds as mostly of Eocene origin. As part of the Eocene

[20] *Geology,* Volume I (Western Engraver Co., 1866), Chapter VI, pp. 208-237.
[21] *Op. cit.,* "On the Origin and Formation of Prairies," pp. 238-254.
[22] Pages 323 ff.
[23] See George P. Merrill's *The First One Hundred Years of American Geology* (New Haven: Yale University Press, 1924), pp. 381-383. Concerning Lesquereux and his work and views, see also pp. 515, 518-519, 580-581, 585, 586, 587-588, 720, 721.
See also "On the Origin and Formation of Prairies," *American Journal of Science and Arts,* XXXIX (May 1865), for a letter by Lesquereux on the subject of prairies of recent origin around middle western lakes and rivers.

sequence were studied Green River station materials. Dr. Merrill, in his comprehensive study, *The First One Hundred Years of American Geology*,[24] said Lesquereux "referred to the Lower American Eocene all the coal strata of the Raton Mountains; those of the Canon City coal basin; those of Colorado Springs; those of the whole basin of central and north Colorado extending from Platte River or from the Pinery divide to south of Cheyenne, including Golden, Marshall, Boulder Valley, Sand Creek, etc.; and in Wyoming, the Black Butte, Hallville, and Rock Spring coal. He considered as American Upper Eocene or Lower Miocene the coal strata of Evanston, and from identity of the characters of the flora, those six miles above Spring Canyon near Fort Ellis, those of the locality marked near Yellowstone Lake among basaltic rocks, and those of Troublesome Creek, Mount Brosse, and Elk Creek, Colorado. The coal from Bellingham Bay, in Washington, he also referred to the same horizon. To the Middle Miocene he referred the coal basin of Carbon and those of Medicine Bow, Point of Rocks, and Rock Creek; to the Upper Miocene, the coal of Elko Station, Nevada"; and to the last named Green River might also have been added. ". . . the Flora of Point of Rocks is about the same as that of Black Buttes," said Lesquereux, and Black Buttes "is 3000 feet higher in the measures. The Miocene Flora of Carbon is very closely allied to that of Oregon, Alaska, Greenland, the Baltic, Oeningen, etc., but as yet its types are not clearly defined in other groups of the Lignitic of the Rocky Mountains. . . . We know as yet too little of our Fossil plants and the future will show a great deal more. . . ."

Evidently not until 1883 were descriptions of plants of the Oligocene Epoch made available. During that year Lesquereux made known a portion of this "flora of which little was known before, and which," he said, "is now richly represented by a large number of specimens, especially from Florissant, Colorado," where Miocene beds had been found having plants different from Green River and Alkali Station and Randolph County, Utah. As new discoveries were made, confusions multiplied in the ever-increasing localities, relating mostly to what period or epoch the increasing numbers of plant specimens belonged. By 1883, the year of publication of Part III of his *Contributions*, or volume VIII, Lesquereux estimated 443 species of Cretaceous flora were known, 200 of which were from the Dakota Group—ferns, conifers, a few monocotyledonous plants but by far the greatest number dicotyledonous. Specimens found had relations to those of Greenland, Europe, and many North American localities, relations also to other formations and geo-

[24] *Op. cit.*, pp. 582-583.

logic periods. Indeed, Yellowstone National Park searches, including those at the now famous Amethyst Mountain, would yield specimens believed predecessors of the Sequoias of California, and many of the Lower Pliocene epoch.

A complete elaboration of all controversies that developed in North American paleobotany cannot be made the province of this book. Suffice it to say that in almost every one of consequence Lesquereux wrote as an authority, and deference was paid his voice and pen. In ascertaining the age of a geologic formation, an elusive rivalry developed as to comparative importance of paleontological and paleobotanic materials. Lesquereux knew this, telling Hayden in 1873: "In the lignitic formation, as in the Carboniferous formation, as also in the coal formation of Richmond, etc., botanical paleontology will be always in many points in discordance with animal paleontology; as the one represents land or atmosphere influence which cannot be recognized by the other, and vice versa." To illustrate: There developed the "Laramie Question," in which Lesquereux repeatedly affirmed his view that lignite beds of the Laramie Group—shale, sandstones, and coal beds largely developed in and around Colorado, Utah, and Wyoming—were of Eocene or Miocene origin. Though not denying Cretaceous animal shells and remains were found in lignite strata, he held, supported by Saporta, Heer, and others, that existence of such was really comparatively unimportant "in comparison with the well-marked characters of the flora, characters which have been wholly established by a large number of specimens obtained from all the localities referred to the Lignitic."

Even as he objected to a "separation of the so-called Sub-Carboniferous which though truly Carboniferous has thousands of feet too of strata with Devonian animal fossils overlaying beds of coal and shale with true Carboniferous plants," he fought the Dakota Group's being regarded Tertiary and the "Lignitic" Cretaceous: ". . . the great Lignitic group must be considered as a whole and well-characterized formation, limited at its base by the fucoidal sandstone, at its top by the conglomerate beds; . . . independent from the Cretaceous under it and from the Miocene above it our Lignitic formations represent the American Eocene," although at one time Lesquereux admitted he was disposed to recognize "a lignitic Cretaceous formation."

Animal remains, especially mollusks and invertebrates, should, Lesquereux believed, determine the age of marine formations but as to ages of land formations, plant remains, he said, should be given equal, if not greater, weight. A sensible view for a pioneer to maintain. But the controversy lasted many, many years.

Lesquereux conducted researches everywhere, seemingly. He studied fossil woods from upper coal measures of Missouri—of great interest also to Dawson. He reexamined notes on Tertiary flora of California and Oregon, preparatory to his publication of a *Report on the Fossil Plants of the Auriferous Gravel Deposits of the Sierra Nevada* (1878), based on materials sent in 1872 by Josiah D. Whitney of the Survey of California where Lesquereux had already studied the mosses, and also some materials from Oregon. While the California fossil plants were found by Lesquereux to be Pliocene, the author discussed Tertiary and Cretaceous floras, the latter of which became every day more interesting because of "the beauty, the originality and the varieties of the types."

Of Ohio paleobotany, he wrote:

The Mahoning Sandstone (middle coal measures) in Ohio has in its compounds a mass of silicified trunks of which I have collected hundreds of specimens varying in diameter from a few inches to 2 feet. One specimen weighing more than one ton is with my former collection in the Agassiz's museum of Cambridge measuring 2 feet in diameter and as much in h[e]ight, quite cylindrical and so finely silicified that the texture is discernible to the naked eye. This is the same for all the specimens, of which the silicification is perfect. Now, in all these materials, I scarcely found one or two specimens referable to Sigillaria [an enormous lycopod of the Mississippian and Pennsylvanian periods] or Lycopodiaceous. . . . *No trace of Coniferous wood has been found as yet above the Millstone grit* to my knowledge. But we live to learn and to see. To see and forcibly acknowledge the existence of things which was denied from mere ignorance. At least I speak on my account. We have also never seen neither here nor in Europe any marine plants in the coal measures. I described one from the base of the formation.—Penn[sylvani]a a long time ago and now I have splendid specimens of three species of Fucoids from two localities of Indiana and one of Illinois, the three in the coal measures. We have fruits which I consider of Conifers in all the stages of our Coal measures. Why could or should we not find the wood hereafter in [the] same circumstances? It would be very interesting to know from what geological horizon the specimens lately sent in Limestone have been obtained.

But however much paleobotany interested Lesquereux, he commenced in earnest compilation of the important *Manual of the Mosses of the United States*. He always tried to keep alert to Gray's and Engelmann's activities, following them as closely as means of communication permitted. On October 12, 1876, he wrote Engelmann:

I suppose that you are now returned from your exploration from the Racoon [Roan?] and Lookout M[oun]t[ain]s but do not suppose that you have found much there at this time of year. Those mountains are very rich and beautifull:— April and May at least they were in 1850 when I visited them. There was an abundance of mosses especially Sphagna interspersed with Sarracenia species, etc. But of this you know better than myself. . . .

Lesquereux had continued bryological collecting even after Sullivant's death. On his arrival at Columbus, he had regarded Sullivant's materials as the best he had seen and in 1863 told Lesley: "... we have here all the published books and far richer collections than can be found anywhere in Europe especially for all what concerns the Bryology of this continent." But Sullivant's herbarium had gone to Harvard. At Gray's request, Thomas P. James, then a Cambridge resident, was enlisted to aid Lesquereux. Several times in journals Lesquereux and James together described new moss species and for the *Manual*, Lesquereux not having access to abundant materials, James performed microscopical examinations of "doubtful species."

Lesquereux by now had a much respected name in American botany, although he was not altogether aware of it. Membership in the National Academy of Sciences had come in 1864—however, not so much for work in bryology as paleobotany, and for a meeting in 1865 he prepared a paper on "Fucoids of the Coal Measures." Botany was not largely represented in the academy's early membership. Gray, Torrey, and Engelmann were incorporators and one incorporator—he may have been Sullivant—refused the honor. At any rate, Lesquereux attributed his election to Lesley and probably had much to do with Sullivant's election some years later. With the years Lesquereux's fame spread.

On March 9, 1878, he forwarded to Gray copies "of the pliocene Flora of the auriferous deposits of California" and "of [T]he [T]ertiary Flora" with remarks as to the origin of "predominant types of our present vegetation" as indicated by fossil plants found in pre-glacial strata and as to indicated plant distributions in the earlier geologic periods. Concerning "The Tertiary Flora" he said:

The plan of the work was not fixed from the beginning for I had to determine the materials by series as fast as they were sent to me. I then considered as extraordinary the peculiar development of the dicotyledonous in the Cretaceous which I admitted as a proof of disruption from those of the Juras[s]ic. This idea was confirmed by the preponderance of palms in the lower Eocene while there is none in the Cretaceous. But I found later, types of Cretaceous Conifers and dicotyledonous in the lower tertiary strata of Point of Rocks and after a while I so was forced to abandon preconceived ideas and to do as DeCandolle did for his Geography, to go along recording facts and leaving them to speak for themselves. . . .

"The Tertiary Flora," Part II of Lesquereux's famous *Contributions to the Fossil Flora of the Western Territories*, opened "a page of no less interest and one still more important" than Part I, his "Cretaceous Flora of the Dakota Group," which had stimulated much interest in Europe and America. For more than eighteen years Lesquereux had studied the American Tertiary. In 1859 his publication of Evans's "Oregon & Vancouver [plants had] been received with true delight by Heer and the European palaeontologists since," he told Lesley, they shed "the first ray of light to clear the darkness that has covered till now the American tertiary. Gray also has received it with great pleasure as affording some indication for actual distribution of our species of plants." In 1863 Lesquereux informed Lesley: "Darwin writes me that the few I have already published of the plants of the tertiary is very interesting &c&c." From about 1860, therefore—excluding consideration of the few species gathered by the United States Exploring Expedition in Washington Territory near Fraser River—Lesquereux's studies had gone forward until now, combining his work, Newberry's work, that of Heer, that of Dana, and some undescribed specimens, the known number of North American tertiary had increased to "not far from 500."

Comparing this with the known number of American Cretaceous plant species, the assembling of which had begun by Hayden's "discovery in Nebraska of leaves apparently referable to Sassafras, Liriodendron, Platanus," et cetera, and by 1878 amounting to about 200 "specified forms," mostly Dakota Group species, the advance in American paleobotanic knowledge in a little more than two decades was a great source of meritorious pride to the few systematists of the science.

Whitney's California plants, moreover, pleased Lesquereux. He called them "a blessing" since he could trace "the vegetation from the Cretaceous & the Eocene, the Miocene and the Pliocene, all from the same Region." He commented:

This flora is, up to this time, limited to fifty species. These are related by some identical or closely allied forms to the Miocene, and still more intimately by others to the present flora of the North American continent.

The North American facies is traced by some species to the Miocene, the Eocene, even the Cretaceous of the Western Territories. Hence it is not possible to persist in considering the essential types of the North American flora as derived by migration from Europe or from Asia, either during the prevalence of the Miocene or after it. This flora is connatural and autochthonic.

The relation of the Pliocene plants of Nevada and Tuolumne Counties is with the flora of the Atlantic slope, and not with that of California at the present time. This fact is explained by the influence of glacial action during the prevalence of the ice period, and is even clearly exposed by the distribution of the few Pliocene species remaining in the flora of the Pacific coast. The modification of the characters of the present flora of California have, therefore, to be looked for in climate or other phenomena subsequent to the glacial period. This remarkable fact, so clearly demonstrated by nature, may serve as an exemplification of the causes of the disconnection of some of the other groups of our geological floras. . . .

Lesquereux pleaded for more study of North American Pliocene and post-Pliocene plants, calling attention to Gray's conjecture that much material lay buried awaiting investigation in the lower Ohio and Mississippi river valleys.

Material from Lesquereux's letter and his *Reports* doubtless formed a part of Gray's lecture, "Forest Geography and Archaeology,"[25] delivered April 18, 1878, before the Harvard University Natural History Society and published in *The American Journal of Science and Arts.*[26] There Gray referred to Lesquereux's idea of prairie-origins, accounting, as it did, for the absence of trees on prairies by emphasizing an analogy to peat formation and a consequent chemical soil not conducive to much tree growth. Similarly, Gray considered Lesquereux's materials embodied in his *Report on the Fossil Plants of the Auriferous Gravel Deposits of the Sierra Nevada.* Said Gray:

> The case of the Pacific forest is remarkable and paradoxical. It is, as we know, the sole refuge of the most characteristic and widespread of Miocene Coniferae, the Sequoias; it is rich in coniferous types beyond any country except Japan; in its gold-bearing gravels are indications that it possessed, seemingly down to the very beginning of the Glacial period, Magnolias and Beeches, a true Chestnut, Liquidambar, Elms, and other trees now wholly wanting to that side of the continent,— though common both to Japan and Atlantic North America. Any attempted explanation of this extreme paucity of the usually major constituents of forests, along with a great development of the minor, or coniferous, element, would take us quite too far, and would bring us to mere conjectures . . . [T]he races of trees, like the races of men, have come down to us through a prehistoric (or pre-natural historic) period . . . [T]he explanation of the present condition is to be sought in the past, and traced in vestiges and remains and survivals . . . [F]or the vegetable kingdom also there is a veritable Archaeology. . . .

A year and a half later Gray reviewed for *The Nation* DeSaporta's *Le Monde des Plantes avant l'Apparition de l'Homme*, entitling his article "Plant Archaeology,"[27] and saying, ". . . under [Saporta's] happy exposition, the stony desert is made to rejoice and blossom as the Rose." Again Lesquereux's work in ancient and more recent fossil botany was specifically referred to and praised by Gray. Arguing Saporta to be a "thorough Darwinian," Gray said:

> A vegetable palaeontologist who studies the later geological deposits cannot be otherwise; at least, he must needs be a "transformist." Saporta concludes that palaeontology, if it does not furnish demonstration, yet gives irresistible reasons for a belief in evolution. The ground and the nature of this conviction appear in his rounded statement, that there is not a tree or shrub in Europe, in North America, at the Canaries, in the Mediterranean region, the ancestry of which is not recog-

[25] See Sargent's *Sci. Pap. Asa Gray, op. cit.*, II, p. 204.
[26] XVI (3rd ser.), pp. 85, 183.
[27] See Sargent's *Sci. Pap. Asa Gray, op. cit.*, I, p. 269.

nized, more or less distinctly, in a fossil state. This is too absolutely stated, no doubt, but the qualifications it may need will not invalidate the conclusion. . . .

The general conclusion . . . is that the vegetation of the earth has been continuous through all ages, and that the explanation of the present is found in the past. The history of the genus Sequoia—of the two "big trees of California" . . . is a fair illustration of this. The difference between these two trees is as notable as their resemblance and their isolation. They are the survivors of a numerous family, of wide distribution, which is first recognized in the cretaceous formation, in several species, and which reached its maximum in the middle tertiary, in fourteen recognizable species or forms. Almost from the first these separate into two groups, one foreshadowing the Coast, the other the Sierra, Redwood, yet with various intermediate forms. These intermediate species are extinct, the two extreme forms have survived. . . .[28]

There was no gainsaying the now established fact that plants "are the thermometers of the ages, by which climatic extremes and climates in general through long periods are best measured." Gray was not critical. He did not presume to be a paleontologist. He accepted their data and conclusions eagerly, marshaling his arguments in accordance with their findings.

But Lesquereux, when he reviewed the same work of Saporta, was critical. When, for example, Saporta seemed to maintain that the Lower Lignitic flora of the North American continent should be separated from the Eocene epoch and referred to Saporta's new subdivision, the Paleocene, Lesquereux assailed Saporta's claims with vigor. He said:[29]

. . . [I]t is not merely from the identification of a few plants that a relation between the floras of two epochs should be fixed or admitted, but from the general characters of the vegetation representing the climate, and from the general facies resulting from the progress of the vegetation, in passing from types admittedly inferior to others of a more advanced degree of perfection becoming more predominant. Considered in this way, the vegetation of the Lignitic, taken as a whole, indicates the action of a climate of an average temperature far above that of the Cretaceous Dakota Group, and still higher than that of the Paleocene, where the Oaks predominate and there are scarcely any Palms. The prodigious abundance of remains of Palms at Golden, at the Raton especially, is exactly comparable to that of the sandstone of La Sarthe (Upper Eocene), which, says Saporta, recall, by the beauty and the large size of their fronds, the Sabals of Cuba and Florida. If we cannot refer the whole Lignitic flora to that upper stage of the Eocene, if we find in it some typical affinities with the Paleocene, this results from the great thickness of the formation, which, in its four thousand feet of strata, may represent groups of floras related to two or more of the geological divisions established from separate groups of plants, like those which in Europe are referred to the Paleocene and the Eocene.

[28] Sargent's *Sci. Pap. Asa Gray, op. cit.*, pp. 272, 277.
[29] *The American Journal of Science and Arts*, XVII (3rd ser., 1879), pp. 273 ff.

The Paleocene epoch antedated even the Eocene. To assign part of the North American upper coal bearing flora to the Paleocene—still more, to find an evident relationship with the Paleocene of Europe—seemed to Lesquereux not in conformity with his data and observations. However, with the objectivity characteristic of great men of science, he presented his arguments contrary to Saporta's claim and left to subsequent investigation the task of ascertaining the truth. Time would sift the relevant. Indeed, noticed Lesquereux, vast amounts of collected materials in plant, insect, and animal remains—for examples, collections made at Florissant, Colorado, by the United States Geological and Geographical Survey, and by Princeton University—had not then been completely studied, and "it [was] only recently, or since the publication of the Tertiary flora, that [they had] obtained documents numerous and valuable enough for a future comparison."

From work of Lesquereux, many conclusions of world paleontologists were made possible. Honors came to him; for example, an offer of membership in an honorary Danish scientific society. Without hesitancy, men like Gray resorted to him for confirmation or amplification of views long held or recently maintained. However, the principal agency from which Lesquereux derived materials—the great United States Geological and Geographical Survey of the Territories under Hayden—was discontinued by an Act of Congress in 1879, and the United States Geological Survey under Clarence King organized with Hayden continuing as geologist. No provision was made for future botanical exploration. The search, consequently, for new harvests in botany and also in paleobotany had to be borne for much the greatest part by private finance and effort, particularly in the West.

It was the end of a great period in scientific exploration—for natural history materials and other objectives—by the United States government. Beginning, we may say, with the famous exploring expedition of Major Stephen H. Long in 1819 to the Rocky Mountains; extended to the Pacific with explorations of Frémont and the United States Exploring Expedition; and, after a long period of western interior exploration, southern, central, and northern, among them the famous boundary line surveys, the Pacific Railroad surveys, and many specific reconnaissances; and concluding with the four great surveys—the King expedition, the Hayden surveys, the Wheeler survey, and the Powell surveys—the period was being brought to a close by a series of great accomplishments.

There was yet, nevertheless, much to be done. For example, Lesquereux could seek to increase the effectiveness of the Museum of Comparative Zoology at Cambridge, Massachusetts, in which he had been vitally interested since his years of collaboration with Louis Agassiz. In 1869-1870 Lesquereux had sold to it his entire collection of "Coal or of Devonian plants," numbering among them types of species published by him in various geological reports, including those of the early Pennsylvania survey under Rogers, those of Arkansas under Owen, and those of some of the reports of the Illinois survey. In 1880 Lesquereux sought permission to add to the upper coal specimens at the museum a large number of Tertiary and Cretaceous specimens from the United States geological and geographical surveys.

Nor was the museum at Cambridge the only institution serving as a storehouse for Lesquereux's valuable paleobotanical materials. On December 5, 1879, Lesquereux wrote Spencer Baird:

I congratulate you sincerely on the near completion of the great building of the National Museum. I have been often uneasy about the disposal of the collection of fossil plants made especially by Dr. F. V. Hayden and myself for the survey of the Territories. . . . These collections are composed of a very large number of specimens. More than three thousand are in catalogue and it is only part of the whole.

The great United States National Museum at Washington was becoming the principal repository of materials in paleobotany which Lesquereux had determined.[30]

Lesquereux could and would concentrate more and more of his energies on the North American coal flora—and the completion of his work of the second geological survey of Pennsylvania. Years before, Lesquereux had said that Lesley should be placed in charge of the survey and that a great atlas of the coal flora—involving all the known learning in coal flora of the United States—could be made a part of its published reports. For years Lesquereux had been compiling such an atlas, trying, without success, to get portions or the whole published. He regretted that Pennsylvania in 1873, with many celebrated scientific societies, should be far behind Massachusetts, New York, and even Illinois, and other states, in exploration. On May 28, 1874, he told Lesley: "It would be a shame if the Old state of Penn[sylvani]a which has been enriched by its coal fields and the use of its coal could not pay a small amount for the study of the origin and history of that coal. . . ."

All the while Lesquereux had been working on western paleobotany, he had studied flora of the ancient periods, especially the upper coal bearing beds. When Alabama specimens had arrived, he had found in America a flora that filled the gap of "the so-called subconglomerate

---

[30] For an excellent summary of the work of Lesquereux in Pennsylvania, Illinois, Kentucky, Indiana, the Dakota Group flora, etc.—especially as to Lesquereux's relations with Charles H. Sternberg and R. D. Lacoe and whereabouts of collections today—see George Sarton, "Lesquereux," *Isis*, XXXIV (1942), pp. 97-108.

coal, the Culm of Europe," the lowest coal formation there, which, combined with studies of southern and eastern Ohio coal and his other already immense materials enabled him to formulate five hypothetical stages in the upper carboniferous formations which he called: (1) Catskill; (2) Upper Waverly; (3) subconglomerate; (4) middle coal from the Conglomerate to the Mahoning Sandstone and (5) "the upper which goes higher than the Pittsbourg"—"all positively characterized by differences in their floras." In 1876 he told Lesley:

> Our Carboniferous flora is becoming more and more interesting by the discovery first of a number of types which untill now were considered as limited to formations older or newer than the Coal, as Devonian or Permean or even Triassic types where presence in the Carboniferous seems an anomaly. Beside others I have [types] from Mazon Creek.

Lesquereux fixed stations in various states. He had a station in Alabama "for subcarboniferous plants. One in Illinois for the lower coal, one in Missouri about of the same horizon"; and at least two in Pennsylvania, all of which aided in establishing "the essential identity or contemporaneity of the Mazon Creek Flora of Illinois, the Cannelton flora (Kittanning) of Beaver County, Pennsylvania, and the Mammouth Coal flora of the Southern Anthracite Coal Field of Eastern Pennsylvania."[31] Interesting coal mine owners such as I. F. Mansfield of Cannelton, Pennsylvania, he obtained from them many specimens and aroused their interest in coal floras.

All groups of upper carboniferous plants received attention from Lesquereux, but particularly ferns and Cordaites. In 1877-1878, he wrote:

> I have recently spent a few weeks in working out our Cordaites, the Conifers—Cycades of the Coal from the splendid and numerous specimens of Mansfield. . . . The Cordaites have made a true excitement among phytopalaeontologists, Williamson, Binney, Hooker of England, Saporta, Heer, Schimper, Grand'Eury. . . .

Paleobotany opened a veritable wonderland to Lesquereux. "I live," he said, "among wonders of such an admirable vegetation that nothing, not even a journey along the Rhine, would be comparable to it in splendor (For a fossil botanist)!!!" But problems of nomenclature beset the science. And by 1881 Lesquereux was expressing his views very forcibly on the subject.

Nomenclature in paleobotany, he maintained, above all other branches of botany, must have a flexible system. On April 27, 1881, in a letter to Professor Hagen of Harvard, he explained his reasons why:

> Palæontologists have to determine their fragments of plants and animals as far as possible and to give them a name in order to be understood and quoted. But

[31] G. Sarton, "Lesquereux," op. cit., p. 105, quoting William Culp Darrah.

these names should be so to say provisional and constantly subject to revision and modification until documents are so satisfactory that the relations can be established on a safe ground. In that way the names used for the study of living animals or plants either generic or specific or even of groups and families should be mostly left out or at least hypothetically mentioned. But just the contrary of this is attempted now very often if not generally, not only attempted, but requested of Paleontology. We ha[ve] had already many so called [C]ongress[es] for fixing nomenclature and one is soon to be had for Palæontology in Paris, as well you know. Now they propose to preserve some of the so called classic nomenclature: 1st fix genera and generic character so that nobody is allowed to change them 2nd revise nomenclature according to characters . . . 3[r]d take away from an author the privilege of changing his genera or species or the name etc. etc. Well I would like to know what one has to do when for example he gets some boxes of broken specimens representing leaves, fragments of course, parts of stems etc. and when he is called to determine this trash to name the species and make a report which must be delivered within three to six months. I know this per experience. I know what tribulations, anxieties, trouble I have suffered when working such materials until I could go myself and collect good materials for comparison and study. Of course from these I saw mistakes without numbers and had to recognize that fragments were often far different in aspect from what a whole leaf would show. I changed my names of course also and shall continue to do the same. And for reference to plants of the present times coming to the coal for example we have the numerous determinations of Dawson referring fragments to Conifers (Araucarian) etc etc and establishing a number of generic divisions and now (see the paper I wrote you on Renault['s] anatomical determinations) All these fragments referred to conifers by Dawson and English authors are fragments of Cordaites a genus (sui generis) intermediate to conifer & cycadeae. . . .

Subsequent investigation, especially in anatomy, would reveal much new light on Cordaites and their relationships to other plants or plant groups. Their origin and disappearance became an absorbing problem, and in many respects still is. The original and fundamental discovery and discussions of Lesquereux and Dawson, as well as those of the Europeans, must be accorded recognition. They illustrate not only early problems of research and systematization but anticipation by earlier paleobotanists of the immense and coming investigations in phylogeny to be aided by morphology and anatomy.

More than this, however, Lesquereux's main and ever steady purpose was to publish what he termed an honorable representation of the American coal flora. Lesley and the second Pennsylvania geological survey, "the mother of [Lesquereux's] work" which had "fostered it, prepared it and now helps me," he said, "to put it through to make it good," made this possible. Providence had always taken care of Lesquereux. He believed this so more every day of his life. Financial circumstances induced largely by the Civil War had brought about the dissolution of a partnership with his sons in the jewelry business. Agassiz's employment of him at Harvard as a curator, however, had spared him really knowing want. When Agassiz died, Hayden's employment of his services had saved him once again financially. And on termination of the services with Hayden had come the work for the Pennsylvania survey. Lesquereux planned both a learned and popular manual but the latter was deferred till completion of the former. He worked "very carefully in order to make this work a reliable guide for the future" and in 1879 the Board of Commissioners at Harrisburg published his *Atlas to the Coal Flora of Pennsylvania and of the Carboniferous Formation throughout the United States*, thus adding a monumental work to a number of less important ones, as, for example, a *Catalogue of Fossil Plants of the Coal Measures of Pennsylvania*, published in 1858, and others of still less significance. Nevertheless, much of the survey's work still remained unfinished. And more great publications would be forthcoming.

In the course of the work, however, on October 3, 1882, Sophie Henrietta Lesquereux, his much beloved wife, died. "My wife was half of me," he wrote Lesley, "and the best half in the full sense of the expression. I have always worked alone, for years she has been an invalid." He was not only "crushed but prostrated for a time." Long nights of sleeplessness followed. The loss of his wife, born and raised among European nobility, in whom, as a child, the poet Goethe had taken much interest, who had forsaken riches to marry her amiable and brilliant young teacher, took with her his heart but not the climax of his work. To learn a language he could not hear, but lip read, had been a trial for Lesquereux. Often she had served as his interpreter. With intellectuals he could converse in several languages but scientific exploration had required talking with strangers. Often he wrote what he said. And this was difficult, with coal miners many times impossible. She had helped him endure loneliness and almost poverty. They were proud and had taken nothing from anyone unless they could give something in return. Science had been a "shabby boarder" paying little and the demands of paleobotany "severe and unrewarding." Not until 1885 would he sell most of his collections, some in Europe and others in the United States. Europe accorded him recognition as America's foremost paleobotanist. Recognition, however, did not bring money. Correspondents aided him whenever allowed. He curbed expense, once considering surrendering membership in the National Academy. The extreme simplicity of his home nevertheless did not deprive it of happiness. After his wife's death, Lesquereux's sons and daughters aided him loyally and valiantly.

The Pennsylvania survey work remained. Completion of the Hayden survey was also unfinished. As late in his life as 1887 he would write Lesley: "I am trying to bring to a close the last volume which I may have chance to possibly prepare upon fossil plants, now of the Cretaceous, and I give my whole time to it." This was a flora of the Dakota Group, a work following that of studies incident to work of Hayden's survey, and on this he "worked constantly," intending to continue so doing as long as his enfeebled strength lasted. He realized that during his lifetime the completed coal flora of the United States could not be written. He determined to complete it as far as time and ability permitted.

Lesquereux must have known that his work covering fossil floras from Vermont to California would lead to still greater results. Knowledge of coal, oil, petroleum, natural gas, and other natural resources of great value had been greatly amplified during his life.

Then, too, North American paleobotany was slow in being really recognized. First, the tremendous systematic task of knowing what the materials were had to be reasonably completed before still greater tasks of correlations could be assumed. How could knowledge of past floras be derived until adequate comparison with the present flora was available? Till modern morphology knew developing and interior structure as well as mature and external organs, what could be expected from paleobotany except a study of differences and some affinities in plant groups? Paleobotany had added difficulties. In February 1882, the *Botanical Gazette* commented:[32]

No department of Botany seems to the average botanist so unsatisfactory and perplexing as that of Fossil Botany. We all know how difficult it is to name plants when the specimens are only tolerably complete, but to name them from the merest fragments of stems and leaves is something that must border very closely upon guess work. Such naming too becomes of very great importance when the age of formation rests upon the evidence of fossil plants. . . . Still some splendid work has been done and our countryman, Mr. Lesquereux, has had by no means the least share of it.

On September 30, Lesquereux sent Hayden Part III, "The Cretaceous and Tertiary Floras," published the following year, 1883, at Washington as the third *Contribution to the Fossil Flora of the Western Territories.*[33] Again it was a taxonomic work. Lesquereux, moreover, observed:

I send herewith the manuscript of the eighth volume of the Reports of the United States Geological Survey of the Territories, made under your direction. Besides a short introduction, the volume contains:

1st: A review of the Cretaceous Flora of the Dakota Group, or of what has been

32 VII, Number 2 (February 1882), p. 14.
33 Department of the Interior. *Report of the United States Geological Survey of the Territories,* VIII (1883). xii, 283 pp.

published in volume VI, with descriptions of a large number of new and remarkably interesting species illustrated by 17 plates.

2d: Some remarks on the Flora of the Laramie Group which I consider as Eocene, with descriptions of a few new species, illustrated by 3 plates.

3d: The more valuable part of the volume, viz: the descriptions of the plants of the Oligocene, a flora of which little was known before, and which is now richly represented by a large number of specimens, especially from Florissant, Colorado. This Flora will be quite as well received by paleontologists as has been the Cretaceous Flora of volume VI. It is illustrated by 24½ plates, which are all very finely made.

4th: Half of one plate serves for illustration of a few plants from the oldest Pliocene, or upper Miocene of California.

5th: Descriptions with figures of Miocene plants of the Bad Lands, with 5 plates. The plants, clearly of Miocene type, are very interesting from their relation to species of Arctic Flora.

6th: Descriptions of species of Miocene plants of California and Oregon from specimens pertaining to [those of] the State Museum of Oakland, California. . . .

7th: A short account and description of new species found in a collection of fossil plants made in Alaska by W. H. Dall, of the United States Coast Survey, for the Smithsonian Institution. . . .[34]

It is not necessary to remark that all the plants described in volume VIII are considered in separate groups according to their relation to the age of the formation which they determine. Comparisons are established with the European Floras by tables of distribution, etc. I truly believe that this volume will prove to be a very valuable contribution, not merely to the paleontology but also to the geology of this country.

The value of Lesquereux's earlier works had been quickly recognized in Europe, evoking interest and stimulating discussion. Much could be said for the contention that Lesquereux's work was better known in Europe than in America. The world's ablest paleobotanists—Brongniart, Heer, Count Saporta, Schimper, Grand'Eury, and a few others—were all in Europe. Of the three Americans of the branch of science separable and distinct from general botany and general geology, and yet a part of both—Lesquereux, Dawson, and Newberry—Lesquereux was best known. Discoveries in eastern United States floras of European Liassic genera, his great comparative studies, and much else, naturally made him known. The Cordaites alone would have been enough. And the Western Territories work only increased his fame. Hayden characterized his three volume series as "a grand monument to the industry and fame of the author." But Lesquereux was not satisfied and in his Introduction he challenged science:

Yes, in this case, as in many others, we may collect facts, but the work of nature in its mode of proceeding for the creation or modification of species remains in-

[34] Already referred to in this book. Published in the *Proceedings of the National Museum*, V, pl. vi-x, according to Lesquereux.

scrutable. We may consider the formation of the Dakota Group as produced by a very slow, gradual, prolonged depression of the Western slope of the continent, bringing up from the South or West the invasion of ocean water charged with muddy materials, periodically heaped farther and farther inland by powerful tides. We may suppose, too, the invading flow as bringing with it seeds or fragments of roots of plants derived from a country now covered by the sea, and distributing here and there those germs of vegetable organisms. But all this does not account for much in the solution of the problem; it may explain the distribution; but the first appearance, and it seems the simultaneous multiplication, of the dicotyledonous plants remains a fact inconceivable to reason.

Obviously, Darwin's theory of evolution had been weighed in the course of Lesquereux's investigation. Even the fable of the lost Atlantis, a continent now submerged by the sea, had not escaped his mind. Religious man that he was, Lesquereux had spent much time harmonizing the story of Genesis with the facts of scientific discovery. "*I know a little*, other students of science know each a little, but the whole of what is known is but fragmentary and insignificant—merely a few pebbles picked up along the ocean shore," he sometimes said.[35] Lesquereux knew that paleobotany was immature in its development. In his work on *Specimens of Fossil Plants Collected at Golden, Colorado*, for the Museum of Comparative Zoology at Cambridge, he commented:

Of the species described . . . from Golden, or of those formerly known from the Laramie Group, either by the publications of Dr. Newberry or of my own, none is identified with any of the Middle Cretaceous (Cenomanian) or of the Dakota Group. . . .

Recent explorations have brought on the discovery of a large number of localities rich in remains of fossil plants over the whole extent of the Great Lignitic. The flora of the Laramie Group, which now counts only 250 species, will therefore probably soon become better known, and by the greatly increased number of its species will take an important place in the history of the ancient vegetation of the earth. . . .

In 1879 Lesquereux had regarded the Laramie group flora as representative of a transitional epoch which ushered in a new time.[36]

But such was not the only work demonstrative of Lesquereux's greatness. There were also his famous studies with the coal flora and with mosses. On February 10, 1879, he had written Thomas P. James, "I have no time to give to mosses until I am ready with the text of the Coal flora

[35] See Edward Orton, "Leo Lesquereux," *Ohio Archaeological and Historical Publications*, IV (Columbus, Ohio), p. 289.

[36] *Eleventh Annual Report of the United States Geological and Geographical Survey of the Territories*, embracing Idaho and Wyoming, a report of the progress of exploration for the year 1877 (Washington: Government Printing Office, 1879), p. 366. See also Lester Ward's *Synopsis of the Flora of the Laramie Group*, where, describing the vast area extending from Mexico to Canada and on either side of the Rocky Mountains, and its flora, a new treatment is given in 1886; also his *Types of the Laramie Flora* (1887).

and since the 1st of Jan[uar]y I have been unable to make any progress for that work. Hence I must give all my time now to the preparation of a *mss* which is demanded and must be out soon." Again on April 17 of the next year when he congratulated Engelmann on completion of his *Revision of the Genus Pinus*, Lesquereux wrote, ". . . my researches are too unimportant to merit your attention. . . . The corrections of the proof of the text of the Coal flora takes much time and the work is slow. I shall however soon be out of the descriptive part and then the printer may go on with less trouble than with descriptions and later references." When Engelmann showed, as he always did, great interest in Lesquereux's work by sending him materials, Lesquereux had occasion to congratulate Engelmann on another of his studies, *The Genus Isoetes in North America*, and comment, "All these matters interest me much on account of the light which they may afford for the dark very obscure study of fossil plants. We have a kind of Isoetes very distinctly exposed in the Carboniferous. . . . If you have some point of affinity to offer for comparison to [the] kind of fructification" which Lesquereux described to Engelmann, "you would greatly oblige me to let me know."

On January 21, 1880, Lesquereux transmitted to Professor J. P. Lesley, state geologist of the second geological survey of Pennsylvania, his "Description of the Coal Flora of the Carboniferous Formation in Pennsylvania." He commented in his letter on the need of "some convenient book for the study and determination of fossil plants in the Coal measures," pointing out that Brongniart's *Histoire des Végétaux Fossiles* and "Schimper's Vegetable Paleontology"[37] were the only works available and neither was published in America. The purpose of his work, he said, was to make "a kind of manual, to meet the deficiency of books." He gave more detailed descriptions, either of different parts of such plants as were only partly figured, or of plants not as yet figured at all and said:

I have described all the species of vegetable forms known to me as occurring in the coal measures—not only of Pennsylvania—but of the United States; and I have included among them plants of Carboniferous types discovered in the older or so-called Devonian rocks.

Thus, I trust, the student of fossil botany will find two [volumes I and II] easily accessible books with which alone he can pursue his researches through the whole Carboniferous system from top to bottom.

My materials have been derived from every available source. I have endeavored to see all the accessible localities offering a chance for obtaining specimens. I have examined both private collections and the cabinets of scientific institutions, and

[37] In 1879, Gray wrote: "Any proper enumeration of authorities upon the fossil botany of the later periods should include . . . especially that of Schimper . . . , who, like Lesquereux, has divided his life between bryology and fossil botany, and whose classical 'Traité de Paléontologie Végétale' is a systematic compendium . . . of fossil plants up to the year 1874."

have widely offered my assistance in determining specimens from any who were willing to transmit them for that purpose. This has brought to me a mass of materials which I have put to use in notes or figures.

In 1879 Lesquereux had published an *Atlas to the Coal Flora of Pennsylvania and of the Carboniferous Formation throughout the United States*. Volume I contained (1) Cellular Cryptogamous Plants, Fungi, Thallassophytes, and (2) Vascular Cryptogamous Plants, Calamariae, Filicaceae (Ferns). Volume II contained (1) Lycopodiaceae, (2) Sigillariae, and (3) Gymnosperms. Volume III, published in 1884, contained the cellular cryptogamous plants, marine Algae, and the plates. With the work was accompanied Professors William M. Fontaine's and I. C. White's "The Permian or Upper Carboniferous Flora of West Virginia and Southwestern Pennsylvania." The whole was published as a *Description of the Coal Flora of the Carboniferous Formation of Pennsylvania and throughout the United States*, in three volumes with an atlas. In the second volume was contained a discussion of the "Literature of the United States Coal Flora." Lesquereux had also written on the "Coal and Coal Flora" for the *Encyclopedia of North America*. Without any doubt Lesquereux at this time stood as North America's greatest authority on paleobotany, recognized as such both in America and in Europe.

Dr. Edward Orton, an eminent geological authority of North America in his own right, has said of this work by Lesquereux: "The most valuable single contribution that he has made to paleobotany is unquestionably 'The Coal Flora of Pennsylvania,' published by the Second Geological Survey of that State. There is no other American work on the subject that is even to be named in comparison with it. It was written when the venerable author had long passed his three score years and ten, and while embodying all his knowledge and experience, it [showed] no signs of flagging strength or failing powers...."[38] Lesquereux's labors had covered almost entirely the great Appalachian coal field, of the United States, from the bottom of the series to the summit. Beginning soon after his arrival in the United States, his explorations in paleobotany and botany extended over wide areas in Ohio, Kentucky, Pennsylvania, Tennessee, Alabama, North Carolina, Arkansas, Indiana, Illinois, Missouri, Kansas, Nebraska, and the western Territories and states, as well as other places of the world. In each Lesquereux was interested in investigations which he had commenced in Europe[39]— mosses, especially the sphagnous mosses; peat bogs; the effects of glacia-

[38] *Op. cit.*, p. 288. See also Andrew D. Rodgers, III, *"Noble Fellow" William Starling Sullivant* (New York: G. P. Putnam's Sons, 1940), Chapter XIII and succeeding chapters.

[39] This subject is more thoroughly dealt with in Chapter XIII of Andrew D. Rodgers, III, *"Noble Fellow" William Starling Sullivant, op. cit.*, pp. 198-201, 203 ff.

tion; and the study of all live and fossil flora. As his knowledge increased, Lesquereux's horizon enlarged. And the study of various geologic periods brought with it a study of the early and later coal formations.

In the *Annual Report of the Geological Survey of Pennsylvania for 1885*,[40] Lesquereux announced the completion of his notable theory "On the Vegetable Origin of Coal"; a theory seriously considered at the time and reinforced by much subsequent factual investigation for many years after his death, though later narrowed and limited in its scope and application. Many others had devised theories on the origin and formation of coal;[41] notably H. D. Rogers, who in 1842, with his brother William B. Rogers, had astounded the American Association of Geologists and Naturalists with an important paper on coal bed formation along with a still famous oral pronouncement of the wave-theory of mountain chain formation—a theory accounting for mountain chain elevations, similar to the mode in which waves are raised on a body of water. Both theories were, of course, of enormous complexity but it is with the coal formation theory that we are concerned. In his paleobotanical reports of especially the Illinois geological survey, Lesquereux had announced views, or adherences to views, on coal formation. Rogers believed that the Pittsburgh coal area had been in ancient times an extensive flat near a continent beyond which lay a shallow and open sea. On these flats were enormous peat bogs formed from and supporting an extensive growth of Stigmaria which in turn formed a pulpy peat mass mixed with leaves and other matters—conifers, lycopods, and tree ferns from drier areas, parts of which could have been carried easily over the peat bogs by water on sinking, elevation, or bending of the land surface, there to form the roof slate and other bed strata during a period of tranquillity.[42] Was Lesquereux influenced by Rogers's theory? In 1886 Lesquereux wrote, ". . . in the growth of peat we have a microcosmic but true representation of the formation of the ancient coal."[43] And this, in general, was his theory. Dr. Orton, who knew Lesquereux intimately, said, however:

A passing reference of Brongniart had suggested the view that coal-seams originated under conditions similar to those in which peat bogs are now formed. In the mind of one who knew more of peat bogs than anyone had ever known before, the

[40] Harrisburg: Board of Commissioners for Geological Survey, 1886, pp. 95-121.

[41] See George P. Merrill, "Contributions to the History of American Geology," *Ann. Rep. Board of Regents of Smithsonian Institution* (Washington, 1906), pp. 338, 360, 372, 426, 447, 492, 497.

[42] See George P. Merrill's complete explanation, *op. cit.*, p. 372.

[43] From "On the Vegetable Origin of Coal," *op. cit.*, p. 121.

suggestion took root and expanded into a theory which covers the origin of by far the largest part of our valuable accumulations of coal.[44]

Doubtless Lesquereux was somewhat indebted to H. D. Rogers by virtue of joint efforts in the first Pennsylvania geological survey work. But early in the 1850's a bitter feeling, never mended, developed in a controversy between them, Rogers accusing Lesquereux of breaches of literary faith and even dishonesty, which charges in turn brought equally serious countercharges from Lesquereux. Lesquereux told Lesley: "Neither Rogers nor any other American have a mind for purely scientific researches." And to Lesquereux's aid came Gray, Lesley, Sullivant, Agassiz, and others. Lesquereux's work must be regarded his own, giving presumably final, formal, and complete expression to a theory which met wide favor, and held in 1895 "first place" among theories of coal formation.

In 1884 Lesquereux published with Sereno Watson and Thomas P. James, then dead, the *Manual of the Mosses of North America*, a work begun in collaboration with Sullivant. In this work Lesquereux was not the innovator of theory but the conservative. On September 9, 1882, he wrote Gray:

I have finished the descriptive part of the Synopsis of the American mosses about two weeks ago and already prepared a short introduction. When I have made a kind of key which would be really more useful than a conspectus, I will begin the copy with abbreviations and corrections of the manuscript and hope that the work will then go fast and be easy. For species of Hypnaceae which were uncertain and had not been examined by James, I have found valuable assistance in Europe. Cap[tain] Renault a bryologist of France who has for years studied the most difficult part of the Hypnum, the subgenus Harpidium, has already reviewed most of our specimens which were still undetermined & not satisfactorily analyzed and is still continuing the work. . . .

And on March 3, 1883, he told Gray:

At least, I have finished the mss of the *Synopsis of the North American mosses*. That is a good title, I think, the book describing species of mosses whose habitat ranges from Florida and south California as far North as the Arctic Zone, even Greenland—I have now to review my table of classification or Key which is about like that of Prof[essor] Watson in Bot[any] of the California Survey; but of course of far wider extension. . . .

However, when Watson told Lesquereux of the new systems in cryptogamic nomenclature and systematization, he answered Watson:

That revolutionary system may be good according to some opinions. But I am too old and therefore too conservative to admit it, the less so that I should find as many reasons to admit that of Mitten, or of Muller which also present[s] some

[44] Edward Orton, "Leo Lesquereux," *op. cit.*, p. 288.

difference. The more one studies the mosses the more he finds the impossibility to arrange them in some consistent natural system, and also the difficulty of fixing precise genera & divisions or rather of knowing which of the names given by different authors are the oldest or the most right. Hence I think best to stay with Schimper & Sullivant, Wilson, Muller and other authors, whom I have followed for half a century. I have no time to make a new apprenticeship and I truly believe that the American Bryologists will support my opinion in the matter. At least they have followed Schimper and Sullivant untill now, as you have done yourself. I do not say this to depreciate the works of Lindberg whom I have [held] in high esteem and who is certainly one of the best bryologists now living. . . .

The system which Lesquereux opposed was Braithewaite's,[45] a "disciple of Lindberg."[46] Some European texts were admitting no mention of tribes or other divisions in the textual matter, and some were changing the names of the larger divisions. Lesquereux would have none of either. The following January 8 Lesquereux told Gray:

I never thought of any higher authority on mosses than Schimper and Sullivant. And certainly, the manual of mosses after the addition of the Synopsis of the genera and the Key advised by you, was as good for publication as is the Synopsis of Schimper or the Mosses of U[nited] S[tates] by Sullivant. There is even less attention given to the synonymy in the *Icones* and in Schimper's synopsis than in my mss. . . .

Lesquereux felt, and was, much indebted to Gray and Watson in the completion of the *Manual of the Mosses of North America*. During its preparation Thomas P. James died and in deference to Watson's friendship for James and Lesquereux, Watson was substituted as a co-author. Watson and Gray gave much time to preparation and criticism. Sullivant's herbarium had gone to the Gray Herbarium and much resort had to be made to its materials. Lesquereux's advancing age brought with it failing health, preventing his making frequent trips to Cambridge. When the work was finished, Lesquereux urged Watson many times to receive money for his services but each time Watson generously refused.

Although many times Philadelphia and Cambridge urged Lesquereux to move to one of these cities, he liked Columbus and remained. Mosses had been his first interest but paleobotany his last. Even as to paleobotany, mosses figured. In 1880 he told Lesley, "Mosses and Bamboos did not exist at the Carboniferous epoch. At least no remains of them have been found, though Calamites have been often compared to, even taken for Bamboo."

Lesquereux's last years were spent finishing his works and making his

45 R. Braithwaite, *The British Moss-Flora*.
46 Lesquereux held Watson in high esteem. On April 17, 1884, he wrote Watson: "If you have not used the microscope for the examination of mosses, you have studied them in the books much better than any author, except Lindberg; much better than Schimper, especially."

peace with his God. Toward the Omnipotent One, he liked to be as a little child. Religion was more than undiscovered knowledge, a superstitious explanation of the unknown and unknowable. He told Lesley whom he loved:

The gradual modification or development of the vegetable world is from the oldest Periods of the world remarkably distinct. But that march is here and there modified without apparent natural causes or in an unexplainable way. The lower Cretaceous Flora, for example, even to the end of the Neocomian has not in its compone[nts] any trace of dicotyledonous plant[s]. All the vegetation is still composed of Ferns, Cycadeae, Conifers and the like. At once, in the Cenomanian, the Vegetation is mostly if not altogether of Phaenogamous plants. Not in one or two or few species; but in hundreds of them, not in species of one or a few tribes or families; but representing all the essential divisions of the vegetable reign, as we have them on this Continent. By what kind of law or influence has this change been brought on? just at the epoch preceding the appearance of mammals to which these plants will serve as food. None of the greatest botanists of Europe can explain it. . . .

His point was, there are many things which reason alone cannot explain. His deference to European opinion was no greater than theirs to his. Schimper dedicated his great treatise on paleobotany to four paleobotanists; of them, Lesquereux was the only American. And about 1886 the Academy of Neuchatel, Lesquereux's first real scientific home, elected him an Honorary Professor.

CHAPTER IX

The "New Botany." Completion of Engelmann's Work

URING the 1880's, North American botany began a wonderful era of expansion. A decade had been sufficient to free most botanists from a sole interest in taxonomy. The work of the first great American teachers in the new subjects of morphology, physiology, mycology, "vegetable diseases," anatomy, and the like, was taking effect. Neither Lesquereux nor Tuckerman had medical degrees. Naturally they did not have the same appreciation of the new work's value as had Gray, Goodale, Farlow, Rothrock, Engelmann, and perhaps some others, who were doctors of medicine. To a doctor, the inner workings of the human machine are more important than external appearances. Medicine's contribution to botany and botany's early contribution to medicine are interdependent. Working together, both progressed.

In 1870 the list of professorships of botany in this country was not lengthy—Gray at Harvard; Eaton at Yale; Porter at Lafayette; Albert Nelson Prentiss at Michigan Agricultural, and later, Cornell; and others, for example, Alexander Winchell, at Michigan, who taught other subjects along with botany. Indeed, excepting Gray and Eaton, practically every such professor taught natural sciences, or, along with botany, other branches of science. Furthermore, in botany, their interest was principally taxonomic, along lines of the work of Torrey and Gray. Direct observation of specimens in the field and in mainly a systematic sense, the laboratory, was included in their work.

It was a group of Gray's pupils who took the lead in developing a "new botany" in North America. Gray alone did not teach them the new premises and approach. But he was the American inspiration and the one who had given them sound fundamental training. The Europeans, beginning with the great genius Hofmeister and followed by De Bary, Sachs, Strasburger, and others, led the way. It took almost three decades for their work adequately to reach American shores. Gray never fought the new work. His students say he "saw the whole field and appreciated it." Although in the 1860's, his laboratory work, according to William James Beal, was crude, and only for advanced students, during the first part of the 1870's, according to Charles E. Bessey, Gray required his students to know not only plant "structure, but their morphology also." Beal, Bessey, Coulter, Farlow, Barnes, Penhallow, Bailey,

Trelease, Arthur, and others, who led the way to a "new botany" and "new horticulture" were all at one time or another Gray's pupils, or associated with him, receiving his enthusiasm, inspiration, hospitality, and most of all, learning. Gray never did much in pathology; in fact before 1880, bacteriology and the doctrine of fermentation and disease had had a struggle in Europe. In America, laboratory methods to study fungi and bacteria were being developed by Bessey, Farlow, and Thomas J. Burrill principally but until 1880 most of their work belonged really to mycology, morphology, and physiology. Shown by a paper by Bessey, in 1882, on "Diseases of Plants," there were fifteen treatises on plant pathology in France and Germany, and scarcely any in English. Gray encouraged study. It is told that ten or a dozen botanists were on a train probably en route to Dubuque, Iowa, where Gray delivered in 1872 his famous address, "Sequoia and its History," before the American Association for the Advancement of Science of which he was retiring president. The train was halted by a wreck ahead, somewhere in Iowa. Gray recognizing corn smut in a near by field left the train, cut a stalk, and returned, asking, "Who can tell me what this is? Burrill, keep your mouth shut." Of the group, only one botanist knew. Burrill was definitely a leader.

Taxonomy decidedly was Gray's main subject. And in this most botanists followed him. John Merle Coulter in Indiana was no exception. After his explorations in the Western Territories with Hayden's survey and his publication with Porter of a *Synopsis of the Flora of Colorado*, he had accepted in due course a position as professor of natural science at Hanover College. There in 1875 he established his *Botanical Bulletin*, soon his *Botanical Gazette*. In 1879, however, Wabash College of Crawfordsville made him a professor. The Indiana flora engaged his interest and by 1881 he and his associates, especially his brother Stanley and Charles Reid Barnes, his pupil at Hanover, had studied river valleys, lake borders, prairies, and barrens within the state. In December of that year the "Flora of Indiana" numbered 1,432 species grouped under 577 genera. Near Michigan, where Charles F. Wheeler, Erwin F. Smith, and L. H. Bailey, among others, collected materials (also publishing their findings), these central United States regions were rich fields for rare plants. Indeed, sand dune studies in northwestern Indiana would later influence ecological study.

On October 13, 1881, Coulter wrote Watson:

There is a matter several of us restless fellows out west have been considering, and as things have about come to a head, I propose to lay the matter before you & ask your unbiased opinion. To plunge *in media res*, that great belt of states beginning with Minnesota & Dakota on the north and ending with Louisiana & Texas on the south, is unsupplied with any manual of botany that can be of use in the schools & to ordinary collectors. Why can they not have one? I have been fairly overwhelmed with inquiries from botanists, or would-be botanists, in that region, the burden of their questions being, "Where can we get any manual to work up our flora?" That region is filling up with wonderful rapidity; educational facilities are multiplying & there is already a large demand for suitable books. The demand is such, that I venture the opinion that if a good manual of botany is not soon provided, some wretched makeshift will be thrown on the market. The U[nited] S[tates] seems to divide itself so naturally into 4 or 5 regions, & they are so large, that separate manuals will be necessary for them all. The Appalachian region is already abundantly taken care of by Drs. Gray & Chapman, especially since the latter is soon to republish. The great Mississippi Valley region is another distinct region & for it the manual is yet to be provided, except that those already mentioned cover its eastern slopes. That might stand then, and let the western part be included in a Manual for the Prairie Region, running to the Rocky M[oun]t[ain] region, or about the meridian of 103 or 104°. The [third] great region, needing a manual, will be the Rocky M[oun]t[ain]s; the [fourth], the Pacific Coast. The last region has now been well supplied & we think the time has come to collect the results obtained between the Mississippi River & the Rocky M[oun]t[ain]s. Let such a work be small, compact & cheap, on the plan of Dr. Gray's Manual; a simple *compilation* & not an Elaborate monograph. Hence no original work will be needed to any great extent, simply access to books & herbaria. Of course, there will need to be a little original work in the matter of artificial keys etc., but not much. Dr. Gray or you would be the most suitable persons to do such a work, but you have larger game in view & can't afford to fritter away the time it would take. Why cannot some of the energetic botanists out here undertake a work of the kind, with the understanding that access is to be had to the Cambridge library & herbarium & all questions of nomenclature and classification to be referred to the authorities there, that our work may all be uniform? For instance, we could collect all the information possible here during the winter, & then spend the summer in Cambridge, supplementing our list. Let me once get a complete list of the plants of the region & I ask no better fun than arranging their descriptions & constructing analytical keys. Prof[essor] Bessey is heartily in favor of pushing the work, and also E. J. Hill, of Englewood, & several others, who have all more than once had the idea of going into it themselves. If the work is not published in this way, as it were, under the Cambridge wing, it will be published independently by some one, & then there will be confusion worse confounded. The thing of it is to get a publisher, & if through Cambridge influence that cannot be done, there are several of us out here ready to take the risks. . . .

Charles E. Bessey was at that time professor of botany in the Iowa Agricultural College, having been lately a lecturer in the University of California. No one man in America, other than Coulter, contributed more to the developing transitional period of North American botany than Bessey. Having charge of the botanical department of the *American Naturalist,* Bessey had restored that agency's value, himself publishing a number of creative studies.

Bessey's small pioneer western laboratory at Iowa Agricultural College inaugurated an era in American plant study, and his published letters to William James Beal of Michigan Agricultural College where Bessey graduated, are very interesting. Some may claim that Beal, author in 1879 of the classic lecture and booklet *The New Botany*, Thomas J. Burrill of the University of Illinois, or George Lincoln Goodale of Harvard, established the first botanical experimental laboratory for teaching undergraduate botany. The fact is, however, that all four of these laboratories, begun in the seventies, were established independently of each other and Bessey's was the first, being modeled after Gray's "laboratory," which did not become an experimental laboratory in the complete sense until about the year 1875.[1] Bessey instituted his "Botanical Laboratory" in 1873 after studying with Gray at Harvard.

Of course, from the beginning of American botany, there had been laboratories. Torrey's laboratory, among them, had been for both botany and chemistry. Gray's laboratory under his supervision had conducted some experimental study, but mostly for advanced students. When Goodale took hold of it to teach "vegetable physiology," experimental exercises received more attention. Techniques came from Europe and were furthered by American investigators. Bessey's most influential textbook, *Botany for High Schools and Colleges*,[2] pointed the way to American science generally. Burrill in Illinois, stimulated by German study, "made collections and began an intensive study of plant diseases. In the second term of the year 1869," it is said,[3] "his students were given the opportunity of seeing and studying from actual specimens the characteristics of the mildews and other forms of plant disease." Perhaps, a definition of terms is required. What constitutes a laboratory for experimental purposes in botany? Is a separation of recitation from laboratory work the essential test? Or is equipment and its use? Certainly equipment, room space, schedule, and so forth, which Bessey instituted, were indicia which pointed to the establishment of a laboratory for experimentation to be used by undergraduate students. Professors jeered at him for imitating methods in chemistry—"a mere bit of boasting or buncombe," they said. Nevertheless, Bessey did pioneer work in the history of North American botany—the establishment of at least its first western experimental laboratory equipped with microscope, reagents,

---

[1] See a convincing study by Ernst A. Bessey, "The Teaching of Botany Sixty-five Years Ago," *Iowa State Coll. Journal of Sci.*, IX (2 and 3, 1935), pp. 13-19.

[2] New York: Henry Holt and Co., 1880.

[3] Charles F. Hottes, "Personal Recollections of Thomas J. Burrill and His Work," *Illinois Alumni News* (February 1940), pp. 6, 7. Also, Dr. Hottes's "Changing Emphasis in General Botany and Its Significance," *Ia. St. Coll. Jour. Sci., op. cit.*, pp. 73-76.

specimens, and other material, where students were required to spend an allotted time each week in study. Six years passed and his quarters were enlarged to have a large lecture room, a good and convenient office and study, and a laboratory having eleven microscopes and other apparatus. From this laboratory came his revision of MacNab's *Botany*, his own *Botany for High Schools and Colleges*, and later in 1884 his *Essentials of Botany*. Works such as George Macloskie's *Elementary Botany, with Students' Guide to the Examination and Description of Plants* followed publication of Beal's and Bessey's lectures and treatises and the "new botany" began to become a reality.

Taxonomy, however, remained primary. Accordingly, when Coulter wrote Gray in November 1881, bidding him "speak with the directness of a father & feel sure of causing no offense," Watson and Engelmann approving, Coulter's Rocky Mountain *Manual* was planned for a while to cover Dakota, Minnesota, Nebraska, Iowa, Missouri, Kansas, Indian Territory, Arkansas, Louisiana, and Texas. But Gray defined the range. On January 18, 1882, Coulter wrote him:

> I am more than pleased with the "range" you suggest. I never thought of cutting off New Mexico, & that at once suggested Arizona, and that the Basin; but this arrangement settled all that nicely & I can work from Colorado north in a flora with which I am tolerably well acquainted. I am already at work. . . .

Thus, the Great Basin—Watson's property—and Arizona—Rothrock's—were eliminated and Coulter settled on an area not before academically considered in a manual. Marcus Jones, Coulter heard, was corresponding with a publisher about a Rocky Mountain flora, and the publisher wanted inclusion of the western Mississippi Valley flora, and Coulter to join with him. Coulter refused, as he did not wish to work with Jones. On March 29, 1882, he wrote Watson:

> Have you any influence with the *Torrey Bulletin?* What can it mean to admit new species described by *Marcus E. Jones*, in a magazine too that must be referred to? I confess to having published some of Marcus' "twaddle" but I have turned away his "new species," these very ones among the number, & thus incurred his lasting enmity. His honesty in giving you as differing an opinion is very amusing. But this is no worse than Francis Wolle's work which I also had the privilege of refusing to publish.
>
> Marcus has written to me that he is just about to publish a Rocky M[oun]t[ain] flora, and warns me off *his* ground. I have no doubt but that he will get it into print before I get fairly warm in my work. As I am going to spend next summer in Cambridge, will you please slip a word to that effect into Miss Sweetser's ears? . . .

Parry also had trouble with Jones. Late in February or early March 1882, Parry left Colton and went to San Diego. "Next winter Sonora will be accessible," he wrote Engelmann, "You had better come and winter with us at Hermosillo or Guaymas!!" Pringle was expected every day from Vermont and the San Diego botanists planned an excursion. Parry sent Watson the last installment of his 1881 collections and soon was preparing for a trip to Lower California. The steamer following Parry, however, brought "the irrepressible *Jones* . . . active & lively in his way generally bringing 3 or 4 n[ew] sp[ecies] every trip or 'new to California' *more Jonesii*," said Parry. "We now talk of a trip together into Lower California. [I]f we go far enough we shall get good things. Cleveland is a good useful man, but physically weak." Let it be said Parry always tried to be fair to anyone interested in botany. "Jones expects to go up the Coast in about a month," Parry wrote Watson, "he now thinks he will *not finish up California Nevada Oregon* &c&c&c *this year* but confine himself to California. [H]e is a queer fellow but pretty good company. . . . I am surprized at the improvements here. [I]t makes me feel old to go back 32 years & see the changes. I begin to feel like a Veteran."[4] Parry was much interested in the changes of the region where many years before he had landed when smuggled in, as he characterized his appointment as a botanist of the Mexican Boundary Survey. The railroad from San Diego to Colton would be finished by spring. So he concluded to remain there until summer when he planned to go to San Francisco to keep cool and do work at the Academy, and in the autumn go to Davenport.

On April 6, Cleveland wrote Gray: "Dr. C. C. Parry, Marcus E. Jones & *Mr* Pringle started this morning for Cañon Tantillas, Lower California, where Dr. Palmer got some good things, though he went late in the season. I think that they will get some new things." Pringle and Jones went as far as Ensenada[5] on All Saints Bay, and Parry reported to Engelmann the most remarkable new Rosa find. But the most unfortunate incident occurred. Returning, the party stopped to camp at the hot springs at Tiajuana. The next day, being Sunday, Jones refused to leave, preferring to spend the day in religious meditation. The party left him and next day when John and Charles Orcutt went to get him with a team and wagon Jones ordered them at the point of a revolver from the wagon and drove off. Parry wrote Gray, furiously, as the newspaper at San Diego made much of the story:

. . . no time to speak of our trip to Lower California. Pringle is still out in the Mountains to get a lot of *Pinus Parryana*. [M]ay go to Tantillas Canon. Jones has behaved *shamefully* on the trip capping the climax of his conceit and ignorance by

[4] When in San Diego with the Mexican Boundary Survey, Parry found *Ophioglossum nudicaule* but lost it. In March 1882 when with Cleveland near San Diego, Parry rediscovered the plant, a fern.

[5] The party also went to San Rafael.

drawing a pistol on an inoffensive young man of the party to whom we were under great obligation and who did not choose to be bullied by him. Mr Cleveland says if the transaction had taken place on American soil he would be subject to a criminal prosecution. It has got into the papers and the best thing for Jones would be to sneak out of the country. Of course I have *cut him*, and shall have nothing more to do with him. Pringle who takes the same view will do the same. Jones has intimated that he will "steal a march" on us in the publication of n[ew] sp[ecies] of which I wish to give you warning. [H]e has nothing of any consequence that we have not got *better* specimens of as he is a miserable collector—with this late transaction following his late botanical publication he should go into merited obscurity. . . .

Parry also wrote Engelmann and warned him, calling Jones "a *contemptible puppy*," whom they had left alone "both Sundays to enjoy *his* religion?"

Parry went on a trip to San Bernardino with Cleveland and a hasty "flying trip" to Mojave and Tulare, finding *Yucca brevifolia* "bursting into bloom on every tree!" He collected Ephedra and returned to San Diego to learn that Jones had tried to publish their plants under his names. Parry wrote Engelmann:

I need only say just here that the trip was planned by me, the necessary information procured, and all the details of the journey (ex. Sunday rests) left to my direction. I was the only one of the party that spoke Spanish, and it was through my solicitation that the outfit was secured (including Charley Orcutt the driver) at the nominal price of $1 per day. . . . Jones commenced his unpleasantness by absorbing nearly the whole wagon for his traps, leaving a small corner under one seat for me; abusing the driver, & disgusting Pringle, who fortunately had a separate outfit, and after a few days drew off by himself, and refused to eat with him.—Otherwise I tried to do the best I could and was too glad to leave him to his Sunday meditations at hot springs followed up by his Monday outrage, and subsequent refusal to pay his part of expenses of repairs &c&c. . . . The Rose was *not* first collected by Jones [I]t was growing in thickets by the side of the road when we were all 3 riding. Jones was probably the first out of the wagon, the rest following. [W]hether this entitled him to put a *horrid* name on a beautiful plant I leave you to *decide*. Our wish is that you should give it a good characteristic name with that of the *collectors*, leaving the matter of publication with you, only of course hoping our *darling* may not be prematurely cursed by a *Jones* at the end of it. As to the Cereus (*glomeratus*)? Jones may have seen it first (I was not running races with him or any other *Jackass* at that time.) [H]e admitted himself that he did not know whether it was an Echinocactus, Opuntia or Cereus, till I told him, then expressed his doubts whether there was any good distinction between these genera! Of course I paid no attention but quietly made up my notes, and you can do what you *think best* with them. . . .

Parry wrote both the *Torrey Bulletin* and the *Botanical Gazette* announcing that he and Pringle, who had just left to go on a short trip on the Mojave and Gila rivers, would regard any publication of their joint collection as an "outrage on Science and common decency" and "'a breach of scientific courtesy.'" If, however, Jones named the rose "Rosa *horrida* Jones," Parry told Gray, they would have no objection to the name complete.

The entire matter was unfortunate. Jones had established for himself quite a reputation as a collector of western plants and later became comparatively quite a botanist. He had recently published a list of Utah plants and his paper on his Colorado excursions had been favorably enough received to be translated into a foreign language. But his conduct with Parry and Pringle, not to mention his dealings with Coulter, placed him in disfavor and he left San Diego. Engelmann and Gray both sustained the claims of Parry and Pringle although Engelmann, when written to by Jones, tried to be fair to both sides. Gray, however, rushed Parry's and Pringle's Lower California plants into print—"not troubled with your scruples," Parry told Engelmann.

Parry went north, planning to go to the Redwood and lake country of California while Pringle went into Arizona to collect tree specimens for Sargent. In May, Parry had had some idea of going to the Trinity Mountains by way of Rancho Chico. But after going to San Francisco and back to Colton to pack and ship his materials, going for a day to San Diego and stopping on his return trip at Martinez to see Muir and "'*kiss the baby*,'" also incidentally to eat cherries, he left for Davenport and remained there until the last of September during which time he enjoyed a visit from Engelmann.

Engelmann had gone east to Cambridge and Brookline and, although he had not seen Gray and Watson as the former was in Montreal reading his paper, "The Flora of North America," before the botanists of the American Association for the Advancement of Science, he saw Sargent and Dr. Mohr of Mobile, Alabama, and discussed trees, hearing also of William M. Canby's recent explorations in Montana.[6] "Wood characters will, I suppose, eventually become as essential as leaf and flower characters, especially in plants, like oaks, where the latter show so little," wrote Engelmann to Gray from Brookline. He was at work on a paper on Vitis, and another on the distributions of oaks and conifers. And the gymnosperms of conifers for years had interested Engelmann. His interest in roses had proved more or less temporary and ceased entirely when Watson wrote him asking if he had any intention of working up the genus, and if not, Watson did have. Engelmann left material for Watson in Cambridge and went to Davenport to visit Parry.

---

[6] Canby sent Gray his Montana collections October 19, 1882. Some localities noted were Bull Mountains; Little Missouri, Gallatin rivers; also Huntley.

On September 23, Parry wrote Gray: "I am just getting ready to start back to California, while Mrs. Parry goes East to Conn[ecticut]. I have promised Pringle to go with him to S[an]ta Lucia M[oun]t[ain]s to get *Abies bracteata*, then I shall also run up to Chico & see the agreeable Bidwells. I send before leaving a small parcel of Chorizanthe & Oxytheca for Watson. I do not like to trouble you with odd things, but I am a good deal exercised over Arctostaphylos.[7] . . . My winter plans are not perfected, cannot decide till I get to Cal[iforni]a. I propose to work up a matter in which I may need your assistance, viz, to transfer my *whole collections* books &c to the Cal[iforni]a University at Berkeley. . . ."

At the University of California Eugene W. Hilgard had not only organized the first California class in agricultural instruction but as part of his work he had also had a class in botany. Though an authority on soils and soil technology, he maintained a collecting and systematic interest in botany but, being somewhat in ailing health and having some exploration work to do in Washington Territory, probably for the Northern Transcontinental Survey, he was being aided by Edward Lee Greene.[8] Greene sent on to Gray some of his unusual plants, among them, an Oenothera, a Brodiaea, and an Astragalus. Greene himself had been doing some exploring, working around San Francisco Bay and various points on the California mainland and islands. On July 20, he had thanked Gray for recommending him to an appointment on apparently Sargent's Mount Shasta surveying party, at least, one directed by Sargent: "I cannot find it in me to thrust aside the opportunity for seeing more of that region, and especially the land unknown which lies 'between M[oun]t Shasta & the Coast.' "

Greene evidently had been helping Gray with a manuscript on Grindelia, and with Convolvulus. Continuing to disagree with many of Gray's concepts, that is, calling Gray's attention to alleged overlooked differences of plant-habits and chemical properties indicated by odors, and criticizing his too "tenacious clinging to akenes & pappus" and "*color* of *flowers*" in genera such as Hemizonia, Greene and Gray, notwithstanding, were getting along rather harmoniously. Chemical properties indicated by odors of herbage, he said, are "one of the *very best*" characters available and "you *must not* ignore it. It is *the* character, by which the botanists of the future have *one* good technical distinction." Greene commenced delivering a three months course of lectures on systematic botany at the University of California in September; and

---

[7] "Arctostaphylos," Adanson, *Proc. Davenport Acad. Sci.*, IV (1883), pp. 31-37.

[8] See also "Botanical Work of E. W. Hilgard," by Roland M. Harper, *Bull. Torr. Bot. Club*, XLIII, 7 (July 1916), pp. 389-391. Also Dr. Jepson's articles on Greene.

almost immediately began overhauling the herbarium a little, and correcting some names of plants. On September 19, when he told Gray his appointment might become permanent as an instructor, he said:

> I am sorry that I did not, in the Sept[ember] *Torr*[*ey*] *Bull*[*etin*] more pointedly state, and argue for my conception of the value of mode of branching, and inflorescence, combined with that of odor of herbage: (points not generally thought of in Compositae) and the little importance (in *Calycadenia*) of akenes & pappus, which are so much alike in all. . . .

Greene's criticisms to Gray at this juncture were not always direct, but many times sly. Later Greene would become even more direct, however, and shyness and subtleties would turn to bluntness.

Parry went to Berkeley and there saw Hilgard who urged Parry to persuade Engelmann to come west and return through Washington, Oregon, Idaho, and Montana when the Northern Pacific Railroad would be completed. Parry wrote Engelmann from San Diego and said, after communicating Hilgard's "grand scheme":

> But now with next full moon I expect to make a trip into Lower Cal[iforni]a, perhaps take Mrs. Parry along to collect fresh [plants] of *Rosa minutifolia* &c. Wright of San Bernardino wants to go with me, also the Orcutts. So we may make up a pleasure party, won't you join and take possession of your blankets?

On February 9, 1883, Parry again wrote Engelmann:

> Yes I am back, and your letter *met me*. [C]onsidering it was midwinter our trip was quite successful—gone 16 days. Mrs Parry and your friend *Miss Smith* accompanied us also Mr Wright of San Bernardino and we had no *Jonesian* explosions. So Sundays and Mondays passed peaceably enough. We extended our trip to the Southern bend of All Saints Bay, and camped on a lagoon abounding with huge Turtles—but we got no soup—Saw extensive tracts of *Agave Shawii*[9] many in full flower. [T]ook some notes. . . . We found *Cereus glomeratus* in fl[ow]er & fr[uit], also *C. guminosus*. Cha[rles] Orcutt found a red fl[ow]ering Cereus of which we took notes & measurements. . . . The *Aesculus Parryi* was in full leaf. . . . We secured 1000 roots of *Rosa minutifolia*. . . . I fo[u]nd the lower country is more accessible than I supposed & may attempt another raid as far as Magdalene [B]ay. The steamer from San Francisco to Guymas stops there going & coming. Tonight is a meeting of the Nat[ural] Hist[ory] Society (of San Diego) where I am expected to hold forth. [P]ity you were not here to help me.

Parry had been corresponding with a Miss Fanny Fish who lived near San Diego and knew much of the territory they explored. Probably they called on her in the course of the journey as on January 3 Parry told Engelmann that the next full moon he proposed "to go down . . . & keep her warm" in botany.

---

[9] Named for Shaw, the founder of Shaw's Garden in St. Louis. Engelmann had written on "The Flowering of *Agave Shawii*" in 1878 for the *Transactions of the Academy of Sciences of St. Louis*, III, pp. 579-582.

Engelmann, however, was too busy to envy Parry's journeys. Parry had written of proposed trips with Pringle into Arizona, with Greene to Napa Valley, of his visits to Rancho Chico where he had found new plants, a "new Bergia" and a "fine Cupressus," and activities around San Diego where he had located the groves of *Pinus Torreyana* again. Watson had written of his journey in Vermont and New Hampshire, going over the summit of Mount Washington. Engelmann must have read of W. W. Bailey's journey to Mount Lafayette, New Hampshire, "with all New England mapped out at [his] feet."[10] The great number of western catalogues must have interested him—from the Parish brothers; from Cusick of Union, Oregon; from Vasey, Pringle, Lemmon (although not well and not doing much in botany),[11] Suksdorf, and Rusby; not to mention Lester Ward's *Field and Closet Notes on the Flora of Washington and Vicinity*, numbering 1,249 species. Letterman, Reverchon,[12] Heinrich Eggert, Garber, and others had been sending him correspondence from the central and southern states. He himself had done some botanizing on the Illinois River. Not only was he busy with all this, he was also studying morphology in conifers—". . . the question now is about the Carpel," he said—but John Merle Coulter arrived seeking Engelmann to do some work for his manual of the Rocky Mountain flora. On January 1, 1883, Coulter reported to Watson, "Dr. Engelmann & M. S. Bebb are working at their specialties, while Eaton's Ferns will furnish plenty of material for compilation." Gray had invited Engelmann to spend the winter with him in Cambridge. But while Engelmann appreciated Gray's hospitality, he had said that his home and business and his herbarium required his staying in St. Louis. Pringle arrived home, to Engelmann's regret, in disgust with working for Sargent who had given him much irksome and laborious work to do. Engelmann regarded Pringle's work in Cupressus as very valuable, and he told Gray so. Engelmann was one of the few men of his day who, like the Hookers of England, really realized the greatness of Asa Gray. He, as they, thought Gray should be freed of a great deal of trifling annoyances to accomplish for botany in the large the achievements which his rare ability and peculiar advantage and experience placed him in a position to do. Watson's younger shoulders, Engelmann thought, should take over the numerous small and bothersome botanical puzzles. Gray should do what his greater ability and experience enabled him to do as no other could. On December 13, 1882, Engelmann wrote Gray:

[10] See *Botanical Gazette*, VII, Numbers 8 and 9 (August and September 1882), p. 108.

[11] In the fall of 1884, however, the Lemmons collected in Arizona and New Mexico.

[12] See Reverchon's "Botanizing on Comanche's Peak, Texas," *Botanical Gazette*, VII, Number 4, p. 47.

Today it is half a century that I landed on this continent and I must not let the day pass without celebrating it at least in this way with my friends, and let them know of this at least to me very eventful anniversary. . . .

Parry and Gray were undoubtedly Engelmann's two greatest friends. As Engelmann drew near the close of his life, he tried to do more and more for both. And in trying to do more and more, to bother them less and less with what he regarded as trifling matters, though indeed what he did was more important than Engelmann realized. He worked with his herbarium but not even to Gray and Parry did he confide information concerning it. The assemblage of Engelmann's herbarium had been commenced alone and under most difficult circumstances. When he had arrived in St. Louis to commence his early struggles as a doctor, the town was then not much more than a frontier outpost of the West. Not far west was the actual frontier and very few exploring parties with accompanying scientific equipment had penetrated the vast unknown regions of the West. John Charles Frémont had been one of the early pathfinders who had commenced the practice of stopping for a while with Engelmann to make certain that all scientific equipment was in sufficient order to withstand the rigor and hardships of cart, horse, or human-footed carriage. Knowledge sufficient for this task required knowing something of the habits of wild beasts and Indians, the fear of which was ever present on every expedition whether with military escort or not.

Botany was then a science only beginning to achieve organization on a continental scale under the leadership of Torrey and Gray. Engelmann, Sullivant, and a few others had begun to send collectors into the West but Engelmann especially directed their courses of exploration beyond the frontier. Some came to Engelmann almost exclusively and some took instructions also from Torrey and Gray. At any rate, Engelmann developed a garden of western plants which was soon recognized by Torrey and others as unique in America. Obviously, his herbarium likewise became increasingly valuable. When George Thurber, one of the most important early authorities on grasses, sought to get his important herbarium in order, composed as it was of Brewer's, Bolander's, Lemmon's, his own, and many other collections, he turned to Engelmann. In all the families and genera which Engelmann made his specialties—the Cactaceae, Euphorbiaceae, Isoetes, Juncaceae, the oaks, the firs, the pines, and others—Engelmann was the North American botanist to whom material was always sent. Particularly was this so in Cactaceae and Agave. Only men like Gray, Parry, and Watson knew the full value of Engelmann's herbarium.

Railroad transportation in the West was growing—the last instance, a line east from Mojave to connect with the Atlantic and Pacific Railroad at the Needles[13] and as this grew, Parry, Gray, and Engelmann realized its importance botanically. Many new fields of exploration would be more easily available and the large North American herbaria would become increasingly important scientifically. The report of the botanical section of the Academy of Natural Sciences at Philadelphia for the year 1882 showed an increase of more than 3,346 species, one third of which were new to the section's collection, with 100 genera not before represented. In 1883, the Academy's herbarium added 2,868 species and claimed to possess probably one half of the known species of plants. If Redfield could accomplish this with the herbarium at Philadelphia, how much more were Gray at Harvard, Vasey at the United States National Herbarium, and Engelmann with his own herbarium accomplishing!

Daniel Cady Eaton was doing as much for the ferns of North America. His great work, *The Ferns of North America*,[14] published in several parts, but now complete in two volumes, gave colored figures and descriptions with synonymy and fern-geographic-distribution of the United States and British North American possessions. His *Systematic Fern-List*:[15] a classified list of the known ferns of the United States, with geographic species-range materials, was also now available to botanists of the world.

Similarly, George E. Davenport was doing very creditable work in ferns, recently having reported on Alaskan ferns, collected on the Island of Unalaska in 1879, 1880, and 1881. His article, "A Bit of Fern History," published in May 1882, in the *Botanical Gazette* showed how widely distributed over North America were fern collectors. One year later in the same journal, John Merle Coulter commented on a paper read by Davenport before the American Philosophical Society of Philadelphia:[16]

Up to the date of publication (February 2, 1883) the entire fern flora of the United States contained 162 or 164 known species. Of the States, New York leads with 52 species, followed by California with 48, Florida and Michigan with 47 each, Arizona with a probable 47, and Vermont with 45. Mr. Davenport thinks that owing to the contiguous unexplored Mexican territory, Arizona will lead all the other States in the wealth of her fern flora. . . . Florida is distinguished in monopolizing all the species we have in six genera; these, of course, being tropical. The only other State which has the monopoly of a genus is New Jersey with its very local Schizaea. . . .

In the spring of 1883, John Macoun published his *Catalogue of Canadian Plants*, Part I: *Polypetalae* at Montreal, Canada. The range—from

13 Now part of the Santa Fe Railroad system.   14 Salem, 1879-1880.
15 New Haven, 1880.
16 *Botanical Gazette*, VIII, Number 5 (May 1883), p. 226.

Newfoundland to Alaska—was a tremendous one. But the list, including 907 species, under 243 genera, was not considered final, inviting as it did the cooperation of all Canadian botanists. In the "Preface" to his great work, Macoun told of the many sources which had been consulted in its preparation, and the vast area over which the determinations extended. Up to this time the principal works, other than a comparatively few scientific Canadian publications, on which Canadian botanists had had to depend were Sir William Jackson Hooker's *Flora Boreali Americana* and the published works of Torrey and Gray. Newfoundland, Nova Scotia, New Brunswick, Quebec, Ontario, Lake Superior, Lake Huron, Manitoba, the Rocky Mountains, Lake of the Woods, British Columbia, and Alaska were now included in the compass of a work to be published in five parts and to extend over nearly a decade's time in preparation. The brave, perilous explorations in the early years of the century by David Douglas and Thomas Drummond were at last culminating in a great publication on the native soil of the specimens. Macoun analyzed thoroughly the contributions made by himself and by other especially able collectors such as Dr. G. M. Dawson and Dr. Robert Bell, who widened the orbit of collections not only in the Canadian mainlands but in many adjoining and nearby islands north to Hudson's Bay and the Arctic regions.

Early in 1881 Macoun sent Watson a package of plants and called especial attention to some new things collected at Cypress Hills.[17] He was delighted with books which Watson had sent him and promised to send soon for an opinion "on the new (new to me) plants" which he was embodying in a report. Almost every autumn after returning from his summer Canadian explorations, Macoun wrote either Watson or Gray. On September 17, 1882, he wrote:

I reached home yesterday and found your kind letter of August 7th and the enclosed list of plants sent in the winter of 1880. I also find the pamphlet you mention. Please accept my grateful acknowledgements for all and several and believe me I am doubly grateful for the names of the plants as I wished you to verify the names I had worked out with such an expense of time and trouble.

As I indicated in my letter written from Percé I have made large collections this season and have many rare things obtained last season and if you were not too busy I would send you in the course of a month another installment of plants not before sent by me. Many of the British Columbia and Rocky Mountain plants are extremely interesting to me and I hope may be likewise to you and Dr. Gray.

Many of the Gaspé plants are specially interesting as they are rarities in most herbaria. . . . My collections of Lichens, Mosses, & Liverworts are very large and likely contain many new things. These will be worked up during the winter.

17 By letter dated December 8, 1883, Macoun explained the "Cypress Hills and Hand Hills are in our Canadian North West. The former . . . about 3600 feet in Lat 50 and Long 107-109."

Tell Dr. Gray that I have fine specimens of *Armeria vulgaris* from the top of the Shickshok Mountains at an altitude of nearly 4000 feet and fully 25 miles from the sea. The flora of the White Mountains was reproduced at this point.

Macoun was now the botanist of the Geological and Natural History Survey of Canada. In the summer of 1881 he had gone on his fifth exploration of the Northwest—to Winnipeg and Portage La Prairie again and from the latter place proceeded by wagon to Totogon. He had been asked to go to Lake Manitoba and Lake Winnipegosis and the rivers entering these lakes, and also Assiniboine River. He went by sailboat, therefore, to the upper part of Lake Manitoba and from there to Deep River and Lake Winnipegosis; and, after reaching the mouth of Swan River, made for Red Deer River which enters the lake at its head. Eventually their explorations returned them to Swan River down which they floated to Livingstone from where they crossed to Fort Pelly and the Assiniboine. The journey from there to Fort Ellice was 300 miles to the east. But in boats and canoes they went down the winding Assiniboine to the Fort and thence proceeded to Brandon and Winnipeg. After this journey, Macoun wrote his notable book on the Northwest, *Manitoba and the Great North West.*

Early the following spring, Macoun and his son, James, started on a trip to western Ontario. "The reason of my going," wrote Macoun, "was that Sir William Hooker wrote me in 1861 that they had less information in England about the flowers that grew at Lake Erie than they did about those that grew beyond the Arctic Circle."[18] Dr. T. J. W. Burgess, who accompanied the party, wrote the *Botanical Gazette*:[19]

Having in the latter part of June made a collecting tour with my friends, Prof[essor] Macoun, Dominion Naturalist; Mr. William Saunders, Editor of the Canadian Entomologist; and Mr. James Macoun, to Point Pelee, Essex Co[unty], Ontario, the most southern point on the mainland in Canada, a list of the rarer plants found there might not be void of interest to some of the readers of the *Gazette.* . . . The large size and plenitude of the Papaw, Mulberry, Blue Ash, and Sour Gum trees clearly show them to be indigenous. . . . I might add that during the week preceding our trip, Prof[essor] Macoun had found along Lake Erie, at Amherstburg, Pelee Island, and in the neighborhood of Port Stanley . . . no less than eight [plants] . . . which for the first time find a place in Canadian Flora.

Returning to his home, Macoun learned that he was to go with Dr. Ellis along the Gaspé coast. Accordingly he and his son James went to Gaspé from where they examined the coast of the St. Lawrence from Gaspé to Little Metis. Ellis collected the geological specimens and Macoun and his son the botanical. When they reached the river Ste. Anne des Monts, the parties separated and A. P. Low and Macoun took

[18] Macoun's *Autobiography, op. cit.,* p. 205.   [19] VII, Numbers 8 and 9, p. 95.

canoes and French boatmen up the river to climb the Shickshock Mountains. "My purpose in going up was to study the flora at the summit as I had never seen a species growing which we called Arctic," wrote Macoun. "Hitherto I had never climbed a mountain, except one in 1875, and knew nothing of the plants to be collected. On reaching the summit, we found an extensive plateau and came on fine specimens of cariboo which gazed at us for a time and then ran off. We spent three days on the summit and I collected a large number of Arctic plants which formed the basis, in later years, for the excursions made in Quebec by Dr. Fernald of Harvard University."[20] Macoun returned, moved his family to Ottawa, and took up his work at the museum of the geological survey. From there he wrote his letter to Watson, commenting on *Armeria vulgaris* found on the top of the Shickshock Mountains.

When during the following winter the Royal Society of Canada was formed, Macoun was selected among the first twenty members of the geological and natural history section. In the spring of 1883 he decided to visit Nova Scotia and Cape Breton. Accordingly, he and his son William went to Nova Scotia in the Annapolis Valley and Macoun ascended Cape Blomidan, going to Yarmouth, Halifax, and then Cape Breton and Louisburg. From there they went to Gaspé and by schooner to Anticosti during the summer, arranging to be called for in late August at North West Point. On the beach at Salt Lake they camped on land like a peat bog and were troubled by black flies. But Macoun and his son gathered plants from great numbers there; and later going to S'West Point, as far up Jupiter River as they could, and to Betsie River, increased their collections which included fossils also. Eventually, however, they reached North West Point and, camping near the lighthouse there, were picked up by the schooner and returned to Gaspé from where they went home to Ottawa. On December 13 Macoun wrote Watson:

I am engaged working up our flora and in my examinations have come across a few things I want your opinion upon. . . .

Dr. Burgess and myself are thankful to you for your trouble with our Nova Scotia plants. The specimens you name *S. graminea* are truly indigenous, as I found it every where in Nova Scotia and also along the Gaspé Coast. . . .

Any changes I make in the Catalogue I shall inform you.

Macoun wrote again on February 18 of the next year[21] saying "that the greater number [of plants] are from Nova Scotia and the Gulf," and

[20] Macoun's *Autobiography, op. cit.,* pp. 206-207.

[21] Macoun wrote: "It is worthy of note that the Light House Keeper at South West Point, Anticosti, found, 'F. Pursh' cut on an old spruce." Later he said: "I have no doubt but [Pursh] found *L. arctica* var. *Purshii* exactly where I found it" on Anticosti.

in June started for Port Arthur and Nipigon to go up the Nipigon River to the lake of the same name. On their return trip, they went by boat from Nipigon to Ross Bay on Lake Superior, and from that point walked along the line of the Canadian Pacific Railroad to Michipicoten[22] where they took a boat to Sault Ste. Marie. On August 26 Macoun wrote Watson:

I have just returned from Lake Superior and delighted to find on my table the new Manual of the Moss flora of North America.

Please accept my grateful thanks for your kindness and believe me when I say that I fully appreciate it. During the coming winter I purpose sending a set of *our* mosses to the herbarium and shall be pleased to send you a set if you are having a separate collection. I have about 500 species.

During the summer I have made a careful examination of the country north of Lake Superior and Lake Nipigon and my son is collecting on the Cypress Hills and Lake Winnipeg so that you may expect many additional things from me this winter. The Second Part of my Catalogue is now going through the press.

Macoun was by now sending Watson and Gray great quantities of plants from the Canadian flora for the Gray Herbarium. Part II, the *Gamopetalae*, of the *Catalogue of Canadian Plants* acknowledged Watson's and Gray's aid:

Prominence should have been given (in the Preface to Part I) to the fact that through the kindness of Dr. Asa Gray and his able assistant, Sereno Watson, Esq., all doubtful species were critically examined and reported on by them. In every case their decision was considered final, except where mention is made of divergence of opinion and the reasons therefor given in the text. For many years these gentlemen have assisted me in determining our difficult phenogams, and much of the real value of the present work is due to them.

The next year, 1885, Macoun hoped to go to the Rocky Mountains again.

Watson's *Contribution to American Botany* to which Macoun referred especially in a letter dated June 13, 1883, was that presented on May 29, 1883, which contained Palmer's southwestern Texas and northern Mexico plants and some ferns, vascular acrogens, mosses and lower cryptogams from Parry's and Palmer's collection. The part that must have interested Macoun principally was the second, where descriptions of some new western species were given. There plants collected by the Howell brothers on Trask and Willamette rivers, Oregon; plants collected by Greene on the Scott Mountains of northern California; and plants collected from San Bernardino to Washington Territory by col-

---

[22] Their permission was to walk along the railroad to Missinabie.

lectors such as Kellogg, Lemmon, Parry, Rattan,[23] and Suksdorf were determined.

Rattan was a young California teacher and a correspondent of Gray's. As early as 1878 he had explored around Eureka, California, and in Humboldt County. Going along a trail to the mouth of Trinity River, he had followed the river for many miles and then crossed over the mountains to the Mad River, later going to Cloverdale in Sonoma County. Plants of a genus published as Newberrya interested him. The following year he had attempted to compile a small flora of California species and planned going to the wild regions of northern California, visiting the Siskiyou Mountains. Having several hundred students in botany each year studying Gray's *How Plants Grow* for structural forms and growth, he saw the need of a small book describing the wild flowers. The survey books were too bulky and inaccessible. So with some promised aid, but planning by himself, he began culling from reports on California botany what material he knew would be useful to his classes. In 1879 he traveled 370 miles on foot, going over rough mountain trails exploring regions near Eureka, Arcata, the Hoopa Valley, Martin's Ferry, Orleans Bar, Happy Camp, Waldo (Oregon), Crescent City, Gold Bluffs, and Trinidad. He went by way of the Klamath to Happy Camp and crossed the mountains by the trail taken by Brewer and Clarence King in 1863. "From Waldo," he said, "I traveled a trail which lies south of the wagon road and thus saw more of the Illinois and Smith Rivers than Brewer did but the forest trees were like dwarfs on the old road; so I had no means of learning any thing new about the possible firs that grow in the dark forests which mantle the snow covered peaks farther south. . . . I found Sarcodes near the Siskiyou Summit. On the Illinois River I found our Darlingtonia bogs, and on Smith River one which cannot be more than ten miles in a straight line from the ocean." Returning to Eureka or near there at Kneeland's Prairie and Humboldt Ridge, Rattan found a new Pentstemon "confined to a locality in the spruce forest near the Prairie."

In May 1882, the *Botanical Gazette* noticed a *California Flora or Manual of Botany for Beginners*[24] containing descriptions of plants with conspicuous flowers, numbering something over 600. This popular book had been first published in 1879 and by 1882 was in its third edition. Apparently Rattan worked almost altogether by himself, the botanists

23 Volney Rattan. See Willis Linn Jepson, *The Botanical Explorers of California*, I, Madroño, I, pp. 168-170 (1928).

24 *A Popular California Flora*, or manual of botany, for beginners, containing descriptions of exogenous plants growing in central California, and westward to the ocean. San Francisco, 1879, 1880, 1882.

of the California Academy of Sciences resenting his correspondence with Gray. At least, Kellogg did, Rattan heard and believed. Bolander had left California and gone to Central America during the years Rattan worked. And a Dr. Gibbons was preparing a California botany which was attracting the more prominent botanists of the state. Rattan never considered himself an author. "I would prefer to be considered a small plant collector who loves nature and wildness rather than cheap notoriety," he told Gray.

On May 9, 1883, Gray presented to the American Academy of Arts and Sciences another *Contribution to North American Botany* having two principal parts: (1) "Characters of new Compositae, with Revisions of certain Genera, and Critical Notes"; and (2) "Miscellaneous Genera and Species." Plant collections ranging from the Pacific Railroad surveys to more recent ones were embodied in the determinations; those of Pringle in southeastern California; Wright, the Parish brothers, Parry, Lemmon, and others on the Mohave Desert and around San Bernardino; Rattan in Humboldt County, California; Lyall, Hall, Watson, Suksdorf, the Howell brothers in Washington Territory; Lemmon in the Santa Catalina Mountains and the canyons near Fort Huachuca;[25] Rothrock in southeastern Arizona; Greene in southwestern New Mexico; Ervendberg in Wartenburg, Mexico; Reverchon in western and north Texas; Schaffner, Parry, and Palmer at San Luis Potosi; Palmer near Saltillo; Rusby in Pinos Altos and Mogollon mountains; Curtiss in Florida; and many others. Eatonella, a new genus, was named for Daniel Cady Eaton who wrote Gray, thanking him heartily and saying he, Eaton, was too timid when working with Watson's Compositae, about new species. The name *Lonicera Sullivantii* was given to the form then appearing in Gray's *Manual* as *L. flava*. Most of the work, however, was as the *Botanical Gazette* commented, "done in the preparation of the forthcoming volume of the Synoptical Flora, the appearance of which all botanists sincerely hope may not be much longer delayed."[26]

That same year the *Botanical Gazette* reviewed Volume I, Part II, *Caprifoliaceae-Compositae* of the *Synoptical Flora of North America*, saying:[27]

This elaboration of some of our very complex genera of Compositae is the result of time, and travel, and severe study, and is the matured, as well as probably the most valuable of the many contributions to North American Botany that have issued from Cambridge. To say that it will enhance a reputation already the greatest in American botany seems superfluous.

25 Arizona.   26 IX, Number 1 (January 1884), p. 15.
  27 IX, Numbers 10 and 11 (October and November), p. 181. *Synoptical Flora of North America*, by Asa Gray, LL.D. New York: Ivison, Blakeman, Taylor & Co., 1884, 1886.

In the same number the *Gazette* commented on Charles E. Bessey's *Essentials of Botany*:[28] "No text book ever gave better promise of meeting a long felt want than this. It will be welcome wherever the aim is to learn from nature herself, and to make the book serve only as a *guide*."

Lesquereux and James's *Manual of the Mosses of North America* had been reviewed in the September number, with a list of mosses added by Eugene A. Rau. Concerning Rau's *Catalogue of North American Musci*, published a few years before, Lesquereux had characterized it as "a very poor unimportant affair." Some bitterness evidently developed and Rau's additions to the *Manual* list were resented by Lesquereux. He wrote Rau October 7 showing that every species named by Rau was either not a North American one or a synonym. Only one species was omitted from the *Manual*, Lesquereux said, and that was *Hypnum occidentale*, Sull. & Lesq., which he himself had originally determined.

Concerning Chapman's supplement to his *Flora of the Southern States*, John Merle Coulter wrote Gray: "I am sorry that Chapman's Flora could not have taken the form of a revision, rather than a simple reprint with a supplement, but I expect the D[octo]r hardly had the courage at his age to undertake it." Chapman, though himself not satisfied with his supplement completely, believed in view of his age he had done what he could and was content to praise others. "In your valuable 'Contributions' you have never published anything so valuable to American Botanists as your notes on Aster & Solidago," he told Gray. "I have studied them like a Classic and find the hitherto dark gropings all disappeared. Thank you everlastingly." Soon he named in Gray's honor a "clean little thing *S.*[29] *Grayii.*" He had done some botanizing in Florida, Georgia, and possibly, the mountains north of there. But not much. In point of age, he was the Dean of American botanists. Though never contented to be without the field in his botany, he was compelled to give more time to only his herbarium.

The climax to Gray's career seems never to have been reached. All during the years after Torrey's death, his life and work seemed to enjoy a perpetual state of climax, with few years superior one to another. On July 19, 1883, Parry had written him from Rancho Chico:

Not a mere "congratulation" but a real *ovation* showering down upon your silvered head should commemorate such a *life-work* accomplished. Well may you take a long deep-drawn inspiration! It seems but yesterday since as a medical student in N[ew] Y[ork] I first secured a copy of N[orth] A[merican] Flora

---

[28] *Ibid.*, p. 184. *The Essentials of Botany*, by Charles E. Bessey, M.Sc., Ph.D. New York: Henry Holt & Co., 1884.
[29] *Solidago?*

Vol[ume] II, but what an interval of hard work and rich discovery has been crowded within those past years 1842-1883!! and how hopeful for the future. . . .

Parry planned to remain in California until the following spring, then return via the Northern Pacific Railroad. Late in February he had taken a trip to Table Mountain, "so conspicuous a land-mark from San Diego," and ascended it. In April he had gone to Mount Diablo and returned in May, writing Engelmann on May 9:

Well at M[oun]t Diablo we had a *funny* time, a 4 horse lumber waggon with *8* rol[l]icking children stowed away among the bed clothes & provisions. [W]e camped under the open sky & slept *13 in a bed*! The *ticks* were not countable. [W]e passed up some magnificent "ravines" and nearly encircled the double peak. Some of the party ascended the summit. I staid on the upper slopes as the season was early for high altitudes. Some rare Compositae, Cruciferae, Phacelia &c&c, *Ptelea angustifolia* abundant. I measured an *Arctostaphylos glauca* with trunk 5 f[ee]t in circumference and 25 f[ee]t high! big enough for Sargent! nothing new or strange in the way of pines or oaks. *Juniperus occidentale* the only cedar. We drove home in a drenching rain and you can imagine the muddle of wet clothes & *damp children*! Still we enjoyed it. . . .

I got back from M[oun]t Diablo in a rain storm to be shocked by the astounding news of your going to Europe—"passage engaged"—pray *keep on* and meet us on the other side. . . .[30]

Parry and Mrs. Parry had planned for some time to go to Europe. But no opportunity had presented itself. And much remained to be done in California. Lemmon was exhausted, physically and mentally, and forbidden to work, although since 1880 he had collected at intervals and written on *Woodsia Plummerae, Ferns of the Pacific Coast*, "Four Rare Trees of Arizona," on an alleged discovery of the potato in Arizona, and other subjects.[31] Lemmon, however, had proved a better collector than author in his early years of botanizing and writing. Pringle was at Tucson and discouraged, for it was dry and Indian troubles were there. Bolander had returned to the United States. But he was in San Antonio, Texas, when last heard of. Palmer wanted to return to New Mexico and old Mexico, and, presumably, California. But he found it would be better for him to remain in the southern states, collecting archaeological material, in Alabama near Blountsville,[32] in Arkansas, and other places. And Muir was, as Parry said, "*immured* on his ranch."

Meehan and, possibly, Canby were coming west for a while, Parry heard. But Parry was growing older and more inclined to go alone

[30] On June 1, Parry wrote Engelmann, saying he had just made a hurried trip to San Diego and expected to go during July to Monterey. The letter was written from San Francisco.

[31] Also, Mr. and Mrs. Lemmon botanized in 1882 in Huachuca Mountains, Arizona.

[32] Palmer went to Greensboro, Tuscaloosa, along the Tombigbee, Alabama, and Mobile rivers; also Mobile harbor; and Georgia and Indiana.

places. He postponed indefinitely his trip to the Santa Lucia Mountains, especially when he heard that Pringle had again gone home in disgust. Greene was busy as a lecturer and working occasionally at the California Academy of Sciences among Veatch's Cedros Island and other collections.[33] Greene, however, wrote Gray on July 14: "No, I am not yet professor; nor am I likely to become such. I doubt if I even obtain the lectureship this year. Bolander has applied, & his friends are working for him! *I* am doing nothing in the way of effort to obtain the appointment: do not *care* much about it. . . . Rusby's plants will speak for themselves. Do not tell me that Rusbya is not a good genus. I will obtain for you, if possible, a complete specimen," he added characteristically. Greene had reached the point where he dreaded and feared Gray; and told Gray he did. Consequently, concerning plants which otherwise he would have already had in print, he had deferred publication.[34] Whether the plants here referred to as Rusby's plants were some collected that year by Rusby is doubtful. Rusby had planned in February to go that season to collect in northern and central Arizona. But he too found the country dry and returned to Franklin, New Jersey, with what he considered "trash." "But do not despair!" he told Davenport, "arrangements are made by which I shall travel in many countries, solely in the interest of Botany and its applications."

As planned, Engelmann went that summer to Europe. "Under the circumstances," he told Gray, "it is a bold undertaking to go to Europe in 4 weeks! But my old plan to spend a week or two before sailing with you and Sargent must be given up. If well enough on my return, it may be attempted then." He sent Gray a package of plants from himself, from Eggert, and from Letterman. "They both refused any compensation, were glad to assist science through you. . . . You will find plenty of specimens of the new black Crataegus." Among Engelmann's last publications were: "The Black-Fruited Crataegi and a New Species," "*Vitis palmata*, Vahl.,"[35] "The True Grape Vines of the United States,"[36] "The Diseases of the Grape Vines," and "The Compass Plant," all comparatively unimportant publications published as notices or short articles. He wanted to publish a Synopsis of Ephedra but concluded to wait till more material accumulated. He sent Gray ten gallons of wine and in June went to New York where he visited with George Thurber before sailing.

[33] Late in the summer of 1883 Greene went to the Sierra and Lake County.

[34] Greene began publishing his "Notulae Californicae." See *Bot. Gaz.*, VII, 8 and 9, p. 93; VIII, 4, p. 203; IX, 3, p. 49.

[35] On June 12, Engelmann wrote Gray: "Have I ever mentioned the rediscovery by Eggert of Michaux *Vitis rubra.* . . . I have written an account."

[36] Early in May Engelmann went to the country and watched the grape vines developing.

So alarmed was Thurber about his condition, however, he wrote Gray: "I fear that our wonderfully acute friend has done his botanical work." Engelmann had heart disease and knew it. Thurber bade him good-by with the feeling it would be the last time. Accompanied by his son and daughter-in-law, Engelmann went to Berlin where he saw Eichler and then had a few days with De Bary in Strasburg. But his strength was limited and soon they were sailing from Antwerp back to America.

On September 21 he wrote Gray: "I can not say much of my botanical studies in Europe—these were scarcely any thing, my health and spirit being so miserable, and in this respect the whole trip was a failure." Nevertheless Engelmann came home expecting to do much. Letters were to be answered, specimens to be examined and named, and an article on metamorphosis of the carpellary scale of conifers to be completed. He had gathered some information in Europe. One point was that he disagreed with Eichler on this subject. Another was contained in a letter to Gray dated December 13, 1883:

I had to say something about hybrids because in France they, especially [one] of Bordeaux, sees hybrids in every deviation from the assumed typical form, crediting nothing to innate variability of species. He sent me his last publication (with nice plates) but his position is shocking: he very often discovers 3 and I believe even 4 parents (grandparents) in certain cultivated American forms. And somebody told me that another Vitis man in France has found over 2 or 300 forms in what they get as *V. riparia*.

How similar this attitude was to opinions of older American botanists such as Lesquereux and Tuckerman! On September 17, 1884, Lesquereux, thanking Gray for his review of the *Manual of the Mosses of North America*, commented:[37]

I return today with best thanks the no. of the *Nation* where I read the notice of Prof[essor] Gray on the Manual of the mosses. The notice is very kind and good. I think, however, that Prof[essor] Gray is mistaken in supposing that Bryologists of the next generation may wish to reduce the number of the genera and orders and thus increase their weight and value. The tendency is the contrary way. That, you may see by Mitten, Lindberg, Braithewaite etc. The inflorescence of the mosses is of the same character for all except perhaps for the Sphagnaceae . . . the Bryologist has to rely for his subdivisions to grouping forms. . . . After a long and tedious study even of a single order of the mosses, the Bryologist is forced to acknowledge that all divisions and subdivisions are an affair of opinion. . . . I should have much to say on the subject and may prepare a short paper. . . .

Tuckerman, after complaining that the judgments of too many scientists, one especially, were subjective and therefore not scientific, refused to "confess that there is nothing fit to be called Science, beside Anatomy"

[37] The letter was addressed to Watson.

of plants, making taxonomy nothing but a system for tagging specimens. He became more philosophical, however, and on March 17, 1884, wrote:

I am continuing to do my best with the determination, and characterization of our Lichens, and am content therefore to leave the results to future students to make what they can of—sure that whatever mistakes they find, they will themselves be found to have made perhaps as many more.

I sometimes think that a superior being, looking down with competent eye on much of the systematic labours of our time (I mean of course in one little field), would find terms like "Science" & "scientific" inappropriate in such relation; but we must do what we can, and reach our reward in furnishing Indexes, the value of which, to the philosophical investigator of vegetable structure as to the humbler but not less happy student of Habit, and Special Morphology, is alike unquestionable.

But on May 29 of the same year, Tuckerman added:

. . . the time will surely come when the systematic genius of Fries will be more instructive than ever before; and a new "*reformatio*" of the system as the micrologists have left it, make *all things* new. *Quod efficiat Deus!*

On January 23, 1884, Engelmann wrote Gray, saying: ". . . I am so so, short [of] breath, only when moving, therefore difficulty of locomotion, otherwise well." It was his last letter to Gray. On February 4, Engelmann died. Gray wrote: ". . . the lasting impression which he has made upon North American botany is due to his wise habit of studying his subjects in their systematic relations, and of devoting himself to a particular genus or group of plants (generally the more difficult) until he had elucidated it as completely as lay within his power. . . . More than fifty years ago his oldest associates in this country—one of them his survivor—dedicated to him a monotypical genus of plants, a native of the plains over whose borders the young immigrant on his arrival wandered solitary and disheartened. Since then the name of Engelmann has, by his own researches and authorship, become unalterably associated with the Buffalo-grass[38] of the plains, the noblest Conifers of the Rocky Mountains, the most stately Cactus in the world and with most of the associated species, as well as with many other plants of which perhaps only the annals of botany may take account."

Early in the summer of 1883 Parry received a pass on the new Denver and Rio Grande Railroad and, probably in August, returned to Davenport. That year he had published in the *San Diego Union*[39] some "Remarks on Lower California" and in the fall the *Overland Monthly*[40] published his article, "Early Botanical Explorers of the Pacific Coast."

[38] See Gray's review of Engelmann's "Two New Genera of Dioecious Grasses of the United States," in Gray's *Scientific Papers, op. cit.*, I, p. 112, entitled "The Buffalo-Grass," 1859.
[39] February 13, 1883.   [40] October 1883.

He began selecting 1,000 species for John Donnell Smith of Baltimore who late that summer had taken his usual excursion to the Southern Alleghenies, did some work with Juncus, and published in the *Proceedings of the Davenport Academy of Sciences* his "New Plants from Southern and Lower California" and "Arctostaphylos, Adanson."[41] On December 26 he told Engelmann: "I am still puttering away at plants, writing up n[ew] sp[ecies]. [T]oday my manuscript goes to [the] printer and then for a raking down from you? Besides that Arctostaphylos paper in which I am quite *conservative*, even more so than Dr. Gray, I make out description of *Phacelia suffrutescens . . . Ptelea aptera, & Polygala Fishiae!*"

On April 2, 1884, Parry wrote Gray concerning the proposed dispositions of Engelmann's botanical materials:

I have just returned from a week at St. Louis with George Engelmann & look over his father's notes & collections&c, and consult[ation] in regard to their disposal. Everything was just as he left it and I had to gather up the loose notes & specimens scattered over his table, and put them together as near as I could, marking each parcel to be packed away till the final disposal is decided. It seems that the D[octo]r left no definite directions and we can only judge of his wishes in a general way. He had intended in case the Acad[emy] of Science should have been in a proper condition to receive them, to leave a fund of $8000 for their maintenance & preservation in an old cancelled will, but since then the fund from which this was to have been taken has been lost by an unsecured failure. George wishes as far as practicable to carry out his father's wishes but as far as I can see, there is not even a remote possibility of doing anything satisfactory in reference to the St. Louis Acad[emy] of Science, which have neither a place of deposit, available means or even interest in the matter. So too in regard to the Washington University which is in a cramped locality devoted only to ordinary college instruction with no one to take an interest in preserving or taking care of the collection—So as far as St. Louis is concerned, *Shaw's Garden* affords the only available resource, and on many accounts I am inclined to think favorably of it. We called on Mr. Shaw while there and had a free conversation on the subject, which Mr. Shaw seemed to think favorably of. . . . It is his intention to leave *all* his property to maintain the Garden & its accessories, having Kew as his model. The available annual income will be at present about $40,000. . . . The special point of interest to us as botanists is the [B]otanical Curator who is designated as [first] assistant to the Curator to have charge of herbarium, botanical investigation, naming of plants&c&c also to act as Secretary. . . . [H]e should be in his special department independent, and appointed on the recommendation of some of the leading botanists of the country. . . . I think some such plan will also meet your views and as Mr Shaw himself intimated he would be greatly influenced by your advise in the matter. . . .

Now in regard to personal views which I would like you to regard as *confidential*, without wishing to appear as a *place*seeker. I think under proper conditions, something as above specified I would be willing for a few years at least to assume

41 IV (1883), pp. 38-40; IV (1883), pp. 31-37.

the duties of botanical curator, having special reference in the first place to putting the Engel[mann] Herbarium in good shape for use & reference. [I]t seems that from my long & intimate acquaintance with Dr E[ngelmann] that I am real[l]y the proper one to undertake the important work, and I am satisfied that it would coincide with George's views and I think with yours.

Parry urged Gray to recommend to Shaw a course of action. Shaw was disappointed in the amount of use the Garden's herbarium and library had had. He should be impressed, Parry said, with the fact that possession of the Engelmann herbarium would bring visitors from many places of the world. The present herbarium was, according to Parry, "not of special interest, but there is already the *nucleus* of a good botanical library that by the addition of Dr. E[ngelmann]'s will be at once valuable & by a few additions unusually so." Parry planned to sail for Europe in June and by that time have his Chorizanthe paper published. In 1884 the Davenport Academy *Proceedings* published his "Revision of the Genus, and Rearrangement of the Annual Species—with One Exception, all North American," and the following year the *Western American Scientist* presented his "New Genus of Euphorbiaceae from Lower California" and also his "Notes on Chorizanthe Lastarriaea Parry." Reading Parry's letter with interest, Gray, as will be told in the next chapter of this book, went to St. Louis and interviewed Shaw.

Since the middle of the century Engelmann had been aiding Shaw in formulating plans for his "Botanic Garden and Collection, Kew in miniature." The extraordinarily rich Englishman, Shaw, a resident of St. Louis, had corresponded with Sir William Hooker who in turn had referred him to Engelmann. Shaw early showed great veneration for Gray. But Engelmann had complained that "Scientific botany [was] secondary or tertiary with him," that he had "the ornamental as much at heart as the scientific," and that while this was well to popularize the establishment Shaw's lack of "real scientific zeal [and] knowledge" stood in the way of interesting him "in what interests us and seems important to us." Consequently, Engelmann's herbarium had been kept apart from that of the Garden on Gray's advice, although looking "to an eventual combination, either in Shaw's lifetime or soon after." Gray had wanted Engelmann to become "director of the whole concern" but Engelmann had steadfastly refused.[42]

After Engelmann's death, and influenced by his visit with Shaw at St. Louis, Gray commenced quite an extensive correspondence with

[42] A very interesting article, "Formative Days of Mr. Shaw's Garden," has been published by Dr. Clarence E. Kobuski and the Missouri Botanical Garden in *Missouri Botanical Garden Bulletin*, XXX, Number 5 (May 1942), pp. 100-110. Quotations in this paragraph are taken from this article.

Shaw, and with William Trelease collected and published for Shaw *The Botanical Works of George Engelmann*, a classic memorial of the work of the greatest American botanist which middle western United States had yet produced. In 1887 this volume was published and, a copy being sent Sir Joseph D. Hooker, he wrote Shaw June 17, 1888:

I have just received your most handsome present of Engelmann's Botanical Works, edited by our dear late friend, Dr. Gray, and I do thank you most heartily, no less for your kind gift than for the effective service to botany that this most valuable contribution to the science renders. It is indeed a noble tribute to a man whose labors as a most conscientious and painstaking botanist have never been surpassed, and I prize it for the sake of the man whom I knew so well and esteemed so highly. I shall never forget my visit to him and to you and the afternoon I spent in your garden and museum at St. Louis, in company with Dr. and Mrs. Gray.

I have been most interested in all that Dr. Gray told me last year about the noble botanical institution that you have founded and in his hopes that it would be a center of diffusion of knowledge, the influence of which would be felt far and wide. . . .

Great as was the memorial volume of Engelmann's works, the Missouri Botanical Garden established in 1889 under provisions of the will of Shaw was even greater. Gray did not live to see the Garden's formal establishment on a many hundred acre tract devised for the purpose. But he had much to do with the definition of plans. Parry did not become the curator although he was urged by Gray to accept the curatorship with the idea eventually of placing his herbarium with the Engelmann herbarium. The appointment of William Trelease to the Gray professorship in the Shaw School of Botany met with approval from practically everyone. Parry's offers had been made in deference to his long friendship with Engelmann and his desire to go to California for further work brought forth his rejection of the curatorship. That Trelease was selected for the highest office is further proof that Gray held a thorough and ardent sympathy for development of experimental as well as taxonomic botany in North America, further proof that he aimed to carry out the long expressed wishes of Engelmann to shape the new Missouri Botanical Garden into an institution in which scientific botany and horticulture, the former particularly, would be uppermost.

On June 12, 1885, Trelease wrote Gray:

As I wrote you some weeks ago, I visited St. Louis on Mr. Shaw's invitation and passed a couple of days very pleasantly at the Garden, meeting Mr. Eliot for a short time one day. The Directors of the University, as I learn from Mr. Eliot, met on Monday and established the department and have offered me the professorship. While I am uncertain about one or two details, I think there can be little doubt but that I shall fill the very promising position that has been opened to me through your kindness. . . .

And on June 27, Trelease informed him:

The matter is at length settled, and I am to go to St. Louis next fall. I feel sorry to leave Madison for many reasons, not the least of them being that my department here is going to suffer seriously by any change. After giving the subject a good deal of thought, I have recommended Seymour[43] for trial one year as Assistant Professor for I have more confidence in him than in any other available man.

Trelease, apparently, was not aware that he was telling Gray something which Gray already knew, or, at least, expected would take place. Gray undoubtedly recommended Trelease, understood the type of institution Trelease would build at St. Louis. The laboratory of scientific experimentation was not Gray's primary forte. Trelease, however, had studied at Harvard. His completion of his work with Farlow—and with Gray—marks a significant event in American botanical history, since he was to have much to do with advancing the interests of research and to extend study more inclusively into the lower orders of plants. Under his leadership, and that of his successor Dr. George T. Moore—from 1909 to 1912 professor of plant physiology in the Shaw School—the Missouri Botanical Garden, with valuable greenhouses, an excellent library, research facilities, and a large and useful herbarium composed in part of Engelmann's herbarium, has become one of the great botanical institutions of the world. Its school, maintained in connection with Washington University at St. Louis, has exercised a leadership from its inception that has spread its influence far beyond the dreams of Engelmann, Gray, and Shaw. But its establishment came not without a struggle.

Shaw did not take quickly to all of Trelease's plans. By slow and steady persuasion (as, for example, Trelease had Shaw read DeCandolle on Herbaria when he sought a particular arrangement of materials) Trelease accomplished most of his objectives. He told Gray:

I foresee that I shall have to take things very moderately, but by falling back in good order when I see I can't carry a point, and renewing the attack whenever more favorable opportunities occur I think I shall get what I want in the long run. Mr. Shaw certainly wants to do something unusually good, but as you know his idea of botany is different from ours, and I sometimes think a sprinkling of "Gray" in my wig would make my suggestions more convincing. . . .

Obviously, Gray's influence with Shaw persisted to the point almost of direction. There is little doubt that the founder of the great garden of the Central West respected Gray's vision and knowledge as much as, if not more than, he did that of Engelmann. Moreover, Gray's secret objective seems to have been to place in the garden a scholarly institution that would do honor to the vision and labors of George Engelmann —an American botanist whose abilities Gray revered as second to none.

[43] A. B. Seymour, later at Harvard University.

CHAPTER X

American Botanical Laboratories Extended.
Agricultural Experimentation

AFTER completing Volume I, Part II, *Caprifoliaceae—Compositae*, of the *Synoptical Flora of North America*, Gray decided to have "a bit of holiday." With Engelmann, Arnold Guyot, Decaisne, and Darwin all recently gone, and Bentham gradually going—whose life was to Gray "the very ideal of a naturalist's life"—Gray, too, had his warning. Gray was now well past seventy years of age and had to work more cautiously. Accordingly, he evidently grasped the opportunity presented by Parry's letter, and went to St. Louis. Later, on June 9, he wrote Hooker:

I must tell of our two weeks' run, Mrs. Gray and I. We left the too tardy spring here, one evening; were the next noon in Washington, where the spring was in full force and beauty. After two days, left Washington one morning, followed up the Potomac River to its very rise in the Alleghanies, and down on to Mississippi waters before dark; woke near Cincinnati, had a pleasant day's journey to St. Louis, which we reached before sunset. There had five days, rather busy ones; thence a journey of thirty-six hours, over prairies of Illinois and Indiana to Buffalo, and to New York city; there two days, and then home.

St. Gaudens at this time was finishing the bronze bas-relief of Gray placed later in the Herbarium. Gray went to New York for his last sitting. His letter to Hooker continued, telling of the proposed "Mississippian Kew":

You remember Henry Shaw, his park and Missouri botanic garden. The old fellow is now eighty-four. Something induced him to ask my advice, and to let me know the very ample fortune with which he is to endow the garden, when he dies. I was in doubt whether all this was likely to be quite wasted, or was in condition to be turned to good account for botany and horticulture when Mr. Shaw leaves it and his trust comes to be executed. I wished also to see that dear old Engelmann's herbarium should be properly and permanently preserved. So I went on to St. Louis. Mr. Shaw took me into his counsel and, without going here into details, without seeing a chance for doing much while Mr. Shaw lives, which cannot be very long, I see there is a grand opportunity coming, and I think that none of the provisions he has made will hinder the right development of the Mississippian Kew, which will be "Kew in a corner." And if he follows my advice and mends some matters, there will be a grand foundation laid. . . .

Gray's advice must have been to turn the garden "to good account for botany and horticulture." In August 1885, the *Botanical Gazette*[1] announced:

Professor William Trelease, of the University of Wisconsin, will, in September, take charge of the new school of botany, founded by Mr. Shaw, in connection with Washington University at St. Louis. A laboratory is to be equipped at once, and we understand an assistant is also to be appointed. It is probable that the laboratory will, before long, be removed to the splendid gardens which have made Mr. Shaw and the city of St. Louis so well known. These magnificent gardens, together with the extensive arboretum and greenhouses, will offer almost unrivalled facilities for students when a laboratory, library and herbarium are placed in their midst.

Trelease brought with him his excellent cryptogamic collection and this, together with Engelmann's herbarium, the Bernhardi herbarium of perhaps 20,000 species, Riehl's Missouri plants, and about 10,000 European plants from the Joad collection gotten by Gray and regarded by him "a real bonanza," gave the institution facilities for elementary, graduate, and research work.[2]

By the year 1885 there were almost a dozen other creditable American botanic laboratories. Harvard's, established about 1872 for advanced study following a European practice probably observed by Gray in 1869,[3] now included undergraduate work and no longer occupied room space solely at the Herbarium where were the greenhouses and the famous Botanic Garden. On the east in the garden were abundant materials for study, including aquatic and marsh plants and on the west the North American flora with a corner of sub-alpine plants and plats of grasses, cactus beds, and other special families. Separate quarters had been given cryptogamic botany, "a large and well equipped room in the Agassiz Museum," and about 1883 Goodale's main phanerogamic laboratory was placed in Harvard Hall with rooms "plainly furnished and abundantly supplied with instruments and material."[4] Nearly all important laboratories now, taxonomic or otherwise, had simple and compound microscope and microtome facilities. However, Harvard had still the only complete equipment "in this country that [could] pretend to compass the subject"[5] of botanic teaching and research. There were in the East two other very creditable laboratories—one at Cornell and the

---

[1] Volume X, Number 8 (August 1885), p. 327.

[2] See the *Botanical Gazette*, X, Number 12, p. 405.

[3] See R. J. Pool's "Evolution and Differentiation of Laboratory Teaching in the Botanical Sciences," *Symposia Commemorating Six Decades of Botanical Science, Ia. St. Coll. Jour. Sci.,* IX, pp. 21-28.

[4] J. C. Arthur, "Some Botanical Laboratories of the United States," *Bot. Gaz.,* X, Number 12, pp. 395-396.

[5] Articles on laboratories, appliances and courses of instruction, by J. M. Coulter, *Bot. Gaz.,* X, pp. 409-413; 417-421.

other at the University of Pennsylvania. Yale, Princeton, and other schools had some laboratory work but not on the scale of the others.

The West, moreover, was developing remarkably. John Merle Coulter had made of the department at Wabash College a strong and well respected one. Trelease had made his reputation in the department of the University of Wisconsin, having published by 1885 a number of articles on fertilization methods, on certain plant diseases, and on parasitic fungi. Although at first Trelease lacked confidence in his abilities as a systematist, several able taxonomic studies soon removed all doubt as to his superior ability as a botanist; and the Shaw School of Botany laboratory only served to enlarge his scope in morphological and systematic researches. There were also important laboratories at Illinois University, Purdue University, the University of Michigan, Michigan Agricultural College, Iowa Agricultural College, and the University of Nebraska. From the Iowa institution where pioneer laboratory research in plant pathology and other branches of botany had early begun under comparatively meager circumstances, Charles E. Bessey went to the last named, the University of Nebraska, where a large sum for those years, five thousand dollars, was appropriated for laboratory apparatus and equipment and where Bessey became state botanist to study the flora, develop forestry, and have much to do with developing ecology and physiology in North America.

In January 1885 North American botany became stirred by an article appearing in the *American Naturalist*—a part of a well developed controversy concerning botanical instruction waged between two Englishmen, Sir W. Thistleton Dyer and Reverend George Henslow. Henslow had written a book, *The Theory of Evolution of Living Things*, a book on the side of Evolution "considered as illustrative of the wisdom and beneficence of the Almighty," and this in much part had evoked an article by Gray in the January 15, 1874, issue of *The Nation*. Dyer in 1885 succeeded Sir Joseph Hooker as Director of Kew. Bessey, as botanical editor of the *American Naturalist*, wrote a comment viewing the dispute objectively but stirred the *Botanical Gazette*, representing the views of Coulter, Barnes, and Arthur, to a forceful appeal, reading:[6]

That systematic botany is a dried pod, out of which all the seeds have rattled, is a grand mistake, for our material is now but fairly brought together for the work of the monographer to begin. This is no plea for the study of systematic botany as opposed to the structural and physiological, as the writer's own laboratory will abundantly testify. But it is meant to call attention to the fact that the pendulum has swung farther away from the old side than it can stay, and that a study of botany must include the systematic phase. Whether a class should begin with Gray's text-books and manual, and then follow Bessey, or begin with protoplasm and run the whole gamut of tissues and tissue systems, and then study classification, is for the individual teacher to decide, and is as often a question of convenience as anything else.

Basically the controversy was—where was the emphasis in botanical learning to be? On the systematic side, seeking only to gather and organize the plant life of continents according to established practice? Or was the botanist to strike out to new and unexplored fields—the region of the *plant* and study it not only externally but also its internal composition and potentials. Bessey was an American leader who answered the question on the latter score. In every sense favored with ability and vision for original research, he had gone into new fields and, employing new laboratory techniques, enlarged the scope of botanical investigation and increased the usefulness of the botanist and horticulturist to the now thriving and growing West, and Middle West. Farmers' institutes; agricultural education; plant disease study, considering also harmful and useful insects; forestry; medical botany; landscape gardening; and many other research phases issued from his work. His pupil, Joseph Charles Arthur, carried on pathological work in the East. Liberty Hyde Bailey at Cornell, responding to Beal's, Gray's, and Sargent's influence, would do much to develop biogenetics in America. As would also national and state experiment stations, the real stimulus of work of all horticulturists and agriculturists at this time. In 1881, while in Europe and England, Gray had visited the Vilmorins, "dear friends of thirty years," who when first introduced to him in 1851 were studying strawberry varieties. On his 1881 journey, Gray, furthermore, visited Darwin, Backhouse, and others, always taking great interest in horticultural and agricultural work. It is true that most study in these lines was in the garden and field, and done from practical and economic motives. The experimental laboratory had not been elevated to a real place of ascendancy. There had been a few great leaders, scholars, who, like Darwin, stressed arriving at an understanding of biological laws and truths. They were by no means preponderant. Plants were not widely investigated as objects of study revealing life processes and part and parcel of a great scheme of living creation. Plants were units, of scholarly interest for themselves and of service to man and the earth. Plant disease study, for example, was only beginning to become scientific in the sense it became a decade or so later. Still, foundations of scientific interest certainly had had real beginnings in many quarters; and Gray, Beal, Bessey, Coulter, and others, who were not as yet products of European laboratory study but were keeping abreast of literature issuing from laboratory, garden, field, and orchard investigations, must have realized there would be a large American importation of foreign investigation methods and techniques. Certainly American students who went to Europe and England to study returned confident of this belief. Certainly the great forward looking American agriculturists and horticulturists, most of whom were in the membership of the Society for the Promotion of Agricultural Science, realized this. Certainly such horticulturists with vision as Liberty Hyde Bailey, Emmett Stull Goff, Charles Sprague Sargent, and others, who braved conventions, prejudice, oppositions in many directions, were confident a new and enlarged plant science study would be shaped. For evolutionary belief had taken root. Systematics—the identification of plants and their classification—would be basic in all study which would go forward to investigate the plants in all physical and biological relationships, climate, soil, temperature, light, et cetera, as also their chemical and physical constituents and their relations in the plant itself. The functional study of plants, the growing and producing plant and its behavior, in health and disease, and under varied conditions, the possibilities of improving and ameliorating plants, the introduction of foreign varieties, and much else making for an enlarged science of plants, would, with the years, become highly important.

Julius Sachs's *Lehrbuch der Botanik*, Bessey said, "marked an epoch in botany in America." His *Botany for High Schools and Colleges*, an American adaptation of an English version of the work of Sachs, published in 1880, gave Bessey a reputation far beyond the confines of the West. Gray knew his pupil's ability. Gray recognized that no more was botany to be, using Bessey's own words, "a great out-of-doors laboratory to be diligently studied from border to border." Botany was also to become a science in the pure sense—carrying on the work of Torrey, Gray, Vasey, Engelmann, Sullivant, Tuckerman, and others—but establishing therewith great indoor laboratories to study the plants themselves along with books about plants. Acknowledging indebtedness to Asa Gray, Bessey urged the study and teaching of anatomy and physiology —the study of protoplasm; the plant cell, its wall, formation, and product; the tissues and tissue systems; the intercellular spaces and secretion reservoirs; chemical constituents and processes; relations to temperature, light, etc. Gray reviewed the work and instantly recognized its significance, saying:[7]

It speaks well for the progress of science in the United States when a professor in a college in so new a State as Iowa, situated mid-way between the Mississippi

---

[7] *The American Journal of Science and Arts*, XX (3rd ser.), p. 337.

and the Missouri, can produce so creditable a book as this. The work concerns itself throughout with what the Germans call "Scientific Botany,"—largely with vegetable anatomy and development, and with particular attention to the Lower Cryptogamia. The plan in general is that of Sachs' *Lehrbuch*. . . . Professor Bessey's volume is a timely gift to American students of a good manual of vegetable anatomy and of the structure and classification of the lower cryptogamia, which was very much needed. Here at least is a commendable beginning.

The work of Rothrock was typically illustrative. While greatly interested in medical botany, not only could he write on *Eriodictyon glutinosum*, Benth., as illustrating evolution,[8] but he could effectively determine a "List of, and notes upon, the lichens collected by Dr. T. H. Bean in Alaska and the adjacent region in 1880,"[9] write for *Forest Leaves*[10] on "The American, or White Elm," or publish "Vacation Cruising in Chesapeake and Delaware Bays."[11] On February 8, 1885, he wrote Gray: "I propose to put my classes very carefully through your whole 3 volume series of Botany as I have two years with four hours a week to do it in. Of course along with recitation there will be much laboratory work. It is simply astonishing to me what an interest is being manifested in Botany in Philadelphia at last."

Not all textbooks were favorably received. W. A. Kellerman's[12] *The Elements of Botany*, while praised on some grounds, was criticized for attempting to cover too many branches of the science in too little space. While other works, nontextual in nature, were more cordially received. Francis Wolle's *Review of Desmids of the United States and List of American Pediastrums*, published at Bethlehem, Pennsylvania, in 1884, received high praise. Warren Upham's *Catalogue of the Flora of Minnesota* of the same year was regarded "in many respects a model local flora," enumerating 1,650 species and varieties. Lucien M. Underwood's *Descriptive Catalogue of North American Hepaticae North of Mexico*, W. G. Farlow's "Notes on the Cryptogamic Flora of the White Mountains" in *Appalachia* of 1884 and J. B. Ellis's *North American Fungi* were reviewed as among many welcome additions to North American plant literature. Farlow's enumerations or notes on divisions such as Peronosporeae and Ustilagineae were valuable. L. H. Bailey's *Catalogue of North American Carices*, listing 293 species and 84 varieties, Scribner's revision of North American Melicae published by the Philadelphia Academy, and George Vasey's *Descriptive Catalogue of the Grasses of the United States*, published in 1885, were among significant taxonomic

---

8 *Bot. Gaz.*, VIII, Number 3 (March 1883), p. 184.
9 *Proc. U.S. Nat. Mus.*, VII, 1884.   10 V (Philadelphia, 1884), pp. 104-105.
11 Philadelphia (1884), p. 262.
12 Then at Kansas Agricultural College. See *Bot. Gaz.*, IX, Number 2 (February 1884), p. 35.

works of the period. Popular works were not neglected, as, for example, Henry Baldwin's *Orchids of New England*. There were many works of lesser importance. Farlow, Trelease, Britton, Coulter, Hill, Bebb, Barnes, Ridgway, Meehan, Rau, William Boott, Lester Ward, Joseph N. Rose, F. Lamson Scribner, E. Lewis Sturtevant, G. D. Swezey, F. S. Earle, David F. Day, J. Schneck, Thomas Morong, Joseph F. James, A. B. Seymour, Gattinger, Burrill, Chapman, and others, were busy. Early promise as systematists, notably by Coulter and Britton, was being shown.

Plant life history studies were getting under way in America. Students, such as Douglas Houghton Campbell and others, were to start study in European laboratories and on their return to American shores commence studies in phylogeny. The study of gross and minute structures of plants and their relationships, and cytology were to receive immense impetus. The University of Pennsylvania laboratory established by Rothrock to study medical botany would enlarge in scope, typically illustrative of work in many other botanical research centers, and studies of botanists such as John Muirhead MacFarlane would issue therefrom. Bessey, Coulter, Trelease, Bailey, and others would lead the way. But the inspiration would be the remarkable work being produced in laboratories of Strasburger and the Europeans. North America would realize that its primary task of learning the American flora was far advanced. Now they must learn of other lands and of the plants themselves.

Exploration would not cease. Nor publications. Greene's "Botany of the Coronados Islands," John B. Leiberg's "Notes on the Flora of Western Dakota and Eastern Montana Adjacent to the Northern Pacific Railroad," Vasey, Watson, and Gray's "List of Plants Collected in 1882-3, by Lieut. A. W. Greely" near Fort Conger, Grinnell Land, Ellsworth Jerome Hill's "The Menoninee [River] Iron Region and Its Flora,"[13] published in 1885 in the *Gazette*, Lieut. P. H. Ray's *Report of the International Polar Expedition to Point Barrow, Alaska*,[14] and Gray's "Notes upon the Plants Collected on the Commander Islands [Bering and Copper islands] by Leonhard Stejneger" published in 1885 and as a *Proceeding of the United States National Museum*[15] were not among the least important.

Lieut. V. Havard's "Report on the Flora of Western and Southern

[13] The Menominee River is on the boundary line for a distance between Wisconsin and Michigan, and flows into Green Bay. A decidedly Lake Superior aspect to its flora was found, and the area between Chicago and Michigan City was found "remarkable for the variety and number of species."

[14] 1885, botany by Gray, Farlow, etc. Washington: Government Printing Office, 1885.

[15] VII (1884), pp. 527-529.

Texas"[16] and Gerald McCarthy's "A Botanical Tramp in North Carolina"[17] also added to exploration knowledge of the time; as did Gray's own account[18] of his and John Ball's excursion to Roan Mountain in 1884, at which time Gray conducted a small group of botanists and ladies through the mountains of Virginia and Carolina—to Luray Cavern between the Blue Ridge and Alleghenies, a cavern which Gray considered the finest in the world; then to Natural Bridge, Virginia, "grander than I had remembered," Gray said; then to Roan Mountain, "the base and sides richly wooded with large deciduous forest trees in unusual variety even for this country, the ample grassy top . ... fringed with dark firs and spruces, and the open part adorned with thousands of clumps of *Rhododendron catawbiense*, which when there last before, late in June, we saw all loaded with blossoms, while the sides were glorious with three species of Azalea"; then along the upper Kanawha River to a mountain-top lower than Roan; and then returned north. Gray had attended the meeting at Montreal of the British Association for the Advancement of Science and read his paper, "Characteristics of the North American Flora."[19] There an invitation had been given the botanists to accompany the excursion. As important as these matters and their published determinations were, none were, however, more vital to North American botany than Gray's and Watson's individual *Contributions to American Botany*, communicated during these years 1884 and 1885 to the American Academy of Arts and Sciences.

On January 14, 1885, Sereno Watson communicated his *Contribution*: "(1) A History and Revision of the Roses of North America" based on accumulations at the great North American herbaria and from such persons as Engelmann, Redfield, Smith, Gattinger, Mohr, Shriver, Upham, and H. G. Jesup of Hanover, New Hampshire;[20] and (2) further "Descriptions of Some New Species of Plants, Chiefly from Our Western Territories." Included in the latter were plant determinations from collections of Greene in California, Lester F. Ward on the Aquarius Plateau, Utah, Dr. Havard near San Antonio, Texas, Cleveland on the mesa near San Diego, and many others. Mountain exploration, begun more than two decades before, was obviously increasing. Plants were determined from collections by Mrs. P. G. Barrett at Lost Lake on Mount Hood; by Thomas Howell in the Siskiyou Mountains; by Pringle

---

[16] *Proc. U.S. Nat. Mus.* (September 23-30, 1885), p. 85.

[17] Including the Tar River country of North Carolina. See *Bot. Gaz.*, X, Number 11, pp. 384 ff.

[18] See *Letters of Asa Gray, ibid.*, pp. 757-758.

[19] See page 13 of this book.

[20] The historical portions formed an account of our roses from Gosnold's voyage in 1602 to Palmer's discovery in 1881 of *R. Mexicana*.

in the Santa Rita Mountains; by Canby on Mount Helena in Montana; by Orcutt in the Cantillas Mountain, Lower California; by Suksdorf on Mount Adams; by the Lemmons in the San Francisco Mountains near Cliff-Dwellers Ravine and in a ravine at Bill Williams Mountain, Arizona; by Brandegee near the Simcoe Mountains, Washington Territory on mesas bordering Satas Creek; and by other collectors at many localities. A determination of *Tetracoccus Engelmanni* collected at St. Thomas, Lower California, by Parry in February 1883, and probably Engelmann's last botanical work was included in manuscript form in the *Contribution*. After concluding this publication, Sereno Watson went to Guatemala, Central America, to collect, being there from February 25th to April 20th of 1885.

Gray's latest *Contributions to North American Botany* were spread through the year 1884. On May 14 he communicated his "Revision of the North American Species of the Genus Oxytropis," which he had begun some years previous and more than partially completed on his recent visit to Kew in England. The following June he sent "Notes on Some North American Species of Saxifraga." And in October and December his large *Contribution* was communicated, containing "(1) A Revision of Some Borragineous Genera"; "(2) Notes on Some American Species of Utricularia" inspired in large part by Gray's possession[21] of original drawings and papers of Major John LeConte used by him when writing early in the century concerning Utricularia, Viola, and Gratiola; "(3) New Genera of Arizona, California, and their Mexican Borders, and two additional Species of Asclepiadaceae," in which half a dozen new genera were described; one, Veatchia,[22] dedicated to Dr. J. A. Veatch, the discoverer and excepting one army officer the only explorer of Cedros Island, California; another, Lyonothamnus,[23] to William S. Lyon, the first thorough explorer on Santa Catalina Island; another, Pringleophytum,[24] to Pringle emanating from his "very arduous and hazardous excursions made during this year from Arizona into the northwestern borders of Sonora, where no botanist had hitherto penetrated"; and another, Rothrockia, to Rothrock who on February 8, 1885, wrote and thanked Gray for linking his name "to that very conspicuous genus of Southwestern plants," one of Asclepiadaceae; and "(4) Gamopetalae Miscellaneae," among which was a determination of a new species of Schweinitzia (*S. Reynoldsiae*) found and collected by Miss Mary Reynolds near St. Augustine and on the Indian River in Florida.

[21] By loan from I. C. Martindale.   [22] Anacardiacearum.
[23] Rosacearum?   [24] Acanthac-Justicearum.

Early in 1885 Gray decided to go by the Southern Pacific Railway route through southern Arizona to Southern California.[25] He went by way of St. Louis where he conferred with "old Shaw, and heard him read his rearranged will, which is satisfactory," Gray wrote Hooker, "as it will allow his trustees, and the corporation of Washington University there, to turn his bequests to good account for botany; will be an endowment quite large enough for the purpose." From there he and Mrs. Gray went to Mobile and New Orleans to attend the Exposition there and Dr. Farlow joined them. He "brought, to our surprise," wrote Gray, "passes for us to go by the Mexican Central Road to the city of Mexico and back to El Paso (the junction with the road to California), and we decided to undertake it." They went to San Antonio, Texas, and on to El Paso where, crossing to the Mexican side of the Rio Grande del Norte, they took a Pullman sleeper for three nights and two days riding through Chihuahua, Zacatecas, Agua Caliente, and León to Mexico City where they stayed at the Hotel Iturbide.

While at Mexico City, Farlow and Gray went out to Chapultepec where they were much impressed by the Valley of Mexico, the surrounding mountains; and the Cypress, *Schinus molle*, Yucca, and other trees interested them much. "Opuntias of two or three arborescent species, some huge, and other cacti not a few," noted Gray. They looked forward to the time when they should compare Arizona with the plateau of northern Mexico. A bad cough from which Gray was suffering took them on orders of a physician away from Mexico City, however. They went east to Orizaba to get away from the dry and dusty Valley of Mexico to the more moist areas near the Gulf of Mexico. Immediately after climbing the high ascent out of the valley they noticed on the descent numerous flowers—"two species of Baccharis, Eupatoria, *Erigeron mucranatum* . . . Loeseliae species, Arbutus, (Xalapensis) in bud . . ." and others. At Orizaba, Gray wrote Hooker:

Very comfortable hotel here. Botteri left an élève here who knows something of botany, but lives out of reach on a hacienda. We found a garden combined with a small coffee plantation. The proprietor thereof, speaking a little French, has filled his ground with a lot of things that will stand here. It is just *in medias res*, two hours below Tierra Frias, two above (or at Cordoba, only seventeen miles, but 2,000 feet lower) true tropical. Papaya fruits here, also *Persea gratissima*, etc. And the oranges are delicious. I have passed the whole morning with the garden man, while Farlow went up a small steep mountain, and brought back various things. We shall drive this afternoon to the Cascade of Rincon Grande. . . . On the way here had views of Popocatapetl and the more beautiful and diversified Iztaccihuatl

[25] Gray's letters of his journey are fully set forth in his published *Letters of Asa Gray, op. cit.*, pp. 761-773.

from the sides, and wound round the base of Mt. Orizaba. A true Mexican town this. . . .

Late that afternoon, he added:

We went, but saw the falls (very picturesque) in a wet mist, and for botany got a lot of subtropical Mexican plants, the like of which I never saw growing before: among Compositae, Lagascea (large heads), Tree Vernonias of the Scorpioides set, Calea, Andromachia.

From Cordoba, Gray and Farlow drove to the Cascade of Barrio Nuevo, "almost as beautiful as the other," Gray thought, and spent the morning "in clambering and collecting. In the grounds on the way are planted trees of a Bombacea, in flower before the leaf, probably Pachira," wrote Gray to Hooker.

Southern California was Gray's destination and so by "a long circumbendibus" route by railroad they went in the spring to the southernmost town of California—San Diego. Declining an invitation to go to Lower Southern California and finding the coast too cool and damp, they went on to San Bernardino where two nights were spent with Parish and his wife, and from there to Los Angeles and by steamer from San Pedro to Santa Barbara, "the very paradise of California in the eyes of its inhabitants, and indeed of most others," Gray commented. At their "fine watering-place kind of hotel," they were shown to their rooms "all alight and embowered in roses, in variety and superbness such as you never saw the beat of, not to speak of Bougainvilleas, Tacsonias, and passion-flowers, Cape-bulbs in variety, etc., etc., and a full assortment of the wild flowers of the season. Mrs. Gray was fairly taken off her feet," wrote Gray. "During the ten or eleven days we stayed, there were few in which we were not taken on drives, the most pleasant and various. The views, even from our windows, of sea and mountain and green hills (for California is now verdant, except where Eschscholtzia and Bahias and Layia, etc., and Lupines turn it golden or blue) were just enchanting." They visited some near by ranches, driving up canyons of oaks and plane trees with an occasional *Acer macrophyllum* or Alder, and then, hiring a wagon, went over a pass in the Santa Inez Mountains to the Coast near Ventura. One ranch near Santa Barbara "flanked on the windward sides by eucalyptus groves, apricots, almonds, peach-trees, etc., by the dozens of acres" and where the owner sought to enlarge his produce of olives for olive oil, impressed Gray.

From there they proceeded up the broad and long Santa Clara Valley to Newhall where they took the Southern Pacific Railroad to San Francisco, stopping at the Lick House. At San Francisco, they went across to San Rafael and next day took a drive up behind Mount Tamalpais to the canyon reservoir, seeing the "huge Madroñ[o] (*Arbutus Menziesii*), like one of those great and wide-spreading oaks you used to admire," Gray told Hooker. Next day spent at Monterey, they left Farlow to algologize at Santa Cruz and were soon at Rancho Chico to visit the Bidwells where they enjoyed almost a week botanizing and eating cherries and strawberries then in season. The Sir Joseph Hooker Oak was still there and the Bidwells as cordial as ever. At Lathrop, California, however, they took the train for the East, going by way of "the wonderful T[e]hachapi Pass" in the early morning after a night's ride up the San Joaquin Valley. At Mohave they boarded the Atlantic and Pacific Railroad (now Atchison, Topeka, and Santa Fe Railroad) to go over the "sandy desert to the Great Colorado" River and on to Peach Springs, Arizona, where Gray hired a "buckboard wagon" and in the morning drove twenty-two miles down a descent of 4,000 feet to a point west of the Grand Canyon of the Colorado but still along the canyon then famous because of Major Powell's comparatively recent explorations. "The cañon trip well repaid the journey and its rough accessories," wrote Gray to Hooker.

After this, the high point of the trip had been reached. They continued on to Flagstaff, hoping for an opportunity to see the ancient cliff dwelling evidences near there but, being disappointed, journeyed to Las Vegas, New Mexico, where laying over one train they visited the Hot Springs. Gray watched with interest as they then went eastward along the Arkansas River—the same route taken when going westward with Hooker several years before to Colorado and California. Finally they reached St. Louis where a few days with George Engelmann, Jr., were planned and then on to Cambridge.

As they passed in sight of the foothills of the Rocky Mountains, and at times in sight of the Rockies themselves, Gray may have remembered John Merle Coulter's *Manual of the Botany (Phaenogamia and Pteridophyta) of the Rocky Mountains Region from New Mexico to the British Boundary*,[26] published early that year in New York City. For the first time the great and increasingly populated district between Dakota and Montana on the north and New Mexico on the south was brought into the compass of botanical learning. With only one predecessor manual, Porter's and Coulter's *Synopsis of the Flora of Colorado* some ten years earlier, the area west of about the one hundredth meridian and including Colorado, Wyoming, Montana, western Dakota, western Nebraska,

---

[26] See Asa Gray's review in *The American Journal of Science and Arts*, XXXI (3rd ser., 1886), p. 76, where the work is characterized as "very well done," especially in view of the fact that the Rocky Mountain flora was then still undisturbed and practically in its natural state.

and western Kansas, also parts of the Indian Territory, northwestern Texas, northern New Mexico, northern Arizona, eastern Utah, and eastern Idaho, now had an up-to-date manual which served them as Gray's *Manual* had served the northeastern states.[27] Gray's and Coulter's course had not of recent years been smooth.

On September 20, 1883, Coulter had written Gray:

> The September *Gazette* was either peculiarly unfortunate or it was "the last straw etc." I appreciate your castigations most sincerely & I hope that I profit by them, but the only thing that struck me unpleasantly this time was the idea of warning Foreign Botanists in the *Am[erican] Jour[nal] [of] Sci[ence]* against the *Gazette*, and so killing by a sentence or two the painful growth of years, when the *Am[erican] Nat[uralist]* & *Torr[ey] Bull[etin]* go scot free although full of Marcus E. Jones' rubbish of new species & much other stuff which I have already had the honor of rejecting. Such a course would have been rather severe & hardly justified by the offense. But it is idle to talk of what you did not do, when I think you did the best thing & I will gladly print what you have written. And now for the questions seriatum—
>
> I. Dr. Torrey's letter has struck some persons differently as I have already rec[eive]d some letters from good botanists expressing great interest in it, & chiefly because it was a youthful letter. A youthful letter, so far from detracting from the reputation of a man of such mark as Dr. Torrey only encourages other young fellows & we all like to catch the masters unawares occasionally. But really in accepting this from J[oseph] F. James[28] I took the lesser of two evils. For various reasons I don't want to offend this young man, & I had the choice between Dr. Torrey's letter & an elaborate(?) study of the genus Asclepias. . . .
>
> As for the characterization of the Minneapolis paper,[29] my initials are attached to that article, & the parts relating to Dr. Sturtevant I must confess were too hastily written. . . .
>
> As for James on Compositae, the *Gazette* did not mean to express an opinion, but simply to state the drift of his paper, which, of course, we all knew was as trite as the catechism. . . .
>
> And now to answer your questions in reference to my own paper. Of course, what you refer to was put in by one of the co-editors, who may have stated it too strongly. I made no conclusions with regard to Compositae, but simply told what I saw in Dandelion, nor did I commit myself to the idea of the Dandelion's ovule being produced from the midrib of a carpellary leaf, but said it looked like it. . . .

Coulter, however, was the sort of man who profited from Gray's severity. Once again in 1884, while Coulter was absent, a blunder or two

[27] See Coulter's "Introduction," January 1, 1885, and publisher's note in *Text-book of Western Botany Consisting of Coulter's Manual of the Rocky Mountains, to which is prefixed Gray's Lessons in Botany.* New York, Cincinnati, Chicago: American Book Co.

[28] Of Cincinnati, Ohio, author in *Bot. Gaz.*, VII, Number 4, p. 41, of "Depauperate Rudbeckia," and other works.

[29] Coulter read a paper on the development of the dandelion flower before the Minneapolis meeting Amer. Asso. Adv. Sci. See *Amer. Nat.*, XVII (1883), pp. 1211-1217.

crept into the pages of the *Gazette*. Coulter wrote and explained them to Gray. And soon all past difficulties were completely eradicated.

Soon after his return from the West in 1885,[30] Gray received a letter from Coulter reading:

> I have just declined the chair of botany at the State University, for the good people here back me up so handsomely & have more money than the Univ[ersity] will ever have, so that I could only stay. But the point is that I am so situated now as to be able to pay my share in any collections that are being made. I have subscribed for everything of our regular collectors, but if you have any plan in [which] you want me to take a share, only provided it will bring me plants & work, I can furnish a moderate supply of "funds."

He asked Gray if Britton had so preempted Cyperus that it would be "unbecoming" of Coulter and Barnes "to work at it. I have lots of material & more desire," Coulter wrote, "but Britton has 'warned us off.'" Soon after the publishing of Coulter's *Manual of the Botany of the Rocky Mountain Region*, it became known that a second edition would soon be required. On May 13, 1886, Coulter wrote Gray: "I am not clear yet as to our southwestern boundary. Would it be practicable to make the 2[d] ed[ition] of the Rocky M[oun]t[ain] Manual include New Mexico & W[estern] Texas (as it should), and so take E[astern] Texas into this one & make a clean sweep of the country?" Already he had Scribner at work on the grasses. L. H. Bailey, Jr., would again do the Carices, and Bebb the genus Salix as he had done in almost every important publication since the publication of the *Botany of California*. Coulter had much state work incident to the state surveys. But with two competent assistants, he was anxious for more fields to conquer. On April 24 he had told Gray that his love for Gray personally would at once check any plan of his which might interfere with any of Gray's plans but added:

> I am young and vigorous, eager to work, & if I have what you kindly call "a knack for putting things in shape," here I am at your service, to do for you whatever I can. . . .
>
> You speak of "breaking out paths that others can walk easily & profitably in," but that is both the fate & glory of such pioneers & masters as you represent, & that is really what you have wanted to do for us. There will never be any lack of appreciation.of the work you have done for us, & no botanist would fail to subscribe to your statement about "younger & not *better* men." I am sure I only aspire to become one of your followers in the paths you have broken so easy for us.
>
> Your plan for endowing the Harvard Herbarium is just the thing, & should not be interfered with, & I am ready to help you in it all I can. . . . The Herbarium is for us all.

[30] In August 1885, Coulter read before the American Association for the Advancement of Science a paper "On the Appearance of the Relation of Ovary and Perianth in the Development of Dicotyledons."

It occurs to me that the proposed Manual should include Dr. Chapman's "baili-wick," for it needs a manual sadly, & it is out of the question for Dr. Chapman to undertake it. He told me when he put out his Supplement that he did it that way because he saw no chance of his rewriting the whole book. He is very old, over 90, & as the proposed Manual could not appear for 2 or 3 years, he would still occupy the field for that length of time. . . .

In addition to work on a "Revision of North American Hypericaceae," Coulter planned bringing out another edition of Gray's *Manual of the Botany of the Northern United States*, to include and be a second edition of his Rocky Mountain *Manual*. Evidently, at first, Gray planned the Mississippi Valley region and part of Coulter's district for his *Manual*. But as the doctor aged and other work increased, Coulter assumed the task to be finished with Sereno Watson after Gray's death along lines smaller than planned.

In August 1887, the *Botanical Gazette*[31] editorialized:

Since the consolidation of the national surveys, the government has done nothing for botanical exploration. Millions have been spent in increasing our knowledge of the other riches of our domain, but the plants have been left to private enterprise. It has been claimed that the botanical exploration of this country has been well-nigh completed, but that can only be said by those ignorant of the facts. . . . Hundreds of new species are being described yearly in this country. . . . Money is appropriated for [geology, anthropology and] economic botany, but our plea is for the botany of North America. There are many localities not reached by collectors, or reached at such expense of time and labor that but scanty collections are made.

It was urged that Congress should make appropriations for explora-tion. Not merely to support remarkable investigations carried on by private institutions but to reinstate scientific exploration to its former place of specific government recognition. Professional collectors were busy. Canada, Mexico, and even South America were being further ex-plored. But the *Gazette* called for further exploration in the United States—in taxonomic and other branches.

In 1888 Dr. Vasey was to initiate work preliminary to the establish-ment of grass experiment stations in the arid and semi-arid western re-gions, particularly the Grass Experiment Station at Garden City, Kan-sas. Vasey, the North American authority on agricultural grasses, sought to find grasses and forage plants adapted to regions where irrigation was believed not practicable west of the one hundredth meridian. Money appropriated to this purpose was in part used to revive scientific collec-tions by the government and included as formerly, during the period prior to the abandonment of natural history collections on government surveys, the employment of professional collectors.

[31] XII, Number 8, p. 197.

In the matter of experiment stations, however, the states may be said to have been soon in advance of the work of the Federal Government, at least, in the matters of scientific fact finding and experimentation. In 1882 the board of trustees of the New York State Agricultural Experiment Station had constituted Dr. Edward Lewis Sturtevant its director and his studies and experiments with farm vegetables, notably corn, peas, and melons, and other farm products, were of immense significance. Likewise, his studies in the history of vegetables. On January 6, 1884, he wrote Bessey:

I know that my views as to agricultural botany are considered by many of the systematic botanists as ultra radicle. So long however as I consider my position as a correct one, I shall continue my endeavors to get a hearing before the scientific public.

I only wish I could show to you the success in classifying for the purpose of identification some of our garden plants—notably the pea. By applying the doctrine of motion(?) we seem to have a key which explains the divergences that occur between varieties, and we find a very great permanency to each type or stage of evolution. Indeed there seems a fixity to all our garden varieties except as disturbances come from hybridization.

Again, on February 22, 1886, he wrote to Bessey:

I wish I could interest botanists in cultivated vegetables. This is a new field which is deserving of explorers, and attractive, yet difficult. In preparing myself for the work I have found it necessary to have for comparison and reference, the drawings which have been made a long time ago, and have therefore had to collect a special library of books not readily obtainable for this purpose.

And on January 12 of the same year he had told Bessey:

The tendency of the public is toward forcing a station into doing a class of work, . . . if sufficient for the vanity of the public, yet is of no real merit, and is at the furtherest pole from experimentation. This station will undoubtedly at some time in its history (I hope not immediately) be forced to abandon, or rather the attempt will be made to force it to abandon all its truly experimental work, and expend its energies on a fertilizer control, together with that class of experimenting known under the name of the plat system. When such a period arrives, it is only by calling attention to what has been written about us in the scientific periodicals that we can hope to avert the danger. I speak not only for this Station, but also for all Stations that may be established. The plea must ever be for work more scientific in its character, and every Station must abandon the crudities forced upon it by the popular prejudice just so rapidly as danger to its maintenance will admit. I have been, at times, ashamed to read notices of German work, some of it so poor as hardly to admit of criticism, and yet all the work of our Stations absolutely ignored. The reception that I have received personally on the part of scientists has been such that no soreness exists, and hence my remarks apply only to the general condition of experiment stations, their needs and dangers.

The state experiment stations did have years of struggle. Although as early as 1883 (when at the Minneapolis meeting the Botanical Club of the American Association for the Advancement of Science was formed), Dr. Sturtevant's work in agricultural botany was accorded special recognition, the work was interpreted as important taxonomically rather than experimentally. Coulter in the *Gazette* stood in awe of the "new world of labor" Sturtevant had opened:

He is bringing to bear upon plants every possible influence that can be made to affect their growth, and really he is seeing incipient species springing up under his own manipulation and can recognize the forces that are effecting the change. Many other experimenters in agriculture are seeing the same results but very few have the acuteness to discuss the causes. The work is but begun, but we look for it to become a source of unlimited material not only for the agriculturist but for the professional botanist. Already has Dr. Sturtevant intimated certain results which will completely overturn and tear up by the roots some of our preconceived notions, and one of these days we may look for something startling.

Bessey with his keen interest in horticulture, which led to an interest in forestry, separating it from horticulture, took an interest in Sturtevant's work. On September 30, 1886, Sturtevant elaborated to Bessey:

Last winter I received from the Smithsonian Institution a few Mexican peppers for determination, grown in Chihuahua. In shape of fruit, many of these peppers can be identified with the peppers described by preLinnean writers, and they seem to furnish a very good illustration of the permanency of varieties under culture. Another curious reflection is how unchanged many of our garden plants are even after centuries of culture, records extending through centuries not indicating any essential change in type, the scorzonera, salsify, alexanders, cabbage, lettuce, etc., showing that some varieties have remained constant notwithstanding influences induced by climate and selection during long periods. I think this fact has received less attention than it deserves. . . . If botanists would only recognize the importance of popularizing their study by applying it to some use, if only they would recognize that botanic gardens might be of distinct practical benefit, and that the service of botanic gardens would necessitate traveling collectors in distant fields under a full public approbation, I think that more work of an agricultural character would be done in our many botanical laboratories.

How prophetic of North American botany's experimental period instituted in full force during the next two decades was this letter written in 1886! How prophetic this was of increasing plant life-history studies extended in academic institutions as well as nonacademic. How amazingly the contents of this letter seem to have anticipated, though not with exactness, the great experimental work of Hugo DeVries which placed the doctrine of the origin of species on an experimental basis. The work of the great Europeans doubtless influenced Sturtevant. Indeed, some believe Sturtevant in his work came near to arriving at Mendel's epoch-making conclusions. Sturtevant then and there must have fully apprehended the possibilities of work such as Luther Burbank brought to the public's notice. Hybridization studies, selection studies, and even more important, the development of the scholarly subjects, plant genetics and plant pathology, were to receive much attention in the national and state experiment stations.

Before establishment of the New York Station, other states had commenced agricultural experiment stations but at first their work seems to have been mostly concerned with chemical research. New York developed its station with chemical, botanical, horticultural, live stock, and crop departments. Here Joseph Charles Arthur pursued his early important work in rust physiology, in certain plant diseases, and in formulating a method to estimate loss from oat smut, the first time a plant disease was described by statistics. The influence of Bessey, Arthur's teacher, was evident in Arthur's work, in fact, Arthur consulted Bessey on all his studies. In 1885 Arthur was appointed botanist of the Minnesota Geological and Natural History Survey and in 1887 he went to Purdue University where a new experiment station was contemplated. At the New York station also began the important horticultural investigations of Emmett Stull Goff, studies issuing from the garden but also including classification of cultivated plants, studies of root systems, researches in most effective methods of spraying to prevent, control, or cure diseases. The investigational work of the agricultural colleges, together with that of the experiment stations, as these began to spread over the nation, and, beginning about 1886, over Canada, was highly significant to American botanical progress even though much of it was denominated "agricultural botany." In the east there developed the work of the Massachusetts station and the Connecticut station; in the central eastern states the work in New Jersey, Pennsylvania, and in New York, also at Cornell University; in the south, work at North Carolina, Tennessee, and Alabama; and in the middle west, work in Michigan, Wisconsin, Iowa, Kansas, and Nebraska, not to exclude the work of the far western station in California. A long list of talents and men of superior ability—investigators very important to agricultural, horticultural, and incidentally, botanical, scientific development—could be named. Such, however, in justice to them and their work, must await the writing of another book. The development of the Illinois station alone is a story of vast pertinence. So is that of Ohio. Agricultural chemistry, it is true, laid foundations for great laboratory progress. This, nevertheless, could not fill the office of studies in the garden, in the orchard, in the farm field, and in the wild places of nature *where plants were grown.* Combined work of laboratory and field, studying the cultivated plant as thoroughly as the plant of nature, both in health and disease, had to be coordinated. Progress was slow. At first, many college experiment stations labored with field and feeding experiments but not many years went by before the state-supported and land-grant college stations were performing valuable service in developing agricultural science carrying with it valuable botanic investigation. As important as was the early work of the Federal Government in experimentation, it was not in type or consequence comparable to that of the state stations. The Federal Government's work in grass and tree investigations became largely directory, although the division of forestry and other divisions did some world recognized experimentation. When Bessey developed in his classes and in the field the notable Nebraska Survey, George Vasey wrote in 1893: "I admire the way you Nebraskans are taking hold of Botany. You are setting an example, which if followed by other States, will soon give us a complete botanical Survey of the Country."

Bessey's editorship and knowledge of foreign language had always enabled him to read, even before English translations, new works of European authors. As much as anything, these facts, together with an uncanny vision and instinct for practice, accounted for his early recognition of the values obtainable in applying the "biological" method to study of floras and plant distribution—the science of ecology, as it became later known—a study of extrinsic and intrinsic factors in plant growth and development from the standpoint of circumscribed areas. Even as Gray had inspired his pupils to a realization of a vision foreseen for the future, for example, in plant physiology, Bessey inspired his students to go forward with the new method of ecological investigation. Conway MacMillan, Roscoe Pound, and Frederic Edward Clements are among those who definitely were motivated by Bessey's instruction. Furthermore, botanists such as Frederick Vernon Coville, William Trelease, Coulter, and others were also quick to respond. Under influence of these men and later students such as Henry Chandler Cowles, the new subject quickly expanded into a new investigation method combining laboratory and field that sent botanists again over all the areas traversed by taxonomists.

The early agricultural experiment stations gave ecological study a prominent place beside an immature plant pathology and "biogenetics," a pre-Mendelian development of the branch of science known today as plant genetics. Since, however, botanical investigation first studied disease-producing organisms, the deleterious fungi and bacteria, before study of the diseases produced by them really got under way, as late as 1887 from thirty or forty agricultural colleges or departments of colleges in the United States, the results of study were characterized as meager. In the *Report of the Commissioner of Agriculture* for the year 1887, F. Lamson Scribner, chief of a newly created section of vegetable pathology, made his first annual report, "being the second report of the section of the Botanical Division devoted to the investigation of the fungus diseases of plants." The year previous, in 1886, the commissioner reported that agricultural experiment stations had been established in nine states: Connecticut[32] and North Carolina in 1877; New Jersey, 1880; New York, Ohio, and Massachusetts, 1882; Wisconsin and Alabama, 1883; and Maine in 1885. Said the commissioner:

These are all distinctly independent institutions, with their own organizations, and supported by State appropriations or special tax. Some, however, are located at State agricultural colleges, and officered by the college professors. These stations differ greatly in their organization, facilities, and work. Some are required to control the business in commercial fertilizers. . . . In New Jersey and North Carolina at least $10,000 is expended yearly, mainly in laboratory work. . . . In several other States there are provisions made for systematic experiment work at the agricultural colleges and State universities by appropriations from college funds and the assignment of professors to this duty. In some cases the results are becoming very valuable, at least locally, while in others the efforts are feeble and uncertain.

Other States and colleges are considering the inauguration of experimental inquiry, and efforts in this direction are apparently limited only by lack of means. . . .

National legislation has been proposed to extend the work of experimental agriculture, establishing it in every State, as well as to strengthen that already in progress, and to make the results of all available to the country at large. Without interfering with the organization and management of State stations, whether at colleges or independent, Federal support may supplement existing agencies, and provide through this Department a certain degree of control to secure co-operation where needed and furnish such a medium of intercommunication and exchange as to greatly facilitate and improve the work as a whole. . . .

In 1886, at least seventeen stations in as many states having been organized, a "central experiment station, with proper accessories, for the investigation of questions affecting large areas, and such as relate to the whole country" was recommended. Congress enacted the historic Hatch Act granting $15,000 per annum to each state or territory which established agricultural colleges or departments of colleges and by 1888 it was reported that the enactment had "led, according to the latest accounts at hand, to the establishment of new stations, or the increased development

[32] The date of establishment of Connecticut's station is usually given as 1875. See A. C. True's article, "Origin and Development of Agricultural Experiment Stations in the United States," *Report of the Commissioner of Agriculture* for 1888 (Washington: Government Printing Office), pp. 541-558, where is included data concerning each station's history.

of stations previously established under State authority, in thirty-seven States and one Territory." In 1887 the object of the stations' work was characterized, "to experiment with seeds, plants, crops, fertilizers, systems of culture, etc., and to determine what is best for their respective State or Territory." However, with enactment of the Hatch Act and establishment of stations numbering in all, including branches, by 1888 not far from fifty, the scope of experimental inquiry was widened to include physiological, pathological, and the whole range of plant and animal scientific research.

It was under such circumstances that horticulture as a branch of science made rapid strides in development. On April 19, 1888, Liberty Hyde Bailey, Jr., wrote Watson: "I have accepted the new chair of horticulture at Cornell," the first college in America that recognized this as a branch distinct from all others. When Coulter had learned of Bailey's earlier acceptance of a position as an instructor of horticulture, he bitterly complained of Bailey's leaving botany and said: "You will never be heard from again." Although the United States Department of Agriculture had established Divisions of Gardens and Grounds, Seeds, Forestry, and Pomology, obviously horticulture was not regarded as important as botany. Matters such as cross-fertilization, pollination, et cetera—research and investigation subjects—belonged primarily to botany. Corn breeding, for example, was not even regarded as essentially of the province of botany, being, it is suggested, as late as 1893 "chiefly in the hands of practical farmers."[33]

Bailey had been born of strong, Puritanic, New England and southern parentage in a Michigan wilderness located near South Haven between Kalamazoo and Lake Michigan. In an age of homespun he had been raised with Pottawatamie Indians and the woods as his companions. Inspired greatly as a child by his mother and interested in natural sciences by a teacher to whom he later dedicated one of his books, he had begun the study of Latin and an *Encyclopedia of Natural History* and acquired a remarkable fund of knowledge before entering Michigan Agricultural College to study under Beal, then professor of botany and horticulture there.

Beal had learned the laboratory methods of Agassiz while at Harvard, had studied under Gray when almost the only two botanists in this country relying on botany for livelihood were Gray and Daniel Cady Eaton, and become one of the first to urge plant study from all stand-

[33] See an interesting address by Hon. H. A. Wallace, "Six Decades of Corn Improvement and the Future Outlook," delivered at the Symposium: Applied Botanical Research on Maize, Iowa State College, *Ia. St. Coll. Jour. Sci.*, *op. cit.*, pp. 347 ff.

points, the systematic to the experimental. In 1877 he had consulted Bessey as to "a college need [of] a botanical laboratory and a room for museum botanical products" and receiving Bessey's reply established an early American laboratory with compound microscopes. The college garden of Michigan Agricultural was one of the first established in the United States. Beal, being the last to die of the leaders of the experimental movement in American botany, wielded a venerable influence in shaping the science's course through its transitional years. One of the first educators to cross corn varieties to increase yield in the 1870's, an early student of the college orchard, an experimenter with buried seeds contained in bottles dug up, one every five years, and germinated, a pioneer advocate of forest preservation, a seed tester for purity and viability, one of the first to insist on advantages of plant selection, and an author of works such as *The Grasses of North America, Seed Dispersal,* and others, Beal has earned an abiding place in North American botanical annals. Surprising as it may be, however, his fame extends only incidentally to horticulture.

To Bailey, American horticulture owes by far its largest early debt. After Bailey's graduation, he was for a time a reporter on the *Morning Monitor* of Springfield, Illinois, reporting the legislature there, but, just at the time of an offer of the paper's city editorship, there arrived a letter forwarded by Beal from Gray wanting a young taxonomist to come to Cambridge to work on the famous Joad collection of European plants including Maroccan plants collected by Ball. Bailey had studied taxonomy under Beal, and become quite a student of Michigan and middle western plants. To him this opportunity was like a dream come true— to learn the flora of Europe as well as study American plants further at Harvard. He went to Cambridge, married, studied in Boston greenhouses along with his work, and arranged the collection into sets, one set of which went to the Missouri Botanical Garden and another to the National Museum. Bailey became intimately acquainted with Gray and was placed at naming up everything in the Cambridge garden and greenhouses, one of which was filled with Acacias from Australia. His return to botany and horticulture was thus complete and he went to Michigan Agricultural to teach, there, principally through his own abilities, conceiving his visions of a "new horticulture," a science that would emphasize discovery "of some of the laws of plant variation and dissemination, especially in relation to climate and latitude" as well as knowledge of how to raise vegetable crops.

On July 1, 1886, continuing relations with Harvard, he wrote Gray that he was going with a botanical party of the Geological Survey of

Minnesota to Vermilion Lake, in northern Minnesota, located about thirty miles from the Canadian boundary, a totally unexplored region. After the survey work was finished, he went on further to the terminus of Hunters Island, escorted by Ojibway Indians and on his return wrote a book, the manuscript of which is still in his possession. Nor was this all of his explorations. In July 1888 he wrote Watson: "I take the liberty to send you by express, prepaid, a small parcel of plants for examination. Dr. Beal, C. F. Wheeler, and myself have made a botanizing trip across Michigan in the pack pine plains region."

Bailey, however, had talent in agricultural science. Beal believed so and commented on it to Bessey. Bailey wrote Bessey, telling of plans to reveal in a lecture to be delivered in Massachusetts some of his ideas for the "new horticulture." He said to Bessey: "I am enthusiastic over botany in the garden. I like that phase of the science." And on February 20, 1887, he wrote Bessey again:

> In the course of time I hope to have ready mss for a textbook of Horticulture, the science of horticulture, giving what is necessary of plant growth and nutrition, and especially enlarging upon such matters as cross fertilization, influence of climate, physiology of budding and grafting etc. etc. Books of a high standard upon kindred subjects will be forthcoming from various authors.

Bailey was right. Books on horticultural subjects did come forward. But few ranked in importance with his. Bailey went to Cornell and his aggressive labors, experimentally and taxonomically, increased in scope and significance; especially in the matters of improvement and amelioration of plants. Bailey's complete work in horticulture must be evaluated in some other work. Suffice it to say now that David Fairchild, himself with no peer in the science, evaluates Bailey's contributions as of inestimable worth. Bailey could write in popular and scholarly style. His very valuable *Cyclopedia of American Agriculture*, undertaken as a "labor of love" with the hope of making "a work that will be a credit to American agriculture" and his many other great horticultural works are themselves testimony to his many accomplishments.

Bailey's equally, if not more, valuable *Cyclopedia of American Horticulture* (1900-1902), like the agricultural cyclopedia (1907-1909) in four volumes, and the first that had appeared in America, it is said, since Henderson's *Handbook of Plants* (1881, 1890), his manuals of practice, his philosophical dissertations and studies concerning the evolution of cultivated plants, his experimental work in plant breeding, his laboratory, garden, and greenhouse investigations, his establishment of "the first distinctively horticultural laboratory in this country" in 1887-1888 at Michigan Agricultural College, especially, his great advocacy of an improved science of agriculture, horticulture, and botany, and much else, constitute him probably the greatest figure that American horticultural science has yet produced. Indeed, the belief that Bailey stands in relation to American horticultural development as Gray does to American botanical progress is not unwarranted. In many respects Bailey's opportunities were greater even than Gray's; as a consequence, in some particulars in the history of plant science study in America, Bailey's contributions have been of wider compass than Gray's. Analogies between the lives and work of two men are always imperfect, and dangerous. There were so many other great men of their periods. Nevertheless, with no qualification, it may be said that American plant science history looms large with the names of Asa Gray and Liberty Hyde Bailey. Bailey was by far the greater character in the history of American agriculture. The nature study movement, the Rural Science Series which he edited, his work as chairman of a Commission on Country Life under appointment of Theodore Roosevelt, his most important educational contributions both as administrator and teacher add great luster to his name as a scientist. In June 1888 he agreed to "attempt the revision of Carex for the much needed [Gray's] Manual." Today he is regarded not only as an authority in the genus Rubus and the palms, but his writings in plant evolution and plant breeding, as well as knowledge of nomenclature, constitute him one of the great characters in the history of American science. In 1895 he edited Gray's *Field, Forest, and Garden Botany*. The new edition was not all Bailey wished it might be. He told Bessey:

It is an almost impossible task to make this book what it should be, from the fact that it includes everything, and yet is not expected to be complete. The publishers desire that the book shall not be much increased in size, and as it is very weak in cultivated plants I fear that I shall not be able to increase it so much in wild plants as it needs. However, I appreciate the difficulty in the west and hope to be able partially to meet it. I have but very little acquaintance with the Flora of the west, and the only way in which I can get in the most important species is for you or some of your students to make me a list of those which should be inserted. . . .

Bailey, however, was glad to honor Gray's work at all times and especially pleased to bring down to date a work of his close teacher and friend.

Nevertheless, when the experimental period in North American botany and allied sciences arrived, Bailey was one of the first to comprehend its significance. He issued a plea for a wider botany,[34] one that would embrace a science of cultivated plants as well as wild plants

---

[34] "Plea for a Broader Botany," *Science*, XX (1892), p. 48.

which botanical exploration had brought to the herbaria. Botany was a science which might reach a fuller measure of use, comprising in its material equipment, laboratories, botanic gardens, green houses, orchards, vegetable and ornamental gardens, "all of which should be maintained for purposes of active investigation rather than as mere collections," said he. Bailey argued:

[M]ycology is making important additions to horticultural practice, but there are greater fields for the application of an exact science of plant physiology, whenever that science shall have reached a proportionate development. In short, the possibilities in horticulture both in science and practice, are just as great as they are in the science of botany upon which it rests. . . . Horticulture belongs to botany rather than to agriculture. . . .

Usefulness! That became an important word in North American science. Was botany to proceed along lines primarily utilitarian? Or was it to develop the techniques of an exact and pure science, leaving the regimen of usefulness to kindred branches of the sciences of plant life? The final answer would not be forthcoming for many years. However, Liberty Hyde Bailey became a leader among the next generation of botanists to include such names as Coulter, Barnes, Bessey, Farlow, MacDougal, Campbell, Robinson, Fernald, Trelease, Britton, Arthur, Rose, Coville, and many others. Botanical exploration continued. But over the continent, new branches of the science steadily developed to reshape it.

Especially in one branch of a developing "pure science" of investigation—plant pathology—would there soon be substantial progress. B. T. Galloway in 1888 succeeded Scribner as chief of the section of vegetable pathology. Inspired by the discoveries of Millardet and advances made by Americans, he was to assume a leadership as a pioneer plant pathologist of the western hemisphere, and gather around him a very able corps of workers—the workers of the federal government seeking to save immense annual crop losses from diseases.

CHAPTER XI

The Controversy of Greene with Gray

OULTER's and Gray's relationship was a far cry from the situation which developed between Gray and Edward Lee Greene. Parry had gone to Europe as planned, spending much time at Kew studying Arctostaphylos, Chorizanthe, Lastarriaea, and other botanic interests. He hoped Gray would "help out the muddle of Arctostaphylos" and was inclined to return to Gray's opinion of *Oxytheca lateola* "and regard it," he said, "as belonging to the bracteata involucral section." In November 1884 Parry had written Gray outlining a proposed trip for him to the West which he evidently intended to follow until Farlow joined the Grays and persuaded them to go to Mexico. Parry said:

> Well if you will go let me sketch out a *good* winter campaign, i.e., Starting as soon in December as you can, via Washington & New Orleans, when in Washington do not fail to see *Mrs Gen[era]l Bidwell.* . . . Then on to New Orleans, taking in the industrial exposition, then via South[ern] Pacific route to El Paso. [T]oo early to botanize except to take in the general features of the scenery, thence through South[er]n Arizona to Los Angeles, stopping over at *Colton* and notifying *Parish* to meet you and put you on the track of anything of interest near San Bernardino including by all means a drive to . . . the foothills of San Bernardino range. Then to *San Diego* for comfortable quarters, which will have just enough botanical interest to keep you pleasantly occupied.

The letter was written from London and told of what plans had been made for a memorial to Bentham:

> No special news here, all very busy, now busy arranging books including Mr Bentham's loose papers. Sir Joseph has been kind enough to select a series for me including some of the older ones. Would it not be well for you to suggest something in the way of distribution in U.S. In looking over some of Mr Bentham's addresses &c I am struck with admiration at his wonderful sagacity. I do not yet know what steps will be taken in regard to a memorial but I think the matter is in progress. You know the English move very slow[ly] & are not demonstrative. Our 1st frost only a few nights ago has spoiled the Garden here but the greenhouse was splendid. . . . I note what you say in reference to Mr Shaw but am ready to watch further developments.

Parry often referred to fellow botanists in California but less in amount and fewer in number became the references to Greene. Greene was a friend of Parry's and, though he wished to be loyal to him, he wanted also to be loyal to Gray. On June 8 of the next year Parry wrote Gray again, telling of the Parrys' delightful visit of three days in Geneva,

Switzerland, with DeCandolle, whom Parry regarded as "a distinguished and refined character"; of their sojourn at Interlaken; and of their plans to return via Lucerne, Paris, and Kew to the United States. Parry had examined the Paris and DeCandolle's and Boissier's herbaria, and hearing of Gray's conferences with Henry Shaw, Parry, realizing that he would not be wanted as curator at the Shaw School of Botany, replied: "St. Louis matter about what I expected. [I]t is more of a *relief* than a disap[p]ointment under the circumstances." On another point, however, Parry was firm. He told Gray:

I am obliged for your frank criticisms, to which on the whole I do not wish to take exceptions. Imagine your disgust of the *Medium* of publication has influenced your disparagement of my paper, in which I aimed to give as plainly as I could such matters of interest to me as might be also interesting to others, but especially to define clearly my views on *Chorizanthe Lastarriaea* and its true systematic character. I still think I have done this fairly well, at least so as to be understood, and am a little pleased with your admission that it is *one* of the two *possible* views that *may* be taken. I say *MUST*![1]

On returning to the United States, Parry wrote Gray on August 28 from Westford, Connecticut:

I want to confer with you *confidentially* on some Californian matters, particularly your relation with Mr Greene. [H]e complains to me of your treatment as not being generous & straightforward,—i.e.—making use of his information & work without proper acknowledgments&c There is a nasty jealousy kept about the S[an] F[rancisco] Acad[emy], more or less against all Eastern workers and you particularly, which I do not like to see encouraged by such a man as Mr Greene.

Parry refused Gray's offer of hospitality while at Cambridge, called Gray's attention to an error in one of Gray's papers. Reminding himself that that day was his sixty-second birthday, he told Gray that "seeing the immense work you have accomplished between 62 & 74 there is some encouragement left," although like Engelmann he realized "the necessity of 'not getting any older.' " Gray and Parry had their discussion and Gray was left obviously disturbed.

On February 22, 1884, Greene had written Gray: "Dr. Engelmann has been gathered to his fathers! Who, we are wondering, will be able to gather up the ends, and go on with cactaceae, oaks, and so many more of his specialties?" That same year Greene had been made curator of the California Academy of Sciences. He had been hailed by the *Botanical Gazette*[2] and other publications as a publisher of new species whose

---

[1] In Volume XXVIII (1884), of *The American Journal of Science and Arts*, 3rd ser., Gray, reviewing "Chorizanthe, R. Brown: Revision of the Genus, a Rearrangement of the Annual Species" by Parry, had said: "We much prefer the ordinary and obvious interpretation and should keep up Lastarriaea." See p. 76.

[2] VII, Numbers 8 and 9, p. 89.

"work among our western plants has been invaluable." And his "Notulae californicae" continued to appear in the *Gazette's* pages in 1883 and 1884. In 1885 the *Gazette*[3] reviewed a *Bulletin of the California Academy of Sciences* in which appeared Greene's "Studies in the Botany of California and Parts Adjacent." Several new genera and many new species were constituted, among them a new Vancouveria, about which Rattan wrote Gray on May 12, 1885:

Greene makes a new species of the yellow flowered Vancouveria specimens, of which I sent you in 1879. Unless my memory is at fault, Vancouveria with white flowers as large as the yellow flowered variety is not uncommon in North Humboldt and Del Norte Counties. At least the flowers are much larger there than in Marin County.

New species in Eschscholtzia, Astragalus, and Hosackia, and two new genera of Cruciferae! And Vancouveria was represented as from Oregon. Greene had warned Gray. On August 18, 1884, he said:

Evax as you have it, will torture us who know some things which you have had no opportunity of learning: and . . . the very best genus of Madiea, i.e. Lagophylla is spoiled by the addition of the Holozonia, whose most important character i.e. the number of flowers, you have suppressed. I believe Lagophylla has always 10 (5 ray.&5 disk) in each head. Its flowers *are* "vespertine"; Holozonia are not so. My luck in naming *L. congesta* amuses me, for I know nothing of your *Hemizonia congesta*, which you cite as synonymous. I [am] very glad to see the reinstatement of Peucephyllum and the erection of *Atrichoseris*. . . .

And on December 12, Greene said:

The diagnosis of a new Avicularia from Mrs. Austin's (*P.*[4] *Austinae*) has led me to overhaul *P. tenue*, out of which I have taken the western plant altogether, distinguishing it very readily from the eastern, and calling it *P. Douglasii*. . . .

Another variety which Greene said he had always held to be distinct was named *P. Engelmanni*. And these were mentioned in the "Studies" of which another was sent Watson in August of the next year when Greene told "of a new-born relative of these plants—*P. Austinae*, which grows in N[orth] E[astern] Cal[ifornia]." Such matters accumulated —differences as well as agreements—and in 1885 Greene was appointed an instructor in the University of California.

That autumn, however, another stormcloud appeared on the western botanic horizon. Gray wrote to Greene and he replied:

As for my "*Eriodictyon Lobbi*," I submit provisionally, and without remonstrance, to your sharp words, until I can look again at my specimens. If I *am* wrong I shall feel badly enough. . . . Now I am glad you take up on the matter of specific names. I have not read the discussions by Bentham and by DeCandolle.

[3] X, Number 4, p. 266.   [4] Polygonum.

They are probably inac[c]essible to me any way. I have simply followed what seems *to me* to be the *only* way, which is not subversive of all rights of priority. Let the Encrypta affair still be used for illustration: and let me ask two questions.

Shall Nuttall be allowed to name an *Encrypta foliosa* when the plant already bears the name *Ellisia chrysanthemifolia?* If you say yes, then I see but an end, a complete giving up of the right of priority in all cases where opinion can vary as to the proper genus.

But will be said "Nuttall was unaware of the identity of his plant with Bentham's"? or that "He knew nothing of the prior specific name?" To allow this to be a reason is plainly to place ignorance above par, is it not? Can science allow this? Here is my dilemma! Are there men who teach us how we can escape both horns of it? I ask for information.

Within three weeks another letter followed:

Notes from you, with questions to be answered, are accumulating on my time, while I am so occupied with my university and parochial work together, that I seem quite unlikely to give you any more "Studies" to criticize, for a long time to come. And here, let me tell you that I am very deeply mortified by *one* of your remarks in the *Amer[ican] Jour[nal]*. My genera I can calmly endure to have you disapprove. Most of them I have proposed in full expectation of all that (I mean also those of Nuttall's whose restoration I have called for) but it does hurt me to see that if you *will* actually sneer at *one* of them, that *one* is to be Athysanus. I hold that to be on a par with Bebbia[5] for genuineness. You do yourself injustice, as it seems to me, in choosing that for your most unreserved disapproval, and then telling the world, in effect, this: that you have to deal with a man who makes genera on the presence or absence of a wing to the pod! A "wingless Thysanocarpus." Because whoever takes the trouble to read my page carefully will see that, unless I have lied, I have been in possession of something *very* unlike what a wingless Thysanocarpus would be. Such a one might excuse you from the blame of misrepresentation, by thinking you had not looked to see just what it was that you were condemning. I, for my part, do not suppose you have ever noticed what I wrote, or ever looked to see how very different are the pods of Thysanocarpus and Athysanus. I am guessing that you have read the name Athysanus and thought *it* told my story, and described my genus. I *did* allow the absence of the wing to name my genus, but not to characterize it. . . .[6]

On February 1, 1886, Greene followed with what seemed a final letter:

Two rather long letters of yours are before me, on the subject of Kumlienia. In the first of them you blame me for not having submitted my paper to you before printing, adding, that the genus "is not likely to be approved." In the second you

[5] Greene and Gray agreed on a genus of Greene's discovery, naming it Bebbia, for Michael Schuck Bebb, the North American authority on willows who was recommended by Gray to the college now Ohio State University. But horticulture was established before botany and Bebb was not appointed to the college's position.

[6] Apparently Gray sent Greene either his article on "Botanical Nomenclature" (1883) or his "Gender of Names of Varieties" (1884), both published in *The American Journal of Sci. and Arts*, 3rd ser., XXVI and XXVII. See Sargent's *Scientific Papers, op. cit.*, I, p. 358; II, p. 257. Greene replied he had read "all" long ago. Furthermore, the reference to "Nuttal's" may be "Nuttali." Greene's writing in places is difficult to decipher.

counsel for the suppression of the page that contains it. . . . Up to a year and a half ago I had almost always submitted my little affairs botanical to you, with an unquestioning faith that you would do what was right, whether you approved, or disapproved. When, at length, I was ready to undertake a little piece of work which I thought would really count for something, I confidingly informed you of my purpose. You replied: "Let me see your new material" & "Let us have a fair understanding as regards genera, then go ahead." I went on, sent you my specimens, indicating the limits, according to my notion, of Plagiobothrys, upon which my mind was already made up: yet not venturing my then unsettled views of other allied genera. You were long silent. Then came this rather surprising remark from you. "Inasmuch as I am responsible for the putting of Plagiobothrys and Krynitzkia into Eritrichium, I ought to take them out." This was followed up not long after by the promise of "proof in a few days." I am doing you, now-a-days, the credit of supposing that you have forgotten just how you treated me. I take it you have destroyed my letters and have forgotten what yours to me contain: else you would not suggest to me again, that I lay my little projects before [you]: for you are not likely to attribute to me such meekness as to remain unaffected by so grave a violation of the very fundamentals of scientific justice and good faith. No; I believe you have forgotten. But, no matter about that. My own interests will forbid my doing again what you wish. I could, I very well know, profit greatly by your advise. It is inevitable that I shall make some mistakes which I could avoid if I had the happiness of a free, confiding correspondence with you who, in point of wide experience and profound scholarship, are without a rival, as I judge, among living botanists.

Now, at last, I have said that which I have long hoped I should never feel called on to say. But I think it was best.

As regards the suppression of the pages upon Kumlienia, I may as well continue, and enunciate some things which you have not taken for granted. I do not look for your approval of anything whatever which I print. He who attempts, as I am now beginning, to set forth now and then, a view at variance from yours, is badly off, if he have not faith and patience to suffer and to wait: I suppose that the final judges between you and me on such questions as those of Ennanus, Diplacus, Kumlienia etc. are to be men of some future generation: men who shall know Pacific American Botany as Linnaeus and Fries knew the Flora of Scandinavia, and as the Browns, Hookers and Benthams, that of the British Islands. I am well aware that your approval would be worth everything if I had no faith in myself, and were working to win the praises of my botanical friends and correspondents. I perceive it is in your power to exalt a man, in the estimation of the scientific public, to very high rank as a botanist; and that you can as easily relegate him to the limbo of conceited "cranks." You appear to have decided how you will dispose of me. . . .

Gray replied immediately:[7]

Your letter of the 1st inst[ant] came this morning, and requires a prompt reply. I cannot make it very detailed, but I mean to make it explicit.

[7] This transcription is taken from a copy of Gray's letter to Greene contained in the Greene letter file at the Gray Herbarium. The evident correspondence with the matter of Greene's letter, it is submitted, authenticates it. But no actual proof that this is a copy of the exact letter sent, exists, so far as I know.

You have brought forth your grievance, which if warranted, should change our relations of correspondence, and which you should have seen to before today.

I knew you did not like my setting at work anew on Borraginaceae when I did. But I did not know that you took me to have committed a "grave violation of the very fundamentals of scientific justice and good faith." I think *you* have forgotten the facts. The essential fact is: that my change of view about the Genera in question, and my determination of my duty to act upon it, antedated your communication of a specimen of the typical Plagiobothrys, which up to then I had never seen. When your specimens came, I promptly informed you that I had it already from Howell. I am confident that my letter at that time stated that.

As it required a very large overhauling of very special work of mine—mostly in reshaping—as I knew I had or could have all the means & appliances for the revision, and you had not, and as I was not put upon the track by you—tho' you came in soon after, it was I think natural that I should apprize you that as I had put Plagiobothrys into Eritrichium myself I proposed to take it out—If after this reclamation, you think I have violated good faith or done you wrong, you could no longer wish—nor I allow—our correspondence to be anything more tha[n] that of strict botanical interchange. If you are convinced that you do me a gross wrong, you will say so, and will also disabuse Parry's mind, if, as I have some reason to think, you have given your impression to him.

In my last letters I ventured in a frank, friendly way, to offer some botanical advice, from the adoption of which I could not possibly be at all advantaged—which any one would see was dictated purely in reference to your own scientific reputation. I thought you would take it in that spirit. I think, on reflection, now that there is of course an end of any free criticism and advice,—that you will feel that I have done you friendly service. I feel that I have never done you any other.

As to my published criticisms—of course I had to say that I did not think you were quite right nor wise. At least one of my correspondents wrote me speaking strongly of the kind almost deferential way in which I expressed my dissent. I know that I took pains to avoid every harsh turn of phrase.

I mean to treat you with complete consideration. I shall save valuable time & your temper, by avoiding for the future all "tiresome discussion." I am convinced that I should have done so long ago. In wishing that you had informed me in advance of publication of your Kumleinia, I thought I might have saved you from what you could have seen—on a survey of the Genus—is ill-judged. But why should you not have your own way unmolested?

One must be fair to Greene. Perhaps as much as anything a lack of comprehensive knowledge of evolutionary theory and a consequent inability to evaluate plant variations as others did, accounted for most of the differences in judgment between Gray and Greene. In addition to expert field knowledge, particularly his early emphasis on the study of plants *in the field*, Greene must be remembered as a keen observer and one who stood fearlessly in defense of the right to practice science as one saw truth to be. No one can deny his studious zeal and relentless endeavor. Some of his best work as a student was done in historical research. As clearly as anyone of his day, he saw the basic principle in botanical nomenclature was use of a universal language, describing plants as briefly as possible but completely—a principle even more important than priority of publication. He is said to have had "an excellent appreciation and often a rare judgment of [plant] relationships," although this ability was not exercised as often as it might have been. Greene, however, sought to impose on the botanic world of that day a territorial autonomy which, although he denied it, was evident in both his actions and correspondence. He was, as Engelmann supposed, interested in making western botany territorially self-sufficient. Only such friends as Parry and a very few others were in any sense his rivals or competitors. Parry, as shown, sympathized with his efforts more than once. But Parry was not an extremist. He was primarily a plant collector interested in the good of a cause, not interested in dominating a region or competing with the "closet" botanist or systematizer on a large scale. With men of the abilities and stature of Gray and Engelmann, and their work, Parry was satisfied. He saw the economies of effort and expense, the saving of waste, the facility and celerity in localizing systematic labors to institutions of acknowledged skill and experience. Time has more than vindicated this position.

Greene was a collector, too. In 1885, during May, he went to the Mexican coast south of San Diego getting what Cleveland called "a fine collection at Guadalupe and Cerros [Cedros] islands, and at San Quentin and Ensenada on the coast of Lower California." In July 1886 he went to Santa Cruz Island and remained among the Santa Barbara Islands until September, returning to suffer "a pretty sharp malarial attack . . . from too much physical work on those wondrous islands" which probably included Santa Rosa Island. He knew Colorado, Wyoming, New Mexico, Arizona, and California botany. During the years 1886 and 1887 the pages of the *Bulletin of the California Academy of Sciences* were filled with Greene's new proposals.[8] And he planned a new botany of the Pacific coast, although when Ivison, Blakeman, & Taylor's agent in San Francisco told him that Coulter "having finished 'Colorado' [was then] engaged on a similar work for Pacific states, or California," Greene offered to Gray to desist at once from his undertaking. Greene offered a friendly spirit and continued his correspondence

---

[8] "Studies in the Botany of California and Parts Adjacent," Parts I-VI, *Bulletin of the California Academy of Sciences*, 1886, 1887; "Some New Species of the Genus Astragalus," *ibid.* (1886), I, pp. 155-158; "New Plants of the Pacific Coast," *ibid.*, I, pp. 7-12; "New Genus of Ranunculaceae," *ibid.*, I, p. 337; also: "Extended Range of Some California Plants," *West. Am. Sci.*, III, pp. 206-207. San Diego, 1887, "Bibliographical notes on well known plants," printed in *Bulletin Torrey Botanical Club*, July, August, September, October, 1887.

with Gray. But he could never yield his convictions, in error though he might be. On September 16, 1886, he wrote Watson:

I am as strongly on the side of the party who are insisting that the very first specific name of a plant is to be used in defiance of every other, save when it is already in use in the genus to which a species is transferred. On this . . . score I drop the Nuttallian specific name from Phaenicaulis while retaining the genus. By the way, I must say I did not think and do not see how pods of such different shape as those of Phaenicaulis & Parrya can be admitted in the same genus of Cruciferae.

Greene had openly confessed to Gray that "ante-Linnean generic names ought *all* to be restored to their proper authors," though he realized that by then such an accomplishment would present insurmountable difficulties. He blamed "a certain British trait which expresses itself," he said, "so strongly in the *Genera Plantarum*. . . . It was bad enough for the brilliant, sprightly and highly gifted Swede to go on so recklessly ignoring, as he often did (whether he meant it or not) his predecessors and contemporaries, some of whom appear to have been, in some points, his betters: but I feel confident," he continued, "these men, in order to save themselves trouble, have done more deliberate and unpardonable injustice to great names, than Linnaeus ever meant to do."

When Greene established his own publication, *Pittonia*, he wrote Gray on February 26, 1887:

I am sending out the second signature of *Pittonia* i, and shall soon be done with the first part. . . .

And two days later:

Now I am confiding in you, as I like to; and beg for a reason. . . . Why you did not, in treating those Eritricheae, so much as mention Lehmann's generic name Cryptantha? I have always wondered if it was from lack of S[outh] Am[erican] specimens to form its identity with Krynitzkia; or whether you wished to have it an "obscure name" (as you have somewhere written) notwithstanding its eight or nine years of priority over Krynitzkia . . . meanwhile [I am] withholding from my printer a considerable lot of manuscript on these plants. . . .[9]

Why *Pittonia*? Well that is a hard question, altho' I was half expecting it from you. I can not answer. It is an easy name to speak and to abbreviate, and that is enough. These *Proc[eedings] Am[erican] Acad[emy]s, Bull[etin] Cal[ifornia] Acad[emy]s* and all that, are cumbersome!

Then again: I believe in the "dark ages," and always did: am not, nor ever was, well disposed toward revolutionists like Luther and Linne: have a sympathy for Tournefort, Plumier etc.etc. and these are about my stock of poor reasons for *Pittonia*.

Greene had chosen a family name of Tournefort—*Pittonia*. The *Botanical Gazette* tried to remain impartial. When Greene sent them his

[9] This letter has been slightly disarranged to present the material more clearly and concisely. Content, however, remains as it was written.

fifth paper, "Studies in the Botany of California and Parts Adjacent," concerned as it was with a study of the genus Brodiaea, they commented that he had observed the species in the field and was therefore more familiar with the genus than most "closet botanists." However, on December 3, 1887, Coulter wrote Gray:

What in the world is going to become of us, with Greene stirring up synonymy as with a pitchfork? His *Pittonia* No. 2 wh[ich] you review in Dec[ember] no. *Am[erican] Jour[nal]* is bad enough, but has *Pittonia* No. 3 come to your hands? It reads like the work of a crazy man, at least one lost to all sense of propriety. Is he not a second Rafinesque?

What is to be your policy with regard to the priority of *specific* names? For instance, I restored *Conioselinum canadense*. Walter had described it as *Apium bipinnatum*, & along comes Britton in last *Torr[ey] Bull[etin]*, & changes it to *Conioselinum bipinnatum* Britton. I rebel against such a raking up of old specific names, but if it is to be done I prefer to do it myself.

I have finished up Peucedanum after the toughest tussle of my life, & am now in the beauties of Cymopterus. I have all the Umbellifers in the country nearly except those of Cambridge. My plans now are to drive away all this year at them, & spend next summer in the Cambridge collection & library.

Coulter and Greene each had waged a controversy with Gray. If any one of them was victorious, the victor was Gray. But in no sense was the loser vanquished in either instance. Both continued their friendship with Gray—the principal difference being, Coulter's friendship was continued heartily. When, on the approach of Gray's seventy-fifth birthday, the younger North American botanists joined together to give Dr. Gray a silver vase, and cards and letters were placed on a silver salver accompanying the gift, bearing the greetings of one hundred and eighty botanists of North America, Coulter was one of the committee representing the Botanical Brotherhood. Not a few of the flowers associated with Gray's name and special studies were so deftly wrought on the vase's surface—*Graya polygaloides, Lilium Grayi, Notholaena Grayi, Aquilegia canadensis, Centaurea americana, Shortia galacifolia, Aster Bigelovii, Solidago serotina,* and Dionaea among them—that Gray was constrained to reply, ". . . 'The art itself is nature;' " and that, ". . . this full flow of benediction, from the whole length and breadth of the land whose flora is a common study and a common delight, was as unexpected as it is touching and memorable."[10] The day was made a happy one for Gray and he promised his brothers and sisters of the science that the gift would preserve its memory to those who came after him and the botanists of his time.

[10] The complete account of this gift may be found in the *Letters of Asa Gray, op. cit.*, pp. 776-778.

Greene, however, struck out militantly to bring about a reform in plant nomenclature. Seeming from the time of his controversy with Gray to have maintained an effortful friendship, he led in a fight for nomenclatural reform which being similarly led by Nathaniel Lord Britton accomplished the establishment of the so-called Rochester Code.[11] Nevertheless, not even this code completely satisfied Greene.[12] Greene continued to argue against some of its provisions. However, notwithstanding his eccentricities, he had a certain genius and before his death had established himself as not only one of America's ablest botanists but one who by sheer persistency and keen erudition and knowledge was an authority in American systematic botany and on botanical nomenclature. Coulter, sensing the impending quarrels, turned to morphological interests after completing several very valuable taxonomic studies. That he did not pursue taxonomy after going to the University of Chicago was largely a matter of circumstances there. He, as head of the department, had first to equip the facilities for morphology and physiology. Always expecting to establish taxonomy on a first-rate scale at the university, Coulter changed his plans as the great herbarium of the Field Museum was enlarged. Coulter left taxonomy, not because he disliked fighting for what he believed right in this branch of the science, but because he became convinced that morphological studies would have to be pursued on the basis of phylogeny and life history studies to aid in the development of true systematization—an adequate taxonomy embracing the results of the coming, already begun, experimentation in botany.

Parry seems to have taken but little interest in any controversy. On March 14, 1886, after receiving Gray's Supplement of that year to the *Synoptical Flora of North America*, Volume II, Part I, Parry commented:

I see you hardly touch upon Greene's *gen[era] nov[ae]* but accept many of his *sp[ecies] nov[ae]*—and refer the rest to. . . .

I have not heard how Greene takes it. . . .

I have just heard from Pringle, about to start for Mexico.—We have about concluded to start from here [Davenport] fore part of April, stop over a few days at St. Paul and down the river by boat to New Orleans—then stop over at El Paso & perhaps run down to See Pringle at Chihuahua, so on to South[ern] Cal[iforni]a and then select some quiet sea side for the summer to return here in the fall. . . .

Greene did not mend his ways, even with Watson. July 14, 1888, Greene wrote Watson, expressing surprise that two specimens he had

---

[11] 1892.

[12] See "Some Fundamentals of Nomenclature," *op. cit.*; also Willis Linn Jepson, "Samuel Bonsall Parish," *op. cit.*, and letters there contained, showing Greene's attempt to persuade Parish to his point of view.

sent had not arrived in time "to be brought into comparison with the Pringle specimens. . . . I assumed," said Greene, "that Pringle's and Palmer's plants would be printed by you, in one and the same 'Contribution,'" referring to Watson's *Contribution*, XV, of the March previous which included Pringle's 1887 Chihuahua collection, but not Palmer's collection of the same year made around Guaymas, Mexico, and in Lower California, which were not published until October in another *Contribution*, XVI; Greene said:

I knew that you had not come to the naming of Palmer's. I was still for having the printing of anything in relation to Palmer's set (which I have with full notes), in deference to you. But your paper on Pringle's collection, it now appears, had been three weeks in print when my letter and specimens reached you; and so I was too late. . . .

On June 27 Greene had told Watson by letter:

Your postal card must needs be answered; for it assumes that I have contributed two plant specimens, this year, to the Gray Herbarium, a thing which I had no intention of doing.

Merely to help you avoid making a synonym or two, I sent those two fragments, at the same time hoping to indicate, faintly perhaps, but not unintelligently, the possibility on my side, of a cessation of certain outward appearances of hostility between us, and of a return to the former better ways of kindly thinking and writing each of the other.

I seem, however, compelled to understand that there exists, on your side, no such possibility, and that you refuse to meet my very first advance. If I am wrong in this, you will surely correct me.

Watson hastened to correct Greene's impression, but more briefly than had Gray:

I have your favor of 27th ult[imo] in which you refer to personal feeling & relationship between us. In sending the plants previously there was no reference made to this matter & I could not infer an implied connection.

As to the "outward appearances of hostility between us" I am not conscious that there has been any upon my side. In all my botanical writing & doings I have always studiously avoided personalities of every kind, in reference to yourself as to every one else. I have said & done nothing that evidenced hostility to you, as I have felt none. And as I have done nothing to provoke the hostility which you have shown so I can do nothing to placate it. I have not noticed it hitherto, nor had any intention of doing so. As it has been wholly unprovoked so its correction must be as purely voluntary on your part. . . .[13]

Greene became exasperated; and in his letter of July 14 persisted:

[13] This letter, like the Gray letter to Greene, is a copy found in the Greene correspondence file at the Gray Herbarium. Its credibility, as the actual letter sent Greene by Watson, cannot be established and its value, therefore, must be weighed by the individual reader. It appears that copies were made, sensing a need for someday meeting the situation when it would be presented historically.

You lay upon me all the responsibility for that lack of cordiality which has existed between us; and that I judge to be wrong, yet, am not going to open a private controversy for attempting to write a word in my own defense, or by making detailed complaint. Your time and your energies and mine, too, may be more profitably occupied.

I do not regret having opened up a line of criticism of the doings of botanists here and there. I began that because I thought the interests of the cause loudly called for it. I shall no doubt continue the work, difficult and unpleasant though it be. But I do regret some of my own too pungent phrase; needlessly, and worse than that, hurtfully irritating expressions. My own apology would be this—I am of strong sympathies, and not careful about the weapons I may employ. . . .

Maybe I shall, henceforth, be able to maintain the better spirit in which I have criticized, as you will see, the Torrey Club Catalogue. . . .

In some remarks, now in manuscript in your "Contr[ibution] XV," (of which I had borrowed a copy) I have made one point in my own defense, which I shall now call your attention to, and then erase it from the printer's copy. It will be the better way, since you allow me to write. It is in reference to Heterodraba: and I am perfectly confident that the injustice you do me is done unconsciously. But if men do such things innocently, so much the more need, as it seems to me, that that kind of misdoing should have its injustice held up to view. It is so common with American botanists that I believe no one thinks it wrong; hence my determination to take, before long, some opportunity of discovering it: other opportunities will be given. The point, as far as [it] relates to your paper, is this: in the plant in question, I discovered and published certain characters, namely, the total indehiscence of the pod, and the nearly obsolete partition. These are my important contribution to the history of the plant. If I named it as a new genus it was on these two grounds, along with its prostrate habit. These three characteristics are entirely ignored by you, and that is ignoring my whole stock of facts. My conclusion, that it is *sui generis*, you are not bound to. That is unimportant. That you are unable to second my view is not my affair at all. I do think, however, that justice to me demands the recognition of my facts, if facts they be, and that justice to *science* requires that you deny them, if you find that what I gave for facts and new facts, are only my imagination. I really think that, if you will look into this matter, you will see that you have exposed yourself to criticism by not writing into the character of your Heterodraba the essential characters which I indicated for it as a genus.

Watson's *Contribution*, XV, presented, as part of its subject matter, revisions of North American species of the genus Draba; and to this evidently Greene addressed his complaint. Also part of the *Contribution* were determinations of "Some New Species of Mexican Plants, chiefly by Mr. C. G. Pringle's collection in the mountains of Chihuahua, in 1887." Greene's and Watson's troubles had been in reference to certain Pringle collections. But more, as time passed, they were with respect to collections of Edward Palmer—emphatically collections made by Palmer for the United States Department of Agriculture in Southern California in 1888.

The story is quite complicated and, entertaining the views that he did, Greene cannot be criticized with severity. Because of his important field knowledge of California plants, Greene made his geographical claim, believing sincerely that he, better than anyone else, should systematize California collections. At all events, against even the collector's wishes, Greene sought to systematize Palmer's California plants collected in the summer of 1888, urgently soliciting Vasey to send them to him before sending to others. What seems to have been a misunderstanding ensued. Watson evidently had heard that Greene sought to determine Palmer's *Mexican* plants collected in 1887 concerning which, when written, Greene confessed he had a set. Ownership of a set would not arouse Watson's opposition but an intention to systematize before anyone else the plants gathered by Palmer in Mexico would have. For, by agreement it was settled that Palmer's Mexican plants were to be studied by the Gray Herbarium. When about 1888 Palmer entered the service of the United States Department of Agriculture as a collector, Vasey was also drawn into the seeming affray, on the side of Watson.

Furthermore, for several years Palmer's collections had been growing increasingly valuable. After collecting in the State of Chihuahua in 1885, Palmer had gone in 1886 to the State of Jalisco, Mexico, near Guadalajara, the capital. From there, on August 11, he wrote Watson:

I am succeeding as wel[l] as could be expected in this thinly set[t]led and difficult country to get about in. The rainey season though the only one in which there are plants, is very difficult to dry them in—I start tomorrow to a very good locality from present indications. [A] fair collection wil[l] be made. [T]he collection wil[l] not be a financial success—there is so many expenses—every time I go out of the City with an outfit I have to pay one dollar and twenty five cents safety tax. It wil[l] be October before I finish and [I] wil[l] need some money to return on. . . .

One half of the species were collected at Rio Blanco where Palmer spent from June to October. Part of June was spent at Barranca; from August 25 to September 5 at Tequilla; and from October 27 to November 3 at Chapala. With the end of the year Palmer returned to Washington and began preparing his plants, a set of which he promised Watson by letter dated January 24. Palmer sent valuable ears of corn to Dr. E. L. Sturtevant and planned to name one "the San Pedro after the Indian village that grew the corn, another the Dr. Palmer and the thir[d] the Baird corn." Gamopetalae were done by Gray who on his last European trip had taken with him some of Palmer's plants for study. Vasey, Eaton, and Britton studied certain groups. The whole became Watson's *Contribution*, XIV, and of the 675 species of Palmer's Guadalajara plants, 120, it is said, proved new. The *Contribution* was presented by Watson in 1887.

In 1887 Palmer changed the locale of his collecting to north and west in Mexico—to Guaymas in the State of Sonora and Lower California. He wrote:

This is a very peculiar country to collect in. [I]t was nearly three years without a good rainey season [A]nd this year instead of raining in [the] latter part of June, did not untill [the] middle of August. [T]hen what vegetation came up matured so rapidly that before you could gather from all desirable places, seeds was ripe and plants dried up by [the] middle of October [S]o that a stranger to the country would ask, has it rained he[re] within a year. I have gathered 345 species. . . . I am now going to the unusual places to gather in the novelties. I think you wil[l] find some good things. Just returned from a visit to the [Island] of San Pedro Martin . . . 18 species only found, some very interesting ones. . . . Expect to go in a day or two to Los Angeles bay.

Immense groves of a curious Cereus and "some curious Malvacious plants" interested him, and, having written before from Guaymas in November, wrote again from the same place in January informing Watson he had collected about 500 species and finished his task in this locality. He said:

[S]ince I wrote you before [I] have visited Los Angeles Bay in Lower California. [T]hough the latter part of November [I] obtained 100 species of plants this was owing to the summer rain not falling at the usual time from July to September but last part of October it rained and in November several showers so that at my visit it looked like spring while arround Guaymas at the same time vegetation had matured its seed for the most part and was going to rest. I think you wil[l] find many novelties.

Palmer asked Watson and Dr. Gray concerning localities "for desirable plants" on the Pacific side of Lower California, and some on the Gulf of California. But soon he went to Washington where at the Department of Agriculture, with some assistance, he put his collections into sets. For reasons of health, he was ordered by doctors to a dry mountainous country—so he chose California. Watson received his plants in March and on October 10, 1888, presented *Contribution* (XVI) *to American Botany*, aided again by Vasey, Eaton, and Britton, and by Daniel Oliver of Kew. While part of the plants had been collected at Muleje and Los Angeles Bay in Lower California, most of them, Watson said, had come from around Guaymas.

Watson commented interestingly on plant distribution of the Guaymas region:

The characteristics of the flora of the region bordering the Gulf of California, so far as shown by this collection, are for the most part those common to the flora of the whole arid region of the interior, from southeastern California, Arizona, and New Mexico southward into Mexico, distinct in a great measure from that of California proper on the one side, and that of the Gulf States on the other. Nearly or quite two thirds of the species range northward beyond the Mexican boundary. In the mountains about Guaymas we find a considerable number that are identical with or allied to species that have recently been collected by Pringle and Palmer in the mountains of Chihuahua.

During June and July of 1888 Palmer collected for the United States Department of Agriculture in California—Kern, Tulare, and San Bernardino counties—proceeding from there to San Diego during the fall or late summer. On August 26 he wrote Watson concerning Greene's purported wish to determine his plants:

Yours of Aug[ust] 2 was forwarded to me from San Bernardino. [A]s I was going with a party on a trip to Catalina Island. [S]o I wrote to Green[e] on my return yesterday stateing that in accordance with an agreement between myself and you and Dr. Gray all the plants collected in Mexico by me should be determined at Cambridge and that I should expect *that agreement* to be complied with. [T]his Green[e] understood previously—pos[s]ibly your informent is mistaken as to Green[e']s naming some of the Guaymas plants.[14] (Maybe he has commented on some of the Guadelajara plants as I am so informed) I cannot think Green[e] would have the audacity to name the Guaymas plants after knowing my wish to the contrary. My health is some better. It is gratifying to know that the Guaymas plants yielded new and interesting specimens. Though to me a financial loss if mankind derive a benefit that[']s a compensation.

What do you think of Colema or San Blass Mexico for winter work [H]ave the plants of the former place been much gathered. [I]t is a little inland from the Gulf of California. [T]he map show[s] Mountains near it—Xantus was consul there. [H]e gathered birds&c. [I] am not certain about plants—As I have not visited Guatemala it has occurred to me it might be best to go to Colema or San Blass, especially the former place, before going to Guatemala. . . .[15]

Dr. Vasey says next season he may have some work in California.

A week later, Vasey wrote Watson:

I have received Palmer's collection in S[outhern] Cal[ifornia] and will send you everything peculiar. Dr. Palmer writes me that Dr. Greene is very anxious to have a set to work up. He seems to think he has a patent right on all Califor[nia] plants. I should think he is verging toward crazy on the subject. The plants mostly look familiar, but some show variations or peculiarities. My boys are helping me work them up.

Greene, it later developed, was always a purchaser of Palmer's plants.

[14] Proof is not strong that Greene sought to determine Palmer's Mexican collection. However, proof is convincing in respect of Palmer's California collection. On November 21, 1888, Vasey wrote Watson: "I sent the specimens to you so that you might give the names, notwithstanding Mr. Greene's earnest solicitations that they should be sent to him first."

[15] Palmer did not go to Guatemala as planned. Evidently he decided more desirable and new plants could be found in the Southwest. Or, he may have been troubled about finances.

Evidently, either wishing to conform to established practice or for fear of synonym making, he tried to consult Watson's publications always—to prevent "cross-firing," as he said. However, Greene did not receive Watson's *Contributions* and, having on one occasion been forced to borrow a copy from Lemmon, he concluded the failure to receive the *Contributions* was Watson's doing and so ceased exchanging *Pittonia* with Watson. "Late in March I was printing an article," wrote Greene to Watson,[16] "had, indeed, read the last proof, when in one of the Journals I read a notice of your paper on Palmer's Guaymas plants, etc. Some of the species had lately been duplicated to me, in Lieut[enant] Pond's collection, and I was publishing them. When I knew your paper was out, and had been for six weeks, I had to telegraph to my printer to stop.... [I]nasmuch as it was the third successive instance of your papers being sent to all California botanists, me alone excepted, self-respect seemed to demand that I discontinue an altogether one-sided exchange. There is a possibility that you have, in each instance, done your part, and that mail agents are at fault." And Watson presumably had. At least Greene appeared to accept his explanation. A month and a half later, Greene wrote: "It is a relief to be able to lay upon Dr. Vasey the blame of all my trouble and annoyance ... but, let me beg that, another time, you send me my copy direct, not trusting Dr. Vasey or any other man." However, there were but few more letters, if any, between the two men who strove to be friends—Greene and Watson.

Greene at the time, it seems, was working on an article concerned with the "distribution of the 'Oaks.'" He needed little, if any, help. From 1889 to 1890 he issued *Illustrations of West American Oaks. From Drawings by the Late Albert Kellogg M.D.*,[17] and was soon to begin publishing *Flora Franciscana. An Attempt to Classify and Describe the Vascular Plants of Middle California;*[18] and in 1894 his *Manual of the Botany of the Region of San Francisco Bay.*[19] In 1895, however, Greene was to leave California and become professor of botany in the Catholic University of America, and later an associate of the Smithsonian Institution. Greene had persuaded himself in 1885 that his ordination as an Episcopal clergyman had been invalid and, as a consequence, had become a Roman Catholic layman. Greene was much impressed by the inner spiritual effects accompanying sacraments of the Roman Catholic church and although during his last years at the University of Califor-

---

[16] Letter dated July 25, 1889.

[17] San Francisco, 1889-1890. 84 pp. Plates.

[18] Parts I-IV, all published. San Francisco, 1891-1897. 480 pp.

[19] A systematic arrangement of the higher plants growing in several near by counties. San Francisco, 1894. 328 pp.

nia he was professor of botany, he left and took his valuable knowledge of North American flora in the field to a chair of an eastern university. *Pittonia*, which continued for several more years, and his *Leaflets, Botanical Observations and Criticisms*, commenced during the early years of the next century, established a confident and authoritative Greene. And he continued to publish on California plants in *Erythea*, a botanical journal published at Berkeley, and other journals.

CHAPTER XII

Mexican, Central, and South American Explorations. Gray's Last Years

O N January 13, 1884, Palmer had written Watson that Parry had informed him that Pringle was going to northern Mexico. "Unless I go to [s]ome other part of Mexico," commented Palmer, "[I] know not what field wil[l] pay to collect plants combined with a general collection of other things." Palmer was then at Tuscaloosa, Alabama, about to complete four years of exploring prehistoric mounds and graves in Tennessee, Arkansas, Indiana, North Carolina, Georgia, and Alabama. When Engelmann died and Parry went to Europe, Palmer had begun to watch with keenest interest the activities of Pringle.

That same year, 1884, Pringle was selected to succeed to a position held by Rusby in a botanical survey of north and northwest portions of Arizona, conducted under auspices of the Smithsonian Institution. His exploration that year, however, was not confined to Arizona. Pringle went on into the northwestern borders of Sonora, Mexico, where no botanist had been, and returned to be honored with the naming of a genus for him, Pringleophytum. In spite of a severe illness suffered, he brought with him a large collection of plants. On November 22, Palmer wrote Watson, from Washington:

So Pringle went from Tucson to the Gulf [A]t what part did he strike the *Gulf of California*. He is fortunate he could go any where without an escort of Soldiers [A]t the time of my visit to that section he travel[l]ed over, you could go no where without an escort of soldiers or citizens.

The March, 1885, issue of the *Botanical Gazette*[1] announced that "Mr. C. G. Pringle has left for a season of collecting along the line of the Mexican Central Rail Road, especially in W[estern] Chihuahua." This year a botanical conquest of Mexico was determined upon and both Palmer and Pringle were called into service. Evidently Palmer's last work in southeastern United States was a trip into Florida to collect invertebrates—corals, echinoderms, mollusca, and the like. On November 6,[2] while in Washington, he had written Watson: "The beginning of October [I] returned from [the] keys of Florida having made a very

[1] X, Number 3, p. 245.
[2] Palmer often failed to date his letters with the year the letter was written. As a consequence, ascertainment with exactness as to his always changing movements is sometimes very difficult.

large collection for [the] New Orleans Exposition of every thing to be had from the variable waters surrounding the keys." And added: "I am desirous of spending the winter in the Gulf of California among some Islands. Shall gather the botany as it may be of interest."

Palmer asked when the dwarf oaks were in bloom near San Diego, as that was evidently his planned destination. Soon he was off again for the Southwest, to collect materials showing the arts of the Cocopa, Pima, and Yuma Indians as well as materials of an ethnobotanic nature, plant-food-lists, medicinal plants, fibrous plants, and plants of a general botanical nature.

The United States National Museum, however, directed him to go to the mountains of southwestern Chihuahua, a part of the western Sierra Madre Mountains of Mexico, where he was to compare the Mexican cave dwellers, the Tarahumara Indians, with the cave dwellers of Arizona and New Mexico. On January 20, 1885, Palmer wrote Watson from Mesa, Arizona, informing him that it made no difference to him whether he went into Mexico or to Guatemala, as Watson evidently had suggested. "[I]t's the making of a useful collection that I desire," said Palmer and gave as his address "Pimo Indian Agency via Casa Grande Pinal County, Arizona." On February 4, he sent Watson a set of plants with a catalogue and pictures of the mountain scenery about Batopilas, southwestern Chihuahua, Mexico, saying that "A specimen of Yerba de la Fletcha is sent . . . which differ[s] from the kind collected by Pringle in Sonora shown me by Dr. Vasey," and forwarded a bundle of ferns for either Eaton or Davenport. Palmer had had to return to Washington, explaining completely to Watson the following May:

> I could not go out to the Bay[3] as Baird had no available funds. After finishing the Indian collection of Southern Arizona, [I] should have gone to Mexico but my health would not admit of it so returned hear [A]m now better [E]xpect to go out among some Western tribes of Indians in two weeks to make collections for Baird. . . . I am interested in introducing the Mexican edible cactus into Florida. Engelmann called it *Tuna cardova* (an *Opuntia*). . . .

However, on June 15, Palmer wrote:

> As the Indians have broke out against the set[t]lers of Arizona and New Mexico, my plans have to be changed. The long deferred Journey to Chihuahua has been decided for me to make as soon as I can get ready. The funds for the trip are limited as Baird have to pay it out of the Smithsonian fund. . . . I understand Pringle is in Mexico [P]lease inform me what parts he has visited and the extent of his intended Journey, so I may not go over his ground—as two collections might not find sale—For Prof[essor] Baird I have to visit the Indians in the mountains and

---

[3] Probably one of the Pacific coast bays. Parry had recommended a trip to Guaymas, across the Gulf to LaPaz, and across land to Magdalene Bay and San Francisco.

valleys to obtain everything from them. Not an easy task—not wishing to be idle wil[l] undertake the Journey—if plants wil[l] help me out. . . .

On July 3, from Washington, he sent Watson a postal card:

In a few days a party start[s] for the Mining section of Mexico known as Batopilas. I wil[l] accompany them by invitation. Shepherd formerly of this place has charge of the Mines. Pringle did not reach there. As you see by the maps it[']s an important botanical locality. After leaving the R[ail] R[oad] there is a stage ride of three days, then 6 days Journey by mule-travel [I]t is rainey season now [W]il[l] try and return by way of the mountains on Pacific side.

Pringle[4] had arrived at Chihuahua on March 5 and after making a return trip to El Paso gone on to Zacatecas where, disappointed by the dormancy of vegetation, he with his aide had continued on to Agua Caliente and again being disappointed had returned to Zacatecas. They went by horse cars from there to Guadalupe, where they spent March 17; then returned to Chihuahua. Some time was spent around Chihuahua visiting such interesting localities as the Santa Eulalia Mountains and Canyon, Bachimba Cañon, the near by foothills and plains, and the mines located near there. On April 26 they moved 328 miles southward to Jimulco, a railroad station in Coahuila where, walking four miles westward across the valley and mesa to a great canyon they found a "rich variety of plants," collecting 800 specimens in two days. Again they made their way north to Chihuahua and El Paso. Around Chihuahua they spent three days and near El Paso went to Fort Bliss and west of the city, one day being spent on the road to Cusihuiriachic. They returned to Jimulco, visiting the spring and canyon near by and then went "up to the great cañon in the southwestern mountains." As they returned to Chihuahua, the engine of their train ran over an ox and was thrown from the track. But they reached their destination and again explored such regions as Bachimba Cañon, the Chihuahua Mountain, and the hills and river and its branches. The Santa Eulalia Mountains were explored once more as well as the canyon and "not waiting to get [their] 151 species up to 200, because of smallpox infesting the country," Pringle and Welcome returned to Charlotte, Vermont. On July 28, they were to return to much the same regions and remain until November 15, even more closely confining their collecting to near Chihuahua, the Santa Eulalia Mountains, Bachimba Cañon, the Chihuahua and Sacramento rivers, and the hills and valleys in the vicinity.

On February 27, 1886, Palmer wrote Watson:

---

[4] Pringle's trips are fully elaborated in his diaries published in Helen Burns Davis, *Life and Work of Cyrus Guernsey Pringle,* published by the University of Vermont, Burlington, May, 1936; the 1885 trip from pp. 19-24; second trip, pp. 25-32. He was accompanied by George H. Welcome. Dr. Francis W. Pennell has also added a valuable supplement to the work.

I wish to advertisse My sets of plants. Will you please send me such a copy of an advertisement as wil[l] be suitable to designate My sets from Pringles—some that have bought his may think mine are the same—You have seen boath sets and can tell best about the Geographical differences. . . .

Dr. Vasey think[s] it would be best the whole collection is published in one paper. If you are of the same opinion he wil[l] send you his grass notes with list when you desire. . . . I shall have twelve sets in a day or two. . . .

Less than two weeks before, Vasey, acknowledging receipt of Gray's Supplement and Indices, had commented to Gray:

We have now in the field such an excellent corps of botanical collectors, and the field of collecting has been so much extended lately, that the quantity of new discoveries is very surprising. The grasses have shared in the same expansion—during the past 5 or 6 years I think over a hundred additional species have to be recognized. I intend that you shall have specimens of every such addition whenever it is possible, and hope soon to send you a package, including those of Dr. Palmer's recent collection.

Parry wrote Watson: "When do you publish *Palmers* & *Pringles* plants (1885)?" And on March 9, Gray wrote Hooker:

Yes, I have got on Ranunculaceae, and have done up to and through Ranunculus, minus the Batrachium set, of which happily we have few in North America, that we know of. But having done some a while ago in the Gamopetalae of Pringle's interesting North Mexican collection, I am now switched off the same in a hurried collection made by Dr. Palmer, in an unvisited part of Chihuahua,[5] in which very much is new. One after another those Mocino and Sessé plants turn up. Also those of Wislizenus, whom the Mexicans for a time interned on the flanks of the Sierra Madre.

We are bound to know the botany of the parts of Mexico on our frontier, and so must even do the work. Pringle goes back there directly, with increased facilities, and will give special attention to the points of territory which I regard as most hopeful. . . .

Four days later Gray presented to the American Academy of Arts and Sciences another *Contribution to American Botany*, divided into two principal parts:[6] "1. A Revision of the North American Ranunculi," and "2. Sertum Chihuahuense," of which he remarked:

Next in interest to our own botany is that of the northern part of Mexico adjacent to the United States, and especially that of the elevated interior region. Two collections have been made . . . during the past year in the Mexican State of Chihuahua; one by Mr. C. G. Pringle, along the line of the Mexican Central Railway, in the spring and in the autumn of 1885; the other by Dr. Edward Palmer, from August to November of the same year, in the Sierra Madre of the southwestern part of that

[5] The particular localities are shown in Watson's "List of Palmer's Plants"—Hacienda San Miguel, Hacienda San Jose, Cumbre, Frayles, Norogachi, Yerba Buena, at altitudes varying from 2,400 feet to 8,850 feet, and all in a radius of 150 miles from Batopilas. Cyperaceae were done by Britton; Gramineae by Vasey; cryptogams by Eaton; some Juncaceae by Britton.

[6] A third part contained "Miscellanea."

State, with headquarters at the mining settlement of Batopilas, in some maps printed Batopolas. Both are collectors of experience in adjacent regions, particularly in Arizona whence Mr. Pringle had in former years penetrated into Sonora, very beneficially for botany, but to the damage of his own health; Dr. Palmer had made two important explorations in more eastward Mexican States, one in the year 1878, in connection with Dr. Parry, with headquarters at San Luis Potosi, the other in 1879 and 1880, in Nuevo Leon and Coahuila. Mr. Pringle is now returning to the promising field, making the town of Chihuahua his starting point. . . . Having now determined the Gamopetalae of these two collections of 1885, I here bring together some account of the new and otherwise noteworthy species.

Watson combined his "List of Plants Collected by Dr. Edward Palmer in Southwestern Chihuahua, Mexico, in 1885," with another, "Descriptions of New Species of Plants," this time from the Pacific states and Chihuahua, and his "Notes upon Plants Collected in the Department of Yzabal, Guatemala, Ranunculaceae to Connaraceae"—communicating them April 14 as another *Contribution to American Botany* from him. With these were "Notes upon some Palms of Guatemala," the more prominent of Watson's Guatemala collection. Of the twenty-five or so species of palms collected by Watson, "most," he said, "still remain undetermined."

Watson's observations were remarkably important, extending as they did the knowledge of North American herbaria to American plants beyond Mexican confines into Central America. Having gone to Guatemala to visit the father-in-law of Thomas P. James, he spent from the last of February to the middle of April making excursions along the lakes and rivers and into the mountains, observing a flora almost critically unknown and collecting about 500 species of plants. The month of March was spent at a plantation on the Chocon River about thirty miles by boat from Livingston situated on the Gulf of Amatique. The plantation was owned by the president of an American concern located in Boston and so Watson had not only the comfortable hospitality of friends but appreciation of the value of the work he was accomplishing. His investigations, however, did not go deep into the interior, being confined to Izabal, a town pleasantly situated on high ground along Lake Izabal and the eastern terminus of the long-used foot and horse highway, the Camino Real, leading westward across the mountains to the city of Guatemala. Of course, Watson made lake, river, valley, and mountain excursions from Izabal—three days being spent in Motagua Valley and ruins of Quirigia, and a journey was made up Polichic River—but malaria fever took hold of him and he was forced to return to the United States.[7]

[7] See Watson's Introductory Remarks to his "Notes," *Proceedings of the American Academy of Arts and Sciences*, XXI, p. 414.

Rusby's journey to Bolivia, South America, of about the same time, sent botanical searches by North American explorers south of both Mexico and Central America. Foreign and some American explorers had, of course, gathered some botany in South America, from which had issued determinations.[8] But few had gone deep in the interior and many large areas still remained totally unexplored. On December 16, 1884, Rusby had written George Davenport that on January 1 or 10 he was going via the Isthmus of Panama to Bolivia "on a botanical expedition in the interest of [a] most liberal and enlightened house"—Parke, Davis & Company, manufacturing chemists of Detroit, Michigan. "I shall be gone a year or more," wrote Rusby, "and shall make my way overland to Para, Brazil. I expect to look especially for ferns."[9] On July 25, 1885, he had written Gray from La Paz, Bolivia, thanking the doctor for a letter of good wishes for the trip, saying he had then collected 400 or 500 species of plants, and information of about 150 plants used as medical remedies:

While these plants[10] have no particular value, a knowledge of them is of great value to me in studying a thing which is to be my special work,—the relation between the botanical and therapeutical groupings of plants. About this center cluster a group of problems, relating perhaps as much to vegetable physiology as to therapy. . . .

Rusby wanted to be made state botanist of Bolivia and sought a recommendation from Gray, and from Watson and Farlow if possible. "Bolivia," said Rusby, "is the only country that remains (almost wholly) unexplored[11] and I should like very much to spend a few years here." On November 12, however, the matter was settled. Rusby wrote that lack of production in Bolivian mines and unfavorable exchange rates had placed the nation in such a wretched financial condition, such an undertaking was impossible and that he was returning "*via* Mapiri and Beni Rivers, to Para." However, he sought to describe the vegetation:

[8] For example, see "List of the Dried Plants Brought from Chile," determined by Gray, in Archibald MacRae, *Report of Journeys Across the Andes and Pampas of the Argentine Provinces.* Washington, Nicholson, 1855, Appendix G. Also, reference should be made to the Cornell Exploring Expedition under Professor A. N. Prentiss to the valley of the Amazon in 1870. This expedition went several hundred miles above Pará along the Amazon as well as to the rivers Xingú and Tapajóz but, while botanical material was collected, it seems geology was more emphasized. See "Albert Nelson Prentiss" by George F. Atkinson, *Botanical Gazette*, XXI (1896), p. 285.

[9] Rusby was sent to study coca, the basis of the little known drug, cocaine.

[10] Rusby mentioned especially *Fabiana imbricata* R. & P.

[11] See Gray's review, "Ball's Flora of the Peruvian Andes," *Am. J. Sci.*, 3rd ser., XXXI (1885), p. 231, where Gray says that vast regions of the Andes "have been visited at very few points and far between. . . . We are now only beginning to reach some conception of the role which the Andes and their prolongation through Mexico have taken in determining the character of no small part of the North American flora." Also in Sargent's *Sci. Pap.*, *op. cit.*, I, p. 384.

The most striking thing that I have seen is the resemblance of the flora of the coast a little south of the Guayaquill River[12] & southwest to that of our So[uth] Western Desert, not only in general features but in genera and even species. It is remarkable, because north of that point it has an aspect of tropical luxuriance. I wished much to collect there, but only had a few hours at the principal parts, and was horribly seasick, even when ashore. Certainly no observer could fail to be reminded of a time when there was a different sort of a connection between the two grand divisions. The abruptness of the change from luxuriance to sterility was also striking. Another very interesting thing was the dwarfer character of the vegetation on the table land between Tacna[13] and La Paz. I had never seen anything like it. . . . A vine—high-climbing—of the lower part of the western slopes became shorter and shorter as we ascended, until it was only a little herb among the stones. Besides this variation there was a distinct dwarf species on the alto, and another, very stout, as we descended the eastern slope. A shrub with flowers . . . was large and stout near the summit, but [proceeding] rapidly became a low mat of thorns. . . . This table-land is walled in on the east by the last Cordillera, a glistening mass of snow and ice, the most wondrously beautiful spectacle that my eyes have ever rested upon; and the imagination cannot surpass it. Illamani on the south, Sorati on the north, and Huaina Potosi in the center, with a dozen lesser but inaccessible peaks, impart to the whole ridge an aspect of unapproachableness. How strange it seems, how hard to realize, that from its summit a cannon ball could be sent tearing among the tropical vegetation upon the other side. We cross it six leagues east of La Paz, and begin our descent into Yungas, the richest province of Bolivia . . . there are so many ridges and deep valleys to cross, that we are two days in reaching the cool fields. But the ride is a delightful one, and as we pass among Fuchsias, Calceolarias, Begonias gorgeous Melastomaceae—the Rhododendrons of the Andes, bamboos and the ferns, with occasional groves of oranges, bananas and coffee plants, each turn bringing into sight—not something new—but a whole panorama of beauties, we really become sated and weary, of scrutinizing a world of which we know absolutely nothing. In Yungas I spent altogether more than two months, and collected some 700 or 800 species, 100 of them accompanied by wood sections. The leading family here is Orchidaceae, of which about 300 species can be found in a square mile. Probably Filices and Piperaceae come next, Piperomia apparently comprising ½ the species of the last mentioned order. Solanaceae is of course a large order. A family of plants (trees) with 3-ribbed leaves, and panicles of delightfully fragrant white (variegated with purple) flowers, is very conspicuous. Rubiaceae is perhaps even ahead of Filices, and all its sections are represented. Oxalis—as over all of Bolivia, is varied. There are few Ericaceae, but high in the mountains it is a very abundant class. Tupas and Tupa-like plants are common. Bigoniaceae are more abundant than Scrophulariaceae. Of course Leguminosae is generous with puzzling forms. Rosaceae are few but Rubus presents a number of curious species. . . . Urticaceae is common and very vicious— mostly tree-forms. Aroides are of course everywhere, the flowers of one deliciously fragrant at night. A few Cactaceae hang upon the trees. Tree composites abound, and I have become much interested in them. Ranunculaceae, Cruciferae and grasses except bamboos and sedges are very rare indeed. The same may be said of Umbelliferae. There are two species of Coniferae. Only a few Palms.

---

[12] Western coast of Ecuador.   [13] Northern part of Chile.

. . . With your sanction I shall hope to publish my notes through the Smithsonian.

Do not attempt to reply, as you cannot catch me. I hope to reach home in May, and to make my pilgrimage to Cambridge in June, at which time I shall hope to meet you.

Rusby, however, did not arrive home as soon as planned.[14] On January 3, 1887, he wrote Gray: "I hasten to announce the safe arrival of myself and my collections on the 27th ult[imo]. I am going to Detroit for two or three weeks and shall then return to New York and spend three to six months in the Meissner collection[15] working at my plants. My *twin*, Dr. Britton, is as eager for the fray as I am, and we anticipate a great feast. I have collected in the neighborhood of 3000 species." However, by the time he was prepared to go to Cambridge, Gray had other plans—Gray was going to Europe. On March 19 from the herbarium of Columbia College, Rusby wrote, "I had no idea that you were going to Europe until informed the other day by Prof[essor] Baird or I should have been to see you. I have planned to visit Cambridge with Dr. Britton during Easter. Now I shall wait until your return. I have some 20 sets of my So[uth] Am[erican] plants for sale, comprising between 2000 and 3000 species. The specimens are much better than what usually come from those countries."

On June 29 of the year before, Gray had written DeCandolle:

In various ways I am convinced that I am on the verge of superannuation. Still I work on; and now, dividing the orders with Mr. Watson (who, though not young, is eight or ten years my junior), we are working away at the Polypetalae of the "Synoptical Flora of North America," with considerable heat and hope. But it is slow work!

Gray had lamented that Tuckerman, "our lichenologist," and the explorers, Charles Wright and Augustus Fendler, were gone. So were the lesser known but nevertheless accomplished botanists George William Clinton and Samuel Botsford Buckley; and H. W. Ravenel was soon to go. Gray apprehended that his own time might not be far away. In the summer of 1886 Dr. and Mrs. Gray enjoyed a pleasant "holiday" time in Oneida County, New York, Gray's "natal soil," as he termed it. And, although in the late summer and early autumn Dr. Goodale took Watson away from the herbarium for a trip to Europe, and Sargent went on a trip to the southern North Carolina Mountains, Gray persisted with the completion of the *Synoptical Flora*. "Now I am going

14 Rusby wrote a book concerning his travels, entitled *Jungle Memories*. (New York and London: Whittlesey House, McGraw-Hill Book Co., 1933.)

15 Part of the Torrey Herbarium now part of the Columbia College Herbarium.

to pitch into Malvaceae.[16] I am quite alone," he told Hooker, who after twenty years of service had several months before resigned the directorship of the Kew Gardens to give his last years to finishing important studies. "If Sereno Watson," said Gray to Hooker on October 31, "will only go on with the Cruciferae, which he has meddled with a deal, and then do the Caryophyllaceae, which are in like case, we may by March 1st have all done up to the Leguminosae."

Gray, keeping track of collecting activities ranging from western Canada and Alaska to Central and South America, had had a busy time of it. It took a mind of considerable proportions to keep acquainted with the ever increasing significant discoveries of collectors such as Pringle, Palmer, Macoun, Rattan, Greene, Smith, Canby, and many others— besides continuing with his major task, the *Synoptical Flora of North America* and the work of the great European correspondents.

Pringle had continued collections in northern Mexico, going three times to Chihuahua and exploring in 1886 important Chihuahua regions: the Sacramento River, the Santa Eulalia Mountains, Mapula, Horcasitas, Samalayuca, Los Medanos, and other localities. While he regarded the season of 1886 as having proved unfavorable, he estimated he had collected at least 1,000 species; and the next year, making two trips —one from March to May and another from August to November—he went beyond Chihuahua to Mexico City including Chapultepec, Ortiz, and numerous Chihuahua areas. Pringle and Palmer seem to have crossed paths but little. At the time of these explorations, Palmer was west and south in the state of Jalisco, collecting in addition to Guadalajara plants seeds of economic plants which he believed valuable for cultivation in arid districts of Texas, New Mexico, and Arizona. As already explained, in 1887 Palmer went into Sonora near Guaymas.

Macoun's Canadian activities, as planned, had reached the Rocky Mountains again in 1885. On November 3 of that year he had written:

As I mentioned in [the] spring, I passed last summer in the Rocky Mountains and collected from the eastern Foothills to the summit of the Selkirk mountains within the "Great Bend" of the Columbia. My collecting ground was on both sides of the Pacific Railway and on an average about Lat[itude] 51°. My collections are simply immense and mostly in fruit and flower.

He had returned to put up sets of Rocky Mountain and Hudson Bay plants, having traveled on the Canadian Pacific Railway from the summit of the Selkirks to Ottawa—for parts of the trip the new railroad's first passengers. During the year 1886 Macoun had been sent as a Canadian representative to the Colonial Exhibition in London, but

[16] Gray had been at work on Portulacaceae, Montia, Claytonia, and Spraguea.

before leaving completed the writing of Part III of the *Catalogue of Canadian Plants*, which, said the *Gazette*, carried it through the Coniferae, and an addendum and index were contemplated bringing the whole *Catalogue* up to date. Macoun, probably partially on recommendations of Gray and Watson, placed many plant groups with United States botanists—Bebb, Boott, Bailey, Vasey, Coulter, and others —availing himself also copiously of the literature of United States botanists, for illustrations, David F. Day's list of species collected on the Canadian side of Niagara River and the shore of Lake Erie and Thomas Meehan's list of species collected by him on the British Columbian and Alaskan coasts in 1883.

Indeed, under Gray's leadership, for the most part United States collectors kept busy, maintaining in spite of numerous minor disagreements and differences the best sort of spirit. In September 1886, the *Gazette* commented:

> It is probably safe to say that the botanists form the best compacted organization of scientific workers in the country. Their work demands the most widespread exchange of facts, and this has led to correspondence which has often ripened into friendship.

In these years, not only was friendship prevalent but brotherhood and a sense of comradeship dominated. Botany was the envy of other sciences. Although in some respects it may be said to have retarded progress, certainly in North American botany's fundamental task—becoming acquainted with its flora—its closely knit organization contributed toward real accomplishment.

A slowly developing spirit of competition was arising. Nathaniel Lord Britton was yearly becoming more aggressive as a plant systematist. Like Edward Lee Greene, he welcomed comers of ability who brought him plants—and with sincere and praiseworthy motive and spirit. Competition does promote efficiency and often hastens the ascertainment of truth. Gray or Engelmann did not fight competition as such. They knew that in their work, since science was combined with the open field and forest, as well as the farm field and garden, science's service was offered to all worthy comers, and no monopoly would ever be possible, nor was it desirable. Gray at Harvard and Engelmann at St. Louis had accumulated, however, through long years of labor the most adequate facilities. The best work in their lines could be accomplished by collaborations with them. This made for efficiency. Neither had sought deliberately to dominate the field exclusively. What domination existed had been placed with them by force of circumstances. And their positions entailed responsibilities, the most important of which was the maintenance of standards.

The years of the future would not deal so much with learning what nature had done—understanding her works—but would progress with a view to learning what nature could be made to perform. In this Gray and Engelmann laid foundations. The real structure would be raised when North American Botany's first great Transition Period would be complete and the great new work in physiology, anatomy, morphology, mycology, pathology, horticulture, pre-Mendelian genetics, the organization of the knowledge and identification of the plants of the world, and in time, utilizing results of paleobotanic research, would go forward. In March 1886, the *Gazette* editorialized:

Botany in America was never in a more flourishing condition than at the present time. American systematic work, especially that emanating from Harvard, has long stood in the front rank, but other departments of the science have not until recently been so assiduously or successfully cultivated. The study of the anatomy, development and habits of plants received a great impulse by the advent of Sachs' Textbook in 1875, and was especially promoted by Bessey's *Botany* in 1880. The latest addition to this line of text-books, Goodale's Physiological Botany, attests its excellence by receiving commendation, not only at home where it was expected, but abroad. A critical review in the *Botanische Centralblatt* speaks of it as marking an important event for American science, and ranks it in some respects above the text-books of German writers. The *Gardeners' Chronicle* of England calls it "one of the most useful summaries yet issued." This may be taken as an index to our advancement in the teacher's sphere. It would not be hard to trace a connection between good didactic works and the increase of original research. In the latter we are surely making notable progress. *Nature* in noticing the Association number of this journal, took occasion to say of the botanical papers presented at Ann Arbor, that "these furnish satisfactory evidence of the good work done in this branch of science on the American continent, and will not suffer from comparison with a similar record at any of the recent meetings of our own (British) Association." Some of the papers are mentioned as "giving especially good evidence of a capacity for original work." American botanists may well feel encouraged at these signs of intellectual prosperity.

What restrained much physiological work was, as the *Gazette* later revealed, the cost of appliances and apparatus. Want of satisfactory and sufficient instrumentation kept systematic botany uppermost or at least of equal importance. Nevertheless, the number of substantial American botanists "who would willingly exchange all [their] chances in systematic work for a good opportunity to follow out [their] physiological bent" was growing. Gray knew this. To bring about an adequate development in this regard in America, he selected that most remarkable teacher—George Lincoln Goodale—whose ability to keep his students alert, instruct and entertain them, and inspire them toward research accomplishments had as much to do with experimental and research development in botany as any single factor.

Similarly, the radiance of Farlow's teaching in America cast a glow that lit up the vast comparatively unexplored and often minute study of the lower plants. Previous to his European studies, it is said, he practically regarded himself as a teaching failure. A none too strong voice reacted unfavorably. However, acquaintance with new laboratory methods and a vast undeveloped subject brought immediate success and he was soon counted among world authorities. Farlow believed that advanced systematic work must be done by qualified experts having access to large collections and libraries. Dr. Arthur has told that Gray and Farlow in earlier years believed that "for the good of American science no naming of phanerogamic or cryptogamic plants, particularly the latter, should be undertaken outside the precincts of Harvard University, because in no other place was there to be found adequate material for comparison."

As years passed, this view was altered and, with the science's enlargement, completely changed. Still, Farlow, as late as 1887, believed physiological work of high grade could be done only at the few well equipped laboratories. To all workers, whatever their locality or equipment, histology, the study of plant life histories, the "art of specimen making," collecting, and the making of field notes were available. As the gradually developing subjects—ecology, pathology, plant breeding, et cetera— emerged from relative obscurity to prominence, these, too, were added, depending on the worker's proximity to sources of authentication. In the late eighties, both colleges and stations were available; as well as institutions and the government services.

For the most part, however, Gray remained an observer rather than participant in activities dissociated from his main task, the *Synoptical Flora*. He prepared numerous systematic notices and revisions and in 1886 completed a Supplement with indexes—the Gamopetalae, being a second edition of Volume I, Part II, and Volume II, Part I, collected. During the last part of the year, Gray and Watson were working at Polypetalae, and all work—even Gray's next *Contribution*, "Revision of Some Polypetalous Genera and Orders Precursory to the Flora of North America; Sertum Chihuahuense;[17] Appendix; and Miscellanea"—were in aid of the work. Orders were divided with Watson. The task was too much for one man.

Matters of importance were developing in all parts of the country.

17 New Gamopetalae of Pringle's collection in the state of Chihuahua in 1886.

Illustrating with developments in one area—California—Gray aided Volney Rattan with publication of his *Analytical Key to West Coast Botany*, which characterized more than 1,600 species, mostly flowering plants, and received high praise from the *Gazette*. In 1886 Parry had gone via the southern route to California and from San Francisco written Gray, telling of the railroad's opening the Shasta country and urging Gray to confer with Senator Stanford concerning a department of science at the newly proposed Leland Stanford University:

I hope you will be able—without going out of your way—when in Washington to interview Senator Stanford—and ascertain his views in reference to the Bot[any] Dep[artmen]t of his University—he seems willing to take advice and is moving deliberately. What you might suggest to him I have no doubt would have weight —especially as he will see that you can have no personal interest. What is particularly desirable is to suggest the remarkable capacity of the location for horticultural enterprize—which if placed on a scientific basis would make it one of the wonders of the world.

On October 30, 1887, three weeks after returning from his last trip abroad, Gray answered Parry:

I have as yet seen only some of Greene's modest doings since I returned home. But I am in no hurry. I experience great relief of mind in no longer feeling it a kind of duty to understand what he is about and to try to restrain his vagaries. I am too old to take all this trouble, and it would be of no use. He will make a deal of trouble—which those who come after him may take their share of. I have done my part.

I suppose I shall have to go to Washington next month. I may meet Gov[ernor] Stanford. But if I can be of any use to him or to you, as to botanical matters, it will have to be by his initiative. I do not feel at liberty to obtrude advice on him —tho' I entertain a good opinion and no small admiration of the man for all he has set about doing.

Your old friends at Kew & elsewhere asked after you. I stayed at Kew very little—Hooker having moved away so far. Nor did I do much Botany.

Horticultural, agricultural, and botanical enterprises were given an early start at Leland Stanford University. On the Palo Alto ranch, now the university campus, was set an arboretum, where with cooperation of prominent botanists, notably Parry, a large variety of foreign and domestic trees and shrubs were planted to test their adaptability to local climate and conditions. Many experiments in agricultural lines, furthermore, not under Stanford's direction or on his own property, were assisted by him.

Whether Gray consulted Stanford is not known. He probably did. And in 1891 when David Starr Jordan was made president, D. H. Campbell went with him from the University of Indiana to take charge of botany, another recognition of a European and American trained investigator who would advance research. Parry in 1887 continued study in California. Greene was, Parry wrote, a "full fledged Prof[essor] of Bot[an]y" at the University of California.

Strictly speaking, Greene in 1887 was an assistant professor. Not until 1893 was he made professor of botany at the Berkeley institution. However, he is regarded as the founder of the department which had its formal beginnings in the college year 1890-1891 when "regular and systematic work upon the University Herbarium was begun as a unit of the Department's activities," when an assistant was provided, and when in the following spring a garden, The Garden of Native Plants, was laid out and filled with flowering plants from various parts of the state.

Gray found he had to go to Europe. He and Mrs. Gray landed in England, proceeding from Liverpool to Sunningdale to visit Sir Joseph and Lady Hooker. On May 1 they went on to London where Gray did some work, going back and forth, at Kew. Lamarck's herbarium was his objective and so they crossed to Paris, and at the Garden of Plants herbarium Gray completed his study of Asters, having examined all of the genus in important world herbaria. From there they went to Vienna, greatly enjoying the trip through Bâle, Zurich, and Salzburg. At Lake Zurich the fruit trees were white with blossoms. The Arlberg Pass from Feldkirk to Innsbruck impressed Gray by its height. The journey through the Lower Tyrol and the Salzburg Salzkammergut Gray described as "exquisite and wild, and in parts grand." Other choice scenery presented itself and although rain spoiled much of the visit at Salzburg and on the Danube, Vienna was enjoyed with visits at the Natural History Museum and the Academy at Schönbrunn. Returning by way of Salzburg and Munich, they crossed Lake Constance to Zurich and went to Geneva where Gray spent an hour with the aged but cheerful De-Candolle, as much as anyone, other than Hooker, Gray's mentor in botany.

For almost half a century, Gray had been reviewing in American periodicals the works of the DeCandolles, father and son. The great *Prodromus Systematis Naturalis Regni Vegetabilis* had been reviewed in the *American Journal of Science and Arts* beginning about 1839 and continuing through 1873. In 1855 had appeared DeCandolle's *Géographie Botanique*[18] which Gray soon after reviewed in the same *Journal*. An article had appeared from Gray's pen, entitled, "Alphonse De Candolle on the Variation and Distribution of Species," in Volume 35, sec-

---

[18] *Géographie Botanique raisonnée, ou Exposition des Faits principaux et des Lois concernant la Distribution Géographique des Plantes de l'Europe Actuelle.* Paris and Geneva.

ond series, and in 1878 there was published Volume I of DeCandolle's *Monographiae Phanerogamarum Prodromi nunc continuatio, nunc revisio* which was also reviewed by Gray. About 1880 appeared Gray's review of *La Phytographie* styled "De Candolle's Phytography"[19] and a few years later Gray's review entitled "De Candolle's Origin of Cultivated Plants."[20] In each, Gray so elaborated the DeCandolle materials that in some the reviews seemed partially like original articles though always remaining faithful to the work at hand. Some of the ablest writing of Asa Gray is contained in his reviews of the works of his great friends across the Atlantic waters, Darwin, Hooker, Bentham, and the DeCandolles. In fact, in his very able article on "Botanical Nomenclature," Gray embodied a large part of his nomenclatural views, basing them on a review of DeCandolle's *Nouvelle Remarques sur la Nomenclature Botanique*, published at Geneva in 1883. To the DeCandolles Gray owed much and of this debt he was always mindful.

From Geneva, the Grays traveled by railroad to Belgium and the Netherlands, going along the Rhine part way and to Brussels, Amsterdam, The Hague, and Antwerp. They then returned to Paris. Decaisne[21] was gone as were other friends of Gray's and as he worked at the Garden of Plants he was reminded of his many associations of years past. The Grays found time for considerable travel around Paris, however, and on June 14 crossed to England where, going to Cambridge with the Hookers and as guests of Mrs. Darwin, Cambridge University honored Dr. Gray with the D.S. degree. Said Dr. Sandys:

And now we are glad to come to the Harvard professor of Natural History, *facile princeps* of transatlantic botanists. Within the period of fifty years, how many books has he written about his fairest science; how rich in learning, how admirable in style! How many times has he crossed the ocean that he might more carefully study European herbaria, and better know the leading men in his own department! In examining, reviewing and sometimes gracefully correcting the labors of others, what a shrewd, honest and urbane critic has he proved himself to be! How cheerfully, many years ago, among his own western countrymen was he the first of all to greet the rising sun of our own Darwin, believing his theory of the origin of various forms of life demanded some First Cause, and was in harmony with a faith in a Deity who has created and governs all things! God grant that it may be allowed such a man at length to carry to a happy completion that great work, which he long

[19] *La Phytographie, ou l'Art de décrire les Végétaux considérés sous différents points de vue.* Paris.

[20] *Origine des Plantes Cultivées.* Paris. See Sargent's *Scientific Papers of Asa Gray, op. cit.,* Volumes I and II, where these publications with references to Gray's reviews are fully set forth as footnotes. References to *Am. Jour. Sci. and Arts* are contained.

[21] Gray had reviewed Decaisne's *Monograph of the Genus Pyrus* in the *American Journal of Science and Arts,* 3rd ser., IV, p. 489; X, p. 481. See Sargent's *Scientific Papers of Asa Gray, op. cit.,* I, p. 186.

ago began, of more accurately describing the flora of North America! Meanwhile, this man who has so long adorned his fair science by his labors and his life, even unto a hoary age, "bearing," as our poet says, "the white blossom of a blameless life," him, I say, we gladly crown, at least with these flowerets of praise, with this corolla of honor (his *saltem laudis flosculis, hac saltem honoris corolla, libenter coronamus*). For many, many years may Asa Gray, the venerable priest of Flora, render more illustrious this academic crown.

On June 22, Oxford University gave Dr. Gray the D.C.L. degree; and after much entertainment and meeting many celebrities, the Grays went to Edinburgh where the University there conferred the LL.D. degree on Dr. Gray. In 1887 four honorary degrees were awarded Gray, the fourth being conferred by the University of Michigan in the United States— totaling in all more than half a dozen honorary degrees[22] from universities in England, Canada, and the United States in a period from 1860 to 1887.

For a while Dr. and Mrs. Gray traveled in England and France, meeting the Hookers at Rouen and separating from them at Mont St. Michel. They continued their journey, going by way of Chartres, Rouen, Amiens, and then returned to England where Gray went to Harpenden to become acquainted with the famous experiments in agriculture at Rothamstead. A trip to Canterbury was made and toward the end of August, Gray went to Manchester where, attending the British Association's meeting, Gray seconded the opening address, and was a guest in a home where De Bary and Saporta[23] were. From Manchester the Grays went to Failand to visit Sir Edward Fry; then to Gloucester; and then returned to Kew with Dr. and Mrs. Oliver.[24]

The visit with the Olivers stirred Gray to return to America and take up the work of writing accounts of older botanists he had met on earlier trips to Europe. To this intention he was also persuaded by H. G. Reichenbach, professor of botany at Hamburg, and an authority on orchids.

Accordingly, after his return to the United States and a journey to Washington to aid as a regent of the Smithsonian Institution in selecting a successor to Spencer Baird as director, who had died, Gray took up several tasks—Vitaceae for the *Synoptical Flora*,[25] a review of Darwin's *Life and Letters*, the writing of the annual necrology for the *American*

[22] The degrees are set forth in the *Letters of Asa Gray, op. cit.*, p. 825, together with a list of the societies of which Dr. Gray was a member. About sixty-six societies are there listed.

[23] Under the title, "Plant Archeology," Gray had written on *Le Monde des Plantes avant l'Apparition de l'Homme*, by le Comte de Saporta, for *The Nation*, Numbers 742 and 743 (September 18 and 25, 1879), and in the course of the article referred to Lesquereux's researches in carboniferous flora and Silurian botany.

[24] The many other activities and visits of the Grays are described in the *Letters of Asa Gray, op. cit.*, pp. 800-810.

[25] Fascicle 1 and 2 of Gray's *Synop. Flora of N.A.*, I, Part I, *Polypetalae*; Ranunculaceae to Polygalaceae; publication edited by B. L. Robinson, New York, 1895-1897.

Journal of Science and Arts, and the planning of historical accounts of older European botanists.

On Thanksgiving Day, he went to Boston for dinner and after receiving a caller concerning a flower of the southern Alleghenies, sat down to write a letter to Nathaniel Britton which was "important and must be written," he said. Coulter had called Gray's attention the previous December to Britton's systematization of *Conioselinum canadense* as "*Conioselinum bipinnatum* . . . a raking up of old specific names" based on priority as shown by an issue of the *Bulletin of the Torrey Botanical Club.* Gray reminded Britton that nomenclatural rules were violated by "giving a superfluous name to a plant, and also," he said, ". . . in all reasonable probability your name is an incorrect one. . . . We look to you and to such as yourself, placed at well-furnished botanical centres, to do your share of conscientious work and to support right doctrines. So I may proceed to say that, upon the recognized principles since the adoption of the Candollian code, your name of *Conioselinum bipinnatum,* even if founded in fact, would be inadmissible and superfluous. By a corollary of the rule that priority of publication fixes the name [as found in the *Flora of North America,* of Torrey and Gray, I, page 619], taken along with the fact a plant-name is of two parts, generic and specific, it follows that in any case, *Conioselinum canadense* is the prior name for those who hold to the genus *Conioselinum.* I have laid down what I take to be the correct view as to this, in my 'Structural Botany,' paragraph 794, where it is supported by the high authority of Bentham. . . . If you like to adopt [a minority practice described] you have at hand a still older, the very oldest, name, namely, *Conioselinum chinense,* for I can certify that the plant we are concerned with is *Athamantha chinensis* of Linnaeus."

Gray, the next morning, suffered a slight shock in the right arm and, although he sent to friends two copies of his "Review of the Life of Darwin," a more severe shock recurred and the next day the capacity for connected speech was gone. The *Gazette* noticed his illness in November 1887. On December 18, Goodale informed Bessey that Gray's attack of hemiplegia on November 28 had left him helpless and almost speechless and that morning he was much weaker. On January 30 Goodale wrote again: "Dr. Gray passed away after a day of suffering, at seven thirty this evening." On February 2, 1888, Gray was buried in Mount Auburn Cemetery, "where a simple stone, bearing a cross, [soon marked] his grave, with his name and the dates 1810-1888."

Sereno Watson, however, was still at the Gray Herbarium to carry on Gray's great work—with hope of carrying it to its completion.

CHAPTER XIII

Another Generation of American Botanists Nears the End

Leo Lesquereux, one of Watson's most intimate friends, achieved, like Asa Gray, botanic individuality before his death. During his last years, one did not refer to the work of "Torrey and Gray" or "Sullivant and Lesquereux"—but to the work of Gray; or to the work of Lesquereux. Gray earned and merited the heritage left him by Torrey. He then built on it and finally put the star of Torrey into partial eclipse. Lesquereux did likewise. Knowledge and the botanic inheritance bequeathed by Sullivant to Lesquereux was built on in *Manual of the Mosses of North America*. But Lesquereux was not primarily known for this. He was known by his own right in paleobotany, having by his researches laid the foundations for one of the most remarkable extensions of knowledge in the science of botany—the knowledge of structures of fossil plants developed by Penhallow, Wieland, Coulter, Bessey, and others in both laboratory and field. June 24, 1886, Lesquereux wrote Watson: "You think perhaps that I am already dead, buried perhaps. Not yet. Always hard at work with little profit either for myself or for others, I am at a new volume of Cretaceous plants for which I have splendid materials. . . ." He had "always plenty of materials on fossil plants to work upon as far as [he was] able" and so continued "burrowing into the dark field of vegetable paleontology."

Lesquereux lived long enough to complete most of *The Flora of the Dakota Group*,[1] a posthumous publication, which brought the known species of the group to 460 and in which Lesquereux expressed his final belief that the explanation for the supposedly sudden appearance of dicotyledons lay in further examination of the Middle Cretaceous—the Cenomanian period. This was not the completion of Lesquereux's work. In 1885 a large amount of fossil plant material, accumulated at the Smithsonian Institution since its founding, was published as a *List of Recently Identified Fossil Plants Belonging to the United States Museum*.[2] F. H. Knowlton took his list, further compiled and prepared it for publication and the *List* was issued from April 25 to May 17, 1887. A similar procedure evidently was followed in the publications of a *List of Fossil Plants Collected by Mr. I. C. Russell, at Black Creek, Near Gads-*

---

[1] Edited by Frank Hall Knowlton. Washington: Government Printing Office, 1891.

[2] *U.S. Nat. Mus. Proc.* (Washington), X (1887), pp. 21-46. Several new species were described.

den, Alabama, with Descriptions of Several New Species,[3] and *Recent Determinations of Fossil Plants from Kentucky, Louisiana, Oregon, California, Alaska, Greenland, etc.*, with descriptions of new species.[4]

With the Museum of Comparative Zoology at Cambridge, Massachusetts, Lesquereux also collaborated. In 1881 had appeared his *Report on the Recent Additions of Fossil Plants to the Museum Collections*[5] and in 1888 the museum published Lesquereux's list of 118 species, 28 new, *Fossil Plants Collected at Golden, Colorado.*[6] With each there were enumerations of numerous localities, the geologic age to which each specimen probably belonged, and the names of the large number of collectors interested in the science.

The work of the United States Geological and Geographical Survey of the Territories was completed and in 1884 John Strong Newberry, and not Lesquereux, was appointed a paleontologist of the United States Geological Survey. Newberry was an authority in mining and metallurgy. And the survey was interested casually in paleobotany. From Newberry came important monographs on fossil fishes in 1888 and 1889, and he prepared two monographs on fossil plants. But Lesquereux's work sustained an equal, if not greater, prominence as to fossil flora. In 1890 there issued Lesquereux's *Remarks on Fossil Remains Considered as Peculiar Kinds of Marine Plants,*[7] emanating in part from remains collected around the shores of Lake Erie. This was another posthumously published work, however.

On March 14, 1888, Sereno Watson communicated to the American Academy the posthumous *Contribution to American Botany* by Gray styled, "Notes upon Some Polypetalous Genera and Orders, Rutaceae, and Vitaceae"; at which same time Watson presented the fifteenth of his *Contributions to American Botany*. Watson's *Contribution*, XV, became notable not only for Pringle's 1887 Chihuahua plants and descriptions of some plants from Guatemala but for the revisions of Lesquerella (Vesicaria) and of the North American species of Draba. To the Draba revision, Greene took heated exception. But to Lesquereux, Lesquerella was like a crowning event of his life. On June 9, 1888, Lesquereux wrote Watson:

You have indeed a peculiar genius for creating fine words, or for transforming into euphonious names such ones which, like my own, are repulsive to ears and

[3] *U.S. Nat. Mus. Proc.* (Washington), II (1888), pp. 83-87. Plates XXIX, issued Nov. 8.
[4] *Ibid.*, pp. 11-38. Plates IV-XVI, issued Nov. 8.
[5] *Bull. Mus. Comp. Zool.* (Harvard University, Cambridge), VII. (Geol. ser., 1, Number 6.)
[6] *Ibid.*, XVI. (Geol. ser., 2, Number 3.)
[7] *U.S. Nat. Mus. Proc.* (Washington), XIII (1890), pp. 5-12. Plates I, issued July 18. Note: Evidently all of these were issued as separate pamphlets or publications, in most instances, the year following publication.

eyes by orthography and pronunciation. I could never read *Lesquereuxia* without shrieking. Now, *Lesquerella* is like one of those delicately flattering italian diminutives not only acceptable but really lovable. I thank you heartily for this appelation and too for the complimentary remark explaining to right of admittance. . . .

Lesquerella is a Cruciferae genus—bladder-pods of the mustard family. In Lesquereux was much of the poet. Although his stanzas were unpublished, he spent many pleasant hours composing poetry, usually writing in French forms. Lesquerella satisfied both his poetic and scientific sense. And coming from Watson, whose other *Contributions* he admired, he was especially pleased.

On July 20, 1887, Lesquereux had written Watson:

The communication of your beautifull and most valuable memmoir on the plants of Jalisco, Mexico, is gratefully received. . . . For myself I remain the same, enjoying the company of my fossil plants and quietly waiting the end of this world's voyage. It has been already very long as you know.

The voyage of Lesquereux ended two years later—at the age of nearly 83 years. The date was October 25, 1889, and Watson lost a firm friend. Paleobotany lost the man who, more than anyone else in North America, laid the foundation structure on which the science built. Indeed, by many, among them, J. Peter Lesley, Lesquereux was regarded as the world's greatest fossil botanist. Much of his work has since been revised but his place in history stands—and a proud place it justly is. Hearing of Lesquereux's death, Lesley commented, "It will be like missing one of the great stars from the sky, Vega, Aldebaran, or Sirius."

The "valuable memmoir on the plants of Jalisco, Mexico," to which Lesquereux in his letter to Watson referred, was Watson's *Contributions*, XIV, presented in 1887, the same year Watson published in the *Torrey Club Bulletin* on the genera Echinocystis, Megarrhiza, and Echinopepons.[8]

Engelmann was gone. Gray was gone. Lesquereux had left the ranks. It is the way of work and life—the young keep alive the work of their elders. More than a month before Gray became ill, Coulter wrote the doctor:

It is good to have you within reach again, for some of us have run rampant, & you will have your hands full suppressing Greene, Coulter, *et id omne genus*. . . . The *Gazette* has kept you posted in botanical news; as for personal, Barnes has gone to Univ[ersity of] Wisconsin, & Arthur succeeds him at Purdue. I have had several good chances for a change of base, but it would always involve the accumulations again of herbarium & library, & I have no time to do that. The *Manual* is progressing steadily, & I will have no trouble in keeping up with the demands. As you know, for a year now I have been eating, drinking, & sleeping

[8] XIV, pp. 155-158.

Umbellifers. They have fairly possessed me. We have finished our Eastern job.
... We have now turned our faces towards N[orth] Am[erican] Umbellifers & are
making good progress. . . . We don't propose to do any more publishing until the
whole thing is done.

The older taxonomists had left systematics in a condition ready for
changes. Greene, undoubtedly, and Coulter, partially, were neither satisfied. Britton, too, was showing evidence of discontent. They were not
of necessity a new school. A new morphology, a new anatomy, a new
physiology—as far as each in its slow progress had gotten—a developing
mycology, an almost wholly new pathology, and the beginnings of a
reduction of the art of plant breeding to a science, together with a sustained interest in study of business and industrial phases pertaining to
the growing, marketing, and distribution of plant products from the
field, farm, garden, orchard, and forest—not excluding realization of
great potentials for progress to be furthered by going to all parts of the
world and securing plants new to American cultivation, or capable of
improving plants already cultivated—widened, intensified, deepened,
and rendered more exact the scope of botanical, indeed all plant science
knowledge. In the laboratories, the microscope had presented new fields
of study in minute structural differences and relationships. The challenge of the great new science of bacteriology had been presented by
Burrill's and Arthur's microscopic analysis of the alarming fruit disease
known as pear blight. At Washington in the United States Department
of Agriculture, under leadership of Erwin Frink Smith and others, this
challenge would be taken up and from his careful, exacting methods of
research would issue a real plant pathology trailing—but of proportionate value, economically, to—human and animal pathology. Such great
work as that of Herbert Hice Whetzel at Cornell and Lewis Ralph
Jones at the University of Wisconsin, as well as other leaders, would
be patterned in large part after the work of Smith. Soils investigation
would become more and more correlated with study of geological formations, plant societies, chemical and physical constituents, and organic,
including disease, elements, of various soil types. The foundations for
developing a scientific forestry would be laid. Research in plants would
enlarge to aid search for knowledge concerning animal origins and
development. As a matter of fact, in arriving at solutions of problems
concerning human beings, plant study would be utilized. As paleobotany added discovery of theretofore unknown plant groups, the orbit
of scientific research in earth materials, their origins, their processes and
schemes of development—in reality, the whole of evolutionary study in
time and space—would be wondrously widened. Suddenly it would

seem as if the science of botany, a branch of a fundamental unity of all scientific study, had expanded manyfold and segregation of divisions within its own sphere become necessary. Experimentation and researches were revealing the conventional need of training two types of workers—the man of practical applications of knowledge and the man concerned with "pure science" investigations of biological laws. Fundamental to each was the work of the taxonomist. Before a plant could be studied, obviously it was necessary to know what plant was being studied and what were its systematic characteristics. However, as knowledge increased concerning the "physiological species," so-called, as distinguished from the more or less arbitrarily defined taxonomic species, it was natural that some dissatisfaction with established procedures in systematics should develop, and new theories and doctrines in taxonomy should be urged. One, such as Greene, who took little interest in evolutionary study, was dissatisfied because he sincerely believed that systematics, as constituted, did not convey the whole of field knowledge. Another, such as Britton, who became considerable of a student of evolution was dissatisfied because he believed, praiseworthily, that the need of future taxonomy was to harmonize methods in all branches of scientific study. Liberty Hyde Bailey became a great student of the botany of the garden. His work did not deal foremost with herbarium material, although he realized its importance and maintained adequate herbarium facilities. Coulter viewed systematics from the point of view of the laboratories which investigated life histories of plants and phylogenies. Gray, however, saw that all the new work was built on the work begun by Torrey and the early students of classification. He must have known quite certainly there would be revisions but, revisions or not, the basic work he had carried on must be finished as far as humanly possible. Lesquereux in paleobotany knew, before his death, much of his work would be changed. A study of the history of the science shows quite definitely in a great number of instances that the great systematists, before their work was completed, realized that progress and scientific advancement would make necessary revisions of much of their cherished and genuinely enjoyed work. As Coulter later said, the self-gratulation of one age is always the wonder of the next.

Before Gray's death, Parry, still on the Pacific Coast, had written:

Mr. Greene keeps on the *un*even tenor of his way "slashing["] right and left. [H]e has fallen out lately with his Academy associates and will publish hereafter in his own publication (*Pittonia*)! and in his *own way*. I think he has lately taken up Polemoniaceae and will of course do some revolutionary work. . . .

Not all were revolutionary like Greene. Coulter stood for an evolutionary development in taxonomy. Britton, although insisting on a few new principles being adopted into practice, was more steady than Greene. Indeed, his "List of Plants Collected by Miss Mary B. Croft at San Diego, Texas, Near the Headwaters of the Rio Dulce,"[9] and other matters, especially those with regard to plant life near New York indicated clearly that Britton was someone of promise.

On November 30, 1887, not knowing that Gray was ill, he had replied to Gray's last letter:

I thank you sincerely for your kindly criticism of my use of a new binomial for *Conioselinum canadense*, Torr. & Gray which, notwithstanding the kind tone of your letter has evidently displeased you, and as to [this], I deeply regret it.

I would not have you think that I made that name simply for the very poor satisfaction of coining a new one. I worked over it a good while and supposed that I was moderately sure of having taken up the oldest specific name for the plant, for I concluded from a careful examination of Walter's book that he must have known the species in question. And as it comes from the Carolina mountains, at all events, this is not so unlikely.

The *Athamantha chinensis* part of it I was not familiar with, though your letter reminds me that you had a note on it not long ago. Had I seen that at the time I should if satisfied with the reference, written *Conioselinum chinense* as you have done in your letter.

There is no doubt that there are, as you say, pitfalls for young botanists, and I presume the proper thing for them to do is to avoid writing anything. Indeed, as you may gather, I am considerably discouraged at my first attempts at changing a name though of the propriety of using the oldest specific name I am not yet decided what ground to take.

I had rather talk with you about this than write. At all events I thank you for the consideration with which you have treated me.

On January 1, 1888, during Gray's critical illness, Britton wrote Watson:

It is impossible that you are having a Happy New Year, though I sincerely wish it were otherwise. Please accept my high regards at all events, and sympathy for you all in this great trial. Is there, perhaps, a favorable turn in Dr. Gray's illness, or is there still no encouragement? . . .

And on April 8, Britton further elaborated to Watson:

I dislike above all things to hurt anyone's feelings, and it has been only after a very careful study of the matter and a survey of the ground covered and the questions involved that I have brought myself to what you designate "the craze." It is no craze with me, but the deliberate result of my judgment (I am imagining what kind of judgment you are assigning me!)

And you shouldn't, in fairness, charge me with making new names for old plants for that is just what I deprecate as much as you possibly can. I am only contending

[9] With Henry H. Rusby, New York, 1888. *Contribution I from Herbarium of Columbia College*. Reprinted from *Trans. N.Y. Acad. Sci.*, VII (1887-1888), pp. 7-14.

for old names, which never ought to have been laid aside, and for the continued use of these first names in whatever genus the plant may be located. This offers the only probability of stability for nomenclature, and with its strict application we can bring our names of flowering plants into harmony with those of the cryptogamists and the zoölogists and all use the same general system. Is not this a result worth trying hard to attain? And I beg you to understand, my dear Doctor, that if I should, perhaps, move a name of your giving there is no personal feeling in any way connected with it, for you, of all men, I have the very highest regard. . . .

Britton did not emphasize his views as Greene did. Coulter, likewise, while much interested in morphological researches, was not yet wholeheartedly arguing for extension of morphological findings to taxonomy beyond structural determinations which Gray and Watson incorporated.[10] What puzzled Britton more at this time were the applications of rules permitting reestablishment of old names—the first specific name of a species or the first name of a genus abandoned or changed without adequate cause or reason—a subject concerning which Greene had also disputed with Gray. Greene went the whole length arguing for the enlargement of taxonomic descriptions to include morphology and, also, to permit the systematizer to reintroduce the oldest specific name or the oldest name of a genus where more thorough research showed clearly the old names should never have been abandoned. In the latter phase, Britton seemed to join with Greene, although he seems never to have expressly admitted it.

On May 16, 1888, Coulter wrote Watson:

Have you looked over the Torrey Catalogue of "Anthophyta" & Pteridophyta?[11] What *can* we do with Britton? He goes slap dash at the thing, without any critical knowledge whatever, & hauls up old specific names which may or may not have been applied to the species. Look at *Conioselinum canadense* called *C. chinensis* on the strength of your synonym *?Athamantha chinensis*L. How did he know it was that any better than you did? If such changes are to be made they should never be made by catalogue makers, but only by monographers, who have the requisite knowledge. I anticipate a sickening confusion of synonymy in the near future. If, however, we hold on to the *manuals*, catalogue makers can have but little influence.

Almost from the beginning of their relationship, Gray had held Coulter in the best sort of regard as well as Coulter's associate, Charles R. Barnes. Barnes had gone to the University of Wisconsin. He was not eligible to a new appointment. But Coulter was, if he would accept one. On April 15, Britton wrote Watson:

[10] An argument, contra, is possible—based on Coulter's *Synopsis of the Pines*, predicated as it was on a study of leaf anatomy. However, this seems to have been more the exception than the rule with Coulter at this time.

[11] With E. E. Sterns and Justus F. Poggenburg and others. *Preliminary Catalogue of Anthophyta and Pteridophyta Reported as Growing Spontaneously within 100 Miles of New York City*, New York, 1888. 90 pages. Map.

Is there really any truth in the story that is going around that Prof[essor] Coulter is to take Dr. Gray's professorship? I had never for the moment thought of any one but yourself in such position but it has been mentioned by some and I believe the newspapers had it. Are you well supplied with plants from Portugal and the Azores? . . .

Coulter must have been at least considered for the position, as Watson was primarily a curator and not teacher and considerably advanced in age, his beginning in botany having commenced well after his middle years. On June 13, after a conference with President Eliot when the matter of his appointment may have been discussed, Coulter wrote Watson:

I have been flying about since my return, but am about ready to settle down. I wanted to see you after I left Pres[iden]t Eliot, but I barely had time to get to Boston, make a few purchases, & catch my train.

The lists you refer to are good & I have all of them. Certainly Prof[essor] Bessey will gladly help, but he will be on his way to Europe by the time this reaches you. But I have correspondents in E[astern] Nebr[aska] & E[astern] Dakota that can be depended upon.

It occurs to me that the best thing to be done is for me to begin at the beginning & make up the copy, adding the species necessary, changing nomenclature, & correcting range. As fast as I get a little done, I will send it on to you. . . .

Coulter and Watson were working on a sixth edition of Gray's *Manual of the Botany of the Northern United States*, of which by October they had a "continuous Mss from Umbelliferae *through* Gamopetalae, excepting (Watson's) Compositae." Besides this work, Coulter was working with Joseph Nelson Rose, with whom he had published in 1886 a "Synopsis of North American Pines Based upon Leaf Anatomy,"[12] commended by Barnes, Arthur, Morong, and others, and was now completing a *Revision of North American Umbelliferae*,[13] a paper offered as a "Contribution from the Gray Herbarium as much as the Hypericum paper. It would be perfectly fair to call it such," said Coulter, "for it most certainly originated there, was constantly assisted from there, & the consultation of types there were like putting the foundation stones under the structure."[14] Coulter, recently President of the Indiana Academy of Sciences had written many notices, completed with Arthur and Barnes a *Handbook of Plant Dissection*, the famous "A.B.C. book" of laboratory practice, and made with Rose the pines synopsis, and other laboratory investigations of an embryological and

[12] *Botanical Gazette*, XI (1886), pp. 256-262, 302-309.
[13] Crawfordsville, 1888. 144 pp. "Umbelliferae of Eastern United States," *Botanical Gazette*, XII (1887), pp. 12-15, 60-63, 73-76, 102-104, 134-138, 157-160, 261-264, 291-295. Plate. "Notes on Western Umbelliferae," *Botanical Gazette*, XIII (1888), pp. 77-81, 141-146, 208-211.
[14] See *Botanical Gazette*, XI, Number 10, p. 275.

anatomical nature. Was Coulter to be Engelmann's successor? Watson seemed to turn to him as Gray had turned to Engelmann. Gray had, also. One might answer that Coulter was the man more than Greene could ever be—and more than Britton would. Coulter was progressive, yet conservative; a student schooled in Gray's tradition, yet one who ever looked to onward progress of the science.

Palmer's Southern California collection, including plants from areas such as Catalina Island, Mount Whitney, and other localities,[15] went to Vasey but turned out to be a small one, "less than 250 species—many only in single specimens." Five sets were distributed to Smith, Canby, Greene, Britton, and another—the United States botanists who usually could be counted on to purchase sets from Palmer's and Pringle's collections. Watson was sent specimens that he might name the new ones. Vasey and Joseph Nelson Rose published a "List of Plants Collected by Dr. Edward Palmer in 1888 in Southern California," the first *Contribution from the United States National Herbarium*. Joined to this was a "List of Plants Collected" by Palmer in 1889, all of which was issued June 13, 1890. Palmer remained in San Diego most of the winter where Orcutt, recently returned from a collecting trip in Lower California, was trying with aid of Cleveland to establish a "museum on a *scale greater* than the National Museum," to be part of the Society of Natural History there. In 1888 Palmer may have gone to Yuma, Arizona, and "to the interesting points about the head of the Gulf of California." Surely he went there in the spring of 1889 since on April 23 he wrote Watson from Yuma, "I have been absent to Lagoon Head, Cedros, San Benito and Guadalupe Islands and on my return to San Diego had to hasten the drying [of] the plants and shipped them to Vasey before calling on Cleveland—Then I came [here] to go to the head of the Gulf so as to get some special plants desired by Vasey. . . . Suppose you received the Palm flowers sent you by mail sometime since. . . . Should the opp[o]rtunity offer [I] may go to Guaymas to visit localities nearby, then wil[l] be in order the plants you desire. I wil[l] return to San Diego after finishing [here]: please inform me where Pringle is working this year as I wish to select a field not visited."

Vasey, the following summer of 1889, came West and on his return to Washington wrote Watson:

I have had an interesting trip across the continent, but not at a good part of the year for botanical collections. Dr. Palmer is still at San Diego, where he has been sick since July 1. He is improving and hopes to be able to resume work soon. We are considering the question of sending him to Australia for a year to collect

15 Collections were made at Long Meadow, Tulare County.

plants, seeds, grasses &c I saw Mr & Mrs Brandegee at San Francisco—Did not see Mr Greene who was in the mountains of Northern California. Mr Brandegee has made a large collection in Lower California. I saw Mr. C. R. Orcutt who has done some botanical work. We have not quite completed the examination of Dr Palmer's lower California plants.

Vasey made no mention of Parry although Parry had written Watson early in March of that year that he was packing for a return to California. September 12 of the year 1888 Watson had heard from Parry announcing:

> After nearly 2 years sojourn, we are now about turning our face eastward. [W]ill leave for *Davenport Iowa* 21st inst[ant] I have already packed and forwarded my bundles of *hay* to add to the old pile! Though tolerably active for an old man I have mainly confined my collection to specialities, giving the past season considerable attention to Ceanothus, of which I have considerable new material not yet communicated to Trelease. . . . I have seen considerable of Greene and his vagaries. [A]s not crossing his path I manage to keep on friendly terms. Lemmon & wife have a salaried position on the State Board of Forestry and are now in the field. Dr. Anderson[16] of S[an]ta Cruz accompanies us on our eastern trip and I presume will put in an appearance at Cambridge this fall. . . .

Whether Parry returned to California in March 1889 is uncertain. If he did, it was only for a short while. Parry had hoped to go East the winter previous and see Watson. And he probably went. In any event, on February 25, 1890, Canby wrote Mrs. Gray: ". . . we have all been saddened by the news of Dr. Parry's death. It seems a very little time since he was here in, apparently, excellent health." Parry's death came as a shock to American botanists. John Muir wrote: "It seems as if all the good flower people, at once great and good, have died now that Parry has gone—Torrey, Gray, Kellogg, and Parry. Plenty more botanists left, but none we have like these. Men more amiable apart from their intellectual power I never knew, so perfectly clean and pure they were—pure as lilies, yet tough and unyielding in mental fibre as live-oaks. Oh, dear, it makes me feel lonesome, though many lovely souls remain. Never shall I forget the charming evenings I spent with Torrey in Yosemite, and with Gray . . . they told me about Parry for the first time. . . . Then more than a week with Parry around Lake Tahoe in a boat; had him all to myself—precious memories. It seems easy to die when such souls go before. And blessed it is to feel that they have indeed gone before to meet us in turn when our own day is done."[17]

Two of Parry's most important botanical publications had been a work

---

[16] In 1883 Parry had requested Gray to associate an Elatine? from near Santa Cruz with Anderson, who was described as a microscopic botanist.

[17] *Life and Letters of John Muir, op. cit.*, II, pp. 242-243.

on "Pacific Coast Alders"[18] and a revision, as planned, of Ceanothus, L.[19] During Parry's life he wrote much of a general nature for newspapers and magazines. One of the most important of these was his article on "Rancho Chico,"[20] the historically famous ranch of the Bidwells and the home of the Sir Joseph Hooker oak, which article Parry published the year before his death. Born in England, educated in America, a medical student of Dr. Torrey, the equal, if not the greatest, of all North American botanical explorers—was Charles Christopher Parry. Unequaled in the possession of a happy and vital nature, he was remembered affectionately by all who had known him. Excepting Gray, certainly no man influenced Engelmann more. Torrey had certainly looked to him more steadily than any other collector of botanical materials, with the possible exception of John Charles Frémont. Parry took up where Frémont left off. The number of really great North American botanical explorers is not many: David Douglas, Thomas Drummond, Thomas Nuttall, John Charles Frémont, and Parry. None remained with the task as did Parry—from young manhood almost to his last day. Parry knew the mountains, the deserts, and the rivers from the Mississippi to the Pacific Coast. He was familiar with the botany of the entire West and was among the first of the North American explorers to enter Mexico. His name was recorded in almost every genus of the time. And in more than one instance he was the discoverer and namer of genera, as, for example, in the instance of Canbya. Today, however, his name is not so prevalent. Systematic revisions and monographs have removed much of his botanical fame. However, *Lilium Parryi* of Southern California, the Lote Bush (*Zizyphus Parryi*) of the Colorado Desert, and the Ensenada Buckeyes (*Aesculus Parryi*), among many others, perpetuate his memory. Parry was married twice, his first wife dying a few years after early marriage. For his second wife, Lemmon, "to honor a noble lady, who has done eminent service for botany," he said, sought to name a new flower for her—*Gilia Parryae*. After Parry's death, she aided in the preparation of an official biography and prepared a list of papers published by Parry.[21]

On January 11, 1890, Palmer wrote Watson, again from Guaymas:

Arrived hear two days since. . . . I am [here] on Vaseys account to get winter

---

[18] *Bulletin of the California Academy of Sciences*, II (1887), pp. 351-354.

[19] Davenport, 1889. Reprint from *Proc. Dav. Acad. Sci.*, pp. 162-174, 185-194.

[20] San Francisco, 1888. Reprint from *Overland Monthly*, June 1888.

[21] C. H. Preston, *Biographical Sketch of C. C. Parry with a List of Papers Published by Dr. Parry, Prepared by Mrs. C. C. Parry*. See also article on "Charles Christopher Parry," *Dictionary of American Biography*, XIV, p. 262 (article prepared by Willis Linn Jepson). Charles Russell Orcutt has also written of Parry in *West Am. Sci.*, VII (1890), pp. 1-5.

material—An unusual amount of rain has fallen this winter followed by extreme cold, north wind drying up most of the scanty vegetation at this season. The length of my stay [here] depends upon the time Dr. Vasey orders me to Arizona for special work.

Have selected several places to visit that could not be reached when [here] before, but am not certain about being able to carry out my plans. . . .[22]

Mr. Brandegee came on the Steamer from San Francisco as far as Magdalene Bay. [H]e wil[l] collect from that place towards the Gulf ending at La Paz.

On February 8 Palmer wrote again:

I returned to this place today from La Paz. [N]ot much to be had at that place in winter but some of it is different from the summer vegetation. [F]ound some good things. A palm I take to be the Guaymas one (you say new) gathered some good flowering specimens with leaves and two young plants that show character— shall send the collection to Vasey as soon as catalogued. [W]il[l] ask him to send you some Palm flowering specimens &c. . . .

Am now arranging to visit some Islands in the Gulf for I fear that as the summer heat of this hot hole come[s] along wil[l] have to quit. . . .

Hope Pringle met with good success last year.

And on April 20:

Since you he[a]rd from me last I have made a collection at Santa Rosalea, Lower California, and at Alamos, Sonora, Mexico. Am of the opinion there are several new plants—I send a package by mail of [a] plant that grow[s] under shelving rocks up in the Mountains of Alamos directed to you—Tomorrow [I] start for the Huashuca Mountains, Arizona. . . . As I am today not sure of my future address, wil[l] you address me a note in care of Dr. Vasey informing me if Pringle wil[l] go this season to Colema. . . .[23]

In May Palmer had to be given medical treatment at Fort Huachuca, in the southernmost part of Arizona, but he was nevertheless determined to complete some collecting in Lower California and then cross over to Colima. He, however, did not get to the last named place for several months. In 1890, three months were spent in Lower California, collecting at La Paz, Santa Rosalia, and Santa Agueda, Raza Island, island of San Pedro Martin, in the Gulf of California, and some other less important places; and three months were spent in southern Arizona, collecting at Camp Huachuca, Fort Apache, and Willow Springs,[24] with two trips to Alamos. On December 1, 1891, he wrote a last letter to Watson:

I have spent some time in the State of Sinaloa. [C]ollected about 250 species

[22] Sometime during 1890 Palmer collected on Carmen Island, Lower California.

[23] Published as a *Contribution from the U.S. Herbarium.* "List of Plants Collected by Dr. Edward Palmer in 1890 in Lower California and Western Mexico." By Vasey and Rose. Volume I, Number 3, issued November 1, 1890. 173 species.

[24] *Contr. from U.S. Nat. Herb.* "List of Plants Collected by Palmer in 1890 in Western Mexico and Arizona." By J. N. Rose. Volume I, Number 4, issued June 30, 1891. 475 plants enumerated with remarks as to soil, locality, size, and other notes concerning species.

during a very dry season. [C]ame to this place (Guaymas) to ship them to Washington. Some good things are among them. [Y]ou wil[l] remember of my sending you leaves of a palm from Batapilas without fruit. I have found the same palm this summer with fruit, it's about 8 foot high, small top, good size fruit, good for cultivation. [B]efore this reaches you I wil[l] leave for Tepic to spend the winter. [G]oing by Steamer to San Blass, then by Stage to Tepic. [I]t wil[l] be a new field for me and I anticipate a good haul. [A]fter which I go to Colema again to visit the mountains during the next rainey season. This is written so you can inform Pringle of my intentions as it would not pay for us to work in the same places. Though it would give me much pleasure to meet him, we have never met. My health is anything but good. [I]n all probability [I] wil[l] not make many more collections.

Palmer, however, did make many more collections. After 1893, when he collected once more in southern Utah, Palmer went again to the more tropical regions—Sinaloa, Colima, Acapulco, and Tepic. Many regions which he had visited, he revisited, such as Coahuila, San Luis Potosi, Chihuahua, and Durango. One cannot say that he took Parry's place as a collector, although pines and oaks were among his specialties in later years. His last trip was made in the vicinity of Tampico, Tamaulipas, in 1910. A collector of more than a thousand new species of flowering plants, and more than two hundred and fifty Compositae, at the time of his death 200 were said to bear his name.[25]

While Palmer was busy in western Mexico collecting for Vasey and the United States Department of Agriculture, Pringle, under appointment as botanical collector for the Gray Herbarium, was also collecting in Mexico.[26] In 1888, Pringle and his assistant, Welcome, arrived at Chihuahua on May 25. After going to Ortiz and the Mexican side of the Rio Grande near El Paso, Pringle and Welcome left for the Monterey country, going by rail and stage by way of Torreon, Jaral, and Saltillo. There they remained some time, exploring the Sierra Madre Mountains and canyons, bluffs, ridges, slopes, and bases of the hills and rivers near by; till on July 26, they left Monterey and rode over "chaparral covered plains" to Laredo en route to Brownsville, Texas, collecting on the way as occasion permitted at Pena, Rio Grande City, San Miguel, Matamoros, Reynosa, and Hidalgo (Edinburgh), Texas. Some time was spent there and on the return journey. But by August 24 they were again in Chihuahua and Pringle took up further exploration of the lands he had explored years previous. On September 17, however, they set out again on a journey to the Sierra Madre west of the city of Monterey, making camp along the way. Cusihuiriachic, the mesa south of Rosario, Arroyo An-

25 See William Edwin Safford, "Edward Palmer," *Popular Science Monthly*, LXXVIII (1911), pp. 341-354; *American Fern Journal*, I (1911), pp. 143-147.

26 See *Life and Work of Cyrus Guernsey Pringle, op. cit.*, pp. 46-59 for diary of 1888 journey.

cho, and numerous mountain, canyon, and plains localities were visited and returning by way of Guerrero and Rosario they reached Chihuahua, "having made the distance of 160 miles from [their] mountain camp in four and a half days." Monterey in Nuevo Leon was never reached. On October 22 they left for a trip by rail to Guadalajara, going en route through Zacatecas. From October 28 to December 15, Pringle was in the vicinity of Guadalajara, spending much time "in the great barranca (cañon of the Rio Santiago)" and at Atemajac and the hills beyond. The last few days were spent in trips to the vicinity of Esperanza and the Rio Blanco. And on Christmas Eve, Pringle and Welcome reached Charlotte, Vermont.

On May 6, 1889, Pringle left his home for another Mexican journey,[27] this time taking with him Charles C. Hammond as assistant and Miner B. Hayward as collector of birds and mammals. Proceeding to Laredo, they took a train for Monterey where Pringle went immediately "to the rocky cañon at the base of the Sierra Madre" and collected a few mosses, lichens, and other plants—for ferns and all orders of plants were objects of his collections. However, not much time was spent there as by May 19 the party reached Mexico City and, after visiting Chapultepec, took the Mexican Central Railroad for Guadalajara where Pringle again went to the canyons about La Esperanza and the barranca. Returning to Monterey a week later, they took up further exploration of Saddle Mountain (Sierra de la Silla) and the Sierra Madre and its canyon and, although Pringle made trips to Garcia, Laredo and another journey to Mexico City and Guadalajara from June 21 to July 8 collecting nearly 2,000 specimens, most of the time until September 17 was spent near Monterey. A trip was made to Lampazos nearly 100 miles to the north and another to Laredo where Pringle was joined by Fred A. Smith for a while. However, not until late in September did Pringle return again to Guadalajara to the several exploration areas near there. On November 9 he climbed "to the summit of one of the most precipitous peaks of Sierra de San Esteban." And, after a journey to El Paso and Chihuahua, and one to El Castillo where he walked to the next station, La Capella, and explored mountains on the shore of Lake Chapala, also going to Escoba and other places, Pringle left Guadalajara and arrived at El Paso on Christmas Day—another immense collection gathered together and ready for systematization.

On June 12, 1889, and on April 8, 1891, Sereno Watson presented new *Contributions to American Botany* treating of Pringle's collections for

---

[27] See *Life and Work of Cyrus Guernsey Pringle*, pp. 59-73 for diary of 1889 journey.

the years 1888, 1889, and 1890,[28] the last of which embraced a wide area between Guadalajara on the west, Tampico on the east, and Patzcuaro and Mexico City on the south; with Guadalajara and San Luis Potosi the most important focal points. On May 5 Pringle left his home in Vermont with Henry Ash as assistant and, reaching Guadalajara May 16, went to Atequiza and the river marshes and fields near there. Returning to Guadalajara where he visited again regions near, he left on May 24 for San Luis and Laredo, stopping en route at Carneros to botanize. Early in June a trip was made to Tampico and Las Palmas and again at points Pringle pursued a practice often used, of botanizing along the railroad right-of-way. Late in June a trip was made to Bocas and Canoas. But each time the party returned to San Luis. Salinas, Tamasopo Cañon, Canoas, Villar, Mexico City, Flor de Maria, among several other places, were visited during July. And in August and September points theretofore explored were gone over again, adding a few new localities such as Rio Hondo and La Honda. In late October and again in November trips were taken into the state of Michoacan and Patzcuaro characterized by Pringle as pleasant country with a lake, wooded hills, many interesting new plants, and noticeable localities for mosses and a "grand fern hunt among the old pollarded oaks of the hills." On December 4, they left Laredo for Vermont from where Pringle would leave again the following May 5 for similar explorations in Mexico.

Among the "Descriptions of New Species of Plants, from Northern Mexico, Collected Chiefly by Mr. C. G. Pringle, in 1888 and 1889," determined by Watson in 1889, the genera, Sargentia for Professor Sargent of the Arnold Arboretum and Rhodosciadium for Joseph Nelson Rose of Washington ("who," said Watson, "with Coulter has done much to elucidate representatives of the order," Peucedanoid Umbelliferae), were established along with Jaliscoa. Watson combined the descriptions of Pringle's northern Mexico collections with "Miscellaneous Notes Upon North American Plants, Chiefly of the United States, with Descriptions of New Species." As, similarly, he did in his 1891 *Contribution to American Botany* consisting of:

1. Descriptions of some new North American species, chiefly of the United States, with a Revision of the American Species of the Genus Erythronium.

2. Descriptions of new Mexican Species collected chiefly by Mr. C. G. Pringle in 1889 and 1890.

With respect to materials of the first set of descriptions and the revision of Erythronium, a genus which reaches "fullest development in the United States," Watson acknowledged aid from Carl Purdy, G. R.

[28] See *Life and Letters of Cyrus Guernsey Pringle*, pp. 73-84 for diary of 1890 journey.

Kleeberger, and Volney Rattan from California; Mrs. P. G. Barrett, Thomas Howell, and W. C. Cusick from Oregon; L. F. Henderson and W. N. Suksdorf from Washington; and John Macoun from the Canadian Geological Survey.

John Macoun had written Watson on March 17, 1888: "I am well pleased with the work of the past year. Including all orders I have collected nearly 60 species new to science." Macoun's son had been exploring islands of James Bay, the southern extension of Hudson Bay, and Macoun had been on Vancouver Island "making a thorough botanical exploration of the southern half." Macoun's collections were always large. When he wrote Watson during the summer on Vancouver Island at Nanaimo, he told him:

Up to the present I have noted and collected over 1100 species on the island. Of these 744 are flowering plants & ferns, the others are Cryptogamic of all orders. I have fine specimens of a terrestrial Isoetes and have found *Phyllospadix Torreyi* or something very like it. . . .

In the same letter he told Watson:

Two days from now I commence the ascent of Mount Arrowsmith 50 miles from here and 6000 feet high. I purpose making a careful examination of it, as fully 1000 feet of it is above the present snow line, and it must have the greater part of the alpine flora of the island on its upper slopes. I am taking an Indian and six days provisions. . . . So far I have found the alpine flora almost identical with that of the coast range but hope to get Alaskan forms when I reach higher altitude. . . .

Macoun and his son William had made numerous excursions in the locality[29]—to Gordon Head, Mount Tolmie, Cedar Hill, Lost Lake, and other places—but none, with the possible exception of a mountain climb "through a forest of beautiful pine and fir in which the underbrush was Salal" up Mount Benson, was more enjoyed than the trip up Mount Arrowsmith. Macoun described it:

We made our bivouac amongst the trees of the eastern side of the mountain and, on our right, was a snow slope which led almost to the summit. We enjoyed the view very much and, as the evening wore on, I went up to the summit and watched the change from bright sunlight to darkness. The evening was unclouded and I could look over the Pacific and, at the same time, turn around and look over the mainland and see the mountains that border the Gulf on that side. There were only a few fires at this time, but I could see almost every fire that was burning on the Island and could detect the slightest fire by its smoke. What I was most interested in was the change from light to darkness. As I stood there, each summit was bathed in sunlight, the mountains on the mainland also shone out, and the mountains to the northward of the Island stood out boldly also. Gradually darkness seemed to walk in the light and put out the light and, as the darkness increased

[29] Macoun's activities from 1887-1893 are completely told in Chapter XVI of the *Autobiography of John Macoun*, op. cit., pp. 247-265.

and rose on the mountain slopes, we were in twilight and, I decided it was time for me to descend.

Other points to which they went that summer were Mount Mark at the southern end of the Beaufort Range and near Horne Lake, and Alberni and Cape Beale—going to the latter point by way of the canal to the open ocean. When they went home that autumn, they went the entire way from Vancouver to Ottawa on the Canadian Pacific Railroad which was completed during the summer. On Christmas Eve, Macoun received a letter appointing him naturalist to the geological survey, and assistant director and botanist, with rank of chief clerk.

The year 1888 took Macoun and members of his family to Prince Edward Island and in March 1889, he returned to Vancouver City to commence work on the natural history survey, accompanied by his son James, who had been surveying down the Athabaska River the summer before. Hastings, Agassiz, Yale, Lytton, Spences Bridge, Kamloops, Sicamous, Lake Okanagan and the country soon known as the "Garden of British Columbia," Shuswap Lake, and the Gold Range[30] of British Columbia were included in their itinerary, collecting plants, birds, mammals, and snakes. On September 16, 1889, Macoun wrote Watson:

During the summer I made very extensive collections and before spring hope to send you down a large parcel of the collections of the last three seasons. Our work has increased so much of late that my son and myself can scarcely keep up with the routine work, let alone the work of distribution.

I added no new forms of roses this season but can send you when you desir[e] a very good set of all our northern forms as we have them now from the Atlantic to the Pacific.

In 1887 I collected on Vancouver Island, last year on Prince Edward Island and this year from the Coast of British Columbia eastward to the Columbia, so you see I have specimens from much new territory. Will you want the Hepaticae as well as the Mosses?

When the year before Macoun had published Part IV—*Endogens*—of the *Catalogue of Canadian Plants,* he acknowledged indebtedness to Watson for "valuable assistance, especially in the Liliaceae and Juncaceae." At the same time he announced that Part V would include "the ferns and their allies with the mosses and liverworts, and it is intended," he said, "in Part VI. to catalogue the lichens, fungi and seaweeds." Part IV included as to *Endogens* the collections of James M. Macoun on the shores and islands of James Bay and Dr. G. M. Dawson's "valuable and interesting notes and collections in that part of the North-West Terri-

---

[30] West of the Columbia River and north of the Canadian Pacific Railway, "Mountains of Cariboo," where Macoun spent ten days and saw many caribou and Rocky Mountain goats—between Latitudes 55°-56°.

tories bordering on Alaska." Concerning the mosses and hepatics, Macoun said in his letter of September 16 to Watson:

The fourth Century of my Canadian Mosses will be ready for distribution shortly and I will send at the same time the three already published. Kindberg has been at work on my mosses for nearly two years. So far (leaving out this years collections) he has enumerated 608 Canadian species and describes 47 new species and 33 new varieties, also credits me with 46 other species not described in James & Lesquereux's *Manual*. I have the manuscript of Pearson's enumerations of the Canadian Hepaticae (167 species) now in the hands of the printer. He only makes one new species but we are publishing 13 plates of new drawings of imperfectly characterized species. After this is far enough advanced I will send to the printer Kindberg's complete notes and descriptions of new forms and hope to have both ready for distribution in [the] spring.

Macoun went again to British Columbia in 1890, this time going to Revelstoke on the Canadian Pacific Railway. With his son James he "went down the Columbia [River] to Deer Park and stopped there a couple of weeks making collections on the Arrow Lakes. We then," wrote Macoun, "went by canoe down the Columbia to Pass Creek, close to Robson" and at Robson "engaged pack animals to carry our stuff across to Nelson. The boys went with the pack horses and I walked across to collect plants." Macoun narrowly escaped being gored by a mad steer and did suffer a sunstroke from the excessive heat. However, they made their way by boat to Kootenay Lake and, returning to Nelson where further collections were made, proceeded back to Revelstoke and Hector, from where they left for Ottawa. On October 22, he wrote Watson:

We gathered many fine things the past summer in the British Columbian Mountains. Some of these are new to our flora and some may be new to science. As soon as my son gets through the Polypetalae we will send you those that are doubtful and others not hitherto sent to your herbarium. . . .

There are many new species of mosses in Canada yet to be collected and described. Kindberg & C. Müller have been at work on my collections the past summer and have worked out 171 species new to science. Prof[essor] Barnes seems to think these of little account but he will open his eyes *wide* when he sees the descriptions and I hope in many cases the figures. When all are examined there will be fully 200 additions to Lesq[uereux] & James['s] *Manual* as I have over 25 new to America though not to science.

And on November 27:

Many of our northern forms do not seem to be identical with the species which passes for them south of Lat[itude] 49°. Where are the types of Torrey and Gray's *Flora* and Nuttall's species? I am becoming satisfied that many of the old names will have to be revived as we have in our collections now many forms that will not fit in with many of the descriptions published in the last 30 years. If you have the types I mention I may go down to Harvard for a couple of weeks.

Whether Macoun can be said to have been giving support to the claims of Greene and Britton with respect to reestablishment of old names in systematic botany is doubtful, although he certainly fell in with a widespread dissatisfaction with existing conditions. An argument for reestablishing names abandoned because of revisions made during the last thirty years was different from Greene's claims and from Britton's claims. Macoun's complaint seems to have been one directed against the numerous revisions of the last three decades, in the work of which both Greene and Britton had had a part. In any event, Watson and Macoun got into no dispute on the subject.

On August 14, 1891, Macoun wrote Watson the last letter of their important correspondence. It was written from Rocky Mountain Park, Banff, and said:

I have made extensive collections here this year but have nothing *new* as far as I can see but I have secured fine specimens of *Arabis Drummondii* both in flower and fruit. The latter at 6900 feet altitude, the former at 8500 feet. . . . I think I have a small Ranunculus not found before in our mountains.

My son is with Dr. Dawson at Ounalaska and the Pribilof Islands in Behring Sea this year and I expect a number of northern things from his collections when he returns. He is a famous collector so I expect some additions to the flora of these islands when he gets back. I will be home in a month.

Macoun planned to have his son go to Harvard to work up his collections. However, during December, Watson was taken ill with influenza, which resulted in dilated heart condition; and on March 9, 1892, he died. A condition, following malaria fever contracted in Guatemala, had so weakened his otherwise strong constitution that he could not withstand the rigors of influenza. His death was a grave loss both to Macoun and to Pringle.

After Watson's death, available resources at the Gray Herbarium were such that Pringle could no longer be employed as collector for the herbarium. Pringle considered selling his own herbarium to the American Museum of Natural History to finance his further Mexican explorations. However, Mrs. Gray loaned him more than one thousand dollars, taking his unsecured note for repayment. Before her death, these notes were burned in Pringle's presence—symbolic of a gift to science in memory of Dr. Gray. And in 1893 Pringle was restored as collector of botany for the Gray Herbarium.

Pringle without doubt did more than any one individual to make the flora of Mexico known to systematists of the United States—and probably more than any other to acquaint taxonomists of the entire world with the floral riches of that historic land, vast stretches of which nevertheless remained yet unexplored. Macoun was honored by Canadians as "the father of exact natural history in Canada." Natural science in the United States has honored most of its first great naturalists, especially those in botany, by naming towering and important mountains after them—Torreys Peak, Grays Peak, Parrys Peak, Mount Engelmann, Mount Agassiz, etc. Likewise, Canada has honored Macoun with a permanent memorial—Mount Macoun near Mount Sir Donald and Bald Mountain. It has been a matter of no little consequence to botany of the United States that the Gray Herbarium had these two valuable and learned correspondents—Pringle in Mexico and Macoun in Canada.

Dr. Benjamin Lincoln Robinson took charge of the herbarium after Watson's death and ushered in the second great period of its history. To Robinson fell the still unfinished task of completing systematization of the *Flora of North America*. In 1897, Robinson, with collaboration of Trelease, Coulter, and Bailey, published a continuation of Caryophyllaceae to Polygalaceae, Volume I, Part I, of the *Synoptical Flora of North America*, describing it as fascicle 2. Robinson's notable monographic work during his incumbency as director of the Gray Herbarium set a high standard for such work in North America, receiving definite world recognition. In taxonomy, monographs of the larger plant groups and families were to aid decidedly in ordering the chaos which slowly increased as increasing literature and materials became available. It seemed at last that systematization had gotten ahead of exploration. But need of systematizing further the available literature was soon also to be seen.

CHAPTER XIV

A New Era in North American Botany Begun

SERENO WATSON's work in a sense was completed and in another not. His Guatemalan collecting had been carried on by John Donnell Smith.[1] In 1890 the sixth edition of Gray's *Manual of the Botany of the Northern United States*, prepared by Coulter and Watson, had been published and some revisions made in 1891. However, the *Synoptical Flora* remained unfinished; and uncompleted and unpublished was the balance of Watson's *Bibliographical Index to North American Botany*, eight parts of which were at the Gray Herbarium in manuscript form.

Gray's many pupils, nevertheless, remained in the science. Generally, botanists were now classed as systematists, physiologists, or anatomists. Dynamic as opposed to static points of view had had possession of European botanic investigation less than half a century. To illustrate, the concept of cell division was less than a half century old. As late as 1846 protoplasm was described and the embryonic vesicle in the embryo sac discovered. Sexual reproduction in lower plants was doubted by authorities as late as 1853. The algal-fungus theory of lichens had risen some time later. Practically the whole of the science's exact knowledge of life histories of higher cryptogams had been obtained during the last half century. Comparative studies of organs *and their development* and also studies of large plant groups and their relationships and schemes of descent were to take possession of American botany. Gray realized this and must have given the "way clear" signal to Farlow, Coulter, Bessey, Campbell, and others. Gray had been instrumental in Farlow's, Rothrock's, and other students' studies abroad. Each time he urged on them the doing of original work—the breaking out of unexplored paths, as he told Coulter, Bessey, and others, whose European studies were deferred for a while but who read zealously the latest European treatises and texts.

Paleobotany had before it a tremendous future. Yet would come the immensely important work of the National Museum, the United States Department of Agriculture, and state experiment stations. Yet would come physiology, an exact and not an observational science; the inclu-

---

[1] Smith took Guatemalan collections to Cambridge in 1888 for study. In 1889 he collected "a considerable number of plants, from Livingston and Yzabal to Panzos," where Turckheim met him, and they journeyed to Coban, Smith returning by the cart road. In 1890 he collected near Guatemala, Amatitlan, Escuitla, Antigua, and slopes of Volcan de Agua, bringing back 600 species. For many years such collections continued by Smith and others.

sions of phylogeny and paleobotany in morphological research; cytology on an enlarged scale; the transition from the "new botany" of Beal, Bessey, Coulter, and others to the era of scientific research. Yet would arise the important study of ecology having as aids all other branches, including anatomy. The extension of phylogeny to anatomical study would facilitate the understanding of plants and their relationships and descent, both from the insights of inner and outer structures. In exploration would come the rise of Nathaniel Lord Britton and his direction of far reaching expeditions to Pacific and Atlantic waters, to Mexico, and Central and South America. The United States Department of Agriculture, the Smithsonian Institution, and the Gray Herbarium would figure prominently, as would exploratory work by all the larger institutions dedicated to systematic and research work.

The passing of Alvan Wentworth Chapman in 1899 in his almost ninetieth year would terminate the work of a generation of North American botanists. Botany would turn to the challenges of Rothrock, Bessey, Coulter, and others, to study plants as living things in field and laboratory: concentrating more on their physiology, stimuli, responses, and processes; anatomy, cells, tissues, protoplasm, chemical composition; pathology, causes, cures, or control methods of diseases, and study of the diseases themselves; morphology, structures present and historical, cross and self fertilization, factors in evolutionary development; ecology; relations to soils, climates, temperatures, light, all the incidents of habitat. Gray's work was the foundation of all. Darwin had broken loose the bands which had held the science in restricted compass. Gray had furthered and aided the work of Darwin to the limit of his ability. Anatomy and physiology had had his aid.

Gray's books did not encompass all that his reviews of European literature did. The new morphology was not that of his *Structural Botany* nor that of his *Elements of Botany* published in 1887, admittedly a "rehash of [his] 'Lessons of Botany,' more condensed, yet fuller, and with a new name."[2] It would be a new morphology, in the furtherance of which Coulter and Bessey would take most prominent parts *as teachers*. Enumeration of the most able investigators would require going to Europe and delineating the accomplishments of a long line of able men beginning with Hofmeister, Strasburger, and many others. Gray was likewise first a teacher and would be allocated to his sovereign place—with Torrey, the establisher of the science of botany in North America. Gray was a man with versatile talents who could fight for causes such as Redwood Park near Santa Cruz when one of California's glories, her

---

[2] See *Letters of Asa Gray, op. cit.,* p. 792.

trees, were threatened. Gray was a man who could be seer and innovator, interpreting Darwin to a continent; who could interpret to the world's greatest systematists, Bentham, Hooker, and the DeCandolles a continent's scheme of geographic plant distribution and plant migration, both studies precursory to ecological investigation. Gray was a man who could more than adequately be the great associate of Torrey in publishing the North American flora from Alaska and Greenland to Mexico, much of South America, and islands of the Atlantic and Pacific oceans. Gray was North America's greatest botanist.

North American botanists welcomed Coulter and Watson's *Manual of the Botany of the Northern States*.[3] Although Greene and Porter severely criticized it, Macoun, Chapman, and the great number of American botanists were impressive in their praise. Coulter wrote Watson on June 12, 1890:

> I suppose we have had the last explosion in reference to the new *Manual*, as I have just read what Greene has said in *Pittonia*. It is the raving of a madman, which I must believe is more literally true than figurative. If he could only quietly pass away in one of his apoplectic fits, how much better for American botany! . . . there is yet the great body of American botanists who do not write, but who approve of the book, will buy it & use it. It has reminded me somewhat of the ranting of Socialists, et id omne genus, who grow red in the face & gesticulate frantically, while the great American public moves along as usual.

On April 16, Coulter had said:

> I suppose you have, ere this, recovered from the shock of Porter's criticism. I know him well, but I never imagined that he w[oul]d let his spleen display itself in such a childish fashion in print. I have met a good many botanists since the review appeared, & have heard from a good many more, & they are universal in their condemnation. I have rec[eive]d Mr. Faxon's bill for the drawing of the Eriogynia plate, but have not rec[eive]d the plate. . . . I have some good things from Nealley's Texas collection to send you soon. A Dep[artmen]t *Bulletin* will soon be issued containing a list of that collection, with notes, descriptions, etc.

Nomenclatural controversies seemed to increase, rather than diminish, with new publications and revisions. Coulter gave up Greene in despair. On October 29, 1891, he wrote Watson:

> I have just been looking over the last *Pittonia*, & have been immensely entertained by Greene's last scheme for stealing species. I had thought that he must be coming to the end of his ancient genera; but he has opened up a new vista of possible renaming by his announcement concerning "revertible" generic names. He is proceeding now to coin a new generic name (& of course rename all the species) for all genera whose names appear more anciently in the synonymy of other genera!! And so Darlingtonia goes, etc., etc. Such vagaries will finally succeed in

[3] The sixth edition of Gray's famous *Manual*.

making him more amusing than troublesome, & I hope he will get wilder with each succeeding year.

As to Britton, however, Coulter was more serious. On November 12, 1888, he wrote Watson: "What think you of Britton's Hicoria? It looks like priority over Carya, but what a fearful tearing up of specific names! And Greene's Unifolium! I am afraid our troubles have only begun." A year later Coulter asked Watson to send him his Cornaceae as he and his associates had several collections—all of the herbaria (even Greene's), except Britton's—for a contemplated revision; and Coulter added:

I see that Sargent (last *Garden & Forest*) has improved on Britton & changed Hicoria to Hicorius & given all the specific names masculine endings! The hickories will need all their strength presently to carry their load of synonymy.

All the growing confusion, nevertheless, led to results. J. C. Arthur wrote Watson on March 10, 1890, advancing the solution which ended in the promulgation of the Rochester Code of 1892.[4] Said Arthur:

It seems to me that advantage might be taken of the World's Fair in 189[3] to secure an international congress of botanists on American soil. Questions of nomenclature, etc. distribution of species etc. etc. might be made the subjects of profitable discussion with a chance of reaching greater uniformity in practice, or at least of a better mutual understanding. The meeting would not fail to be profitable and interesting for botanists from abroad as well as for those at home. It would also, I am certain, have a direct and lasting influence upon the development and standing of American botany. . . .

Meanwhile, sturdy, conservative, and experienced botanists such as Daniel Cady Eaton held the courses straight, as illustrated by his letter of November 18, 1892, to Davenport, saying:

While I do not think it is right to hold the botanists of 75 to 100 years ago up to the present law of conserving the oldest specific name when they transferred a species to some other genus, it is right to hold *modern* botanists to this rule, unless there is some grave fault in the name, or some fraud in the publication. Especially is this so with those American botanists which recognize "the ironclad law of priority."

Eaton had been so insisting for some time.[5] And stability, for which Britton fought as much as anyone, was maintained. Priority of the names of published places, unless incorrect, as well as priority of specific and generic names and their times or dates of discovery and determina-

---

[4] Held at Rochester, New York, American Association for Advancement of Science meeting.

[5] On September 20, 1890, Eaton wrote Watson: "I have received sample pages of Sargent's *Silva of North America*. I can not afford to take it. What a pity it is that he is so bewitched with the priority craze as to seriously write '*Magnolia foetida*!' This crankiness will do much to destroy the value of his work."

tion were preserved as all abler botanists, including Britton and other special protestants, wanted and fought for.

However, despite wishes of botanists such as Canby and others, it was made increasingly clear that no manual of the entire eastern or the entire western states, much less of the entire United States, could be prepared until competent and complete manuals from all sections had appeared—along with completion of revisions in all orders, families, and genera necessary. Torrey's and Gray's *Flora of North America* and Gray's *Synoptical Flora* not only should but would remain for many years the sole, standard, all-inclusive work. But to furtherance of completion of numerous tasks confronting North American botanists, able men such as Coulter and Britton bent their energies. On June 27, 1891, was issued as a *Contribution from the National Herbarium*,[6] Part I, enumerating and describing 270 genera and 761 species of Polypetalae, of the "Manual of the Phanerogams and Pteridophytes of Western Texas" by Coulter. This work, a valuable adjunct to Coulter's *Manual of the Botany of the Rocky Mountain Region* and published in three parts, was completed in the year 1894 and added much to the growing prestige of various *Contributions from the United States National Herbarium* which, together with *Bulletins* and miscellaneous publications of the United States Department of Agriculture, accomplished the realization of the dream of George Vasey before his death in 1893 of a first-class National Herbarium with first-rate works of national importance issuing from it.[7] Coulter would continue furnishing *Contributions from the United States National Herbarium*. As early as 1889, following the example of George Engelmann, he had decided to do more than a synopsis on pines—the new work to be done not on the basis of leaf anatomy but on the basis of established standards of taxonomy, incorporating new material with revision of the old. On March 28, 1889, he had written Watson:

> Trelease says I can have the use of Engelmann's Cacti in the summer. At present, things there are in too chaotic a state. The notes and sketches I can have at any time. In the meantime I have a term's possible work before me, with a capital assistant, & I want to begin. We know all our own Cactaceae "on sight" & have made some sort of a start.

---

[6] II, Number 1.

[7] In the "Report of the Botanist" published in the *Report of the Secretary of Agriculture for 1893*, pp. 235-237 and p. 244, a complete list of publications prior to March 8, 1893, and publications of the year 1893 are contained. A comprehensive idea of the large number of publications by Vasey, or authorized and completed under his supervision, may be gained from an examination of these, together with materials showing the then condition of the herbarium, the forage experiment work at Garden City, Kansas, and other matters. Washington: Government Printing Office, 1894.

Do you think it w[oul]d be well for us to work over the Cambridge collection first & so be better prepared for Engelmann's? Or is that putting the cart before the horse?

And on May 11:

I think the best plan with the Cactaceae will be for me to study first the Engelmann collection, & use that as a base line. Trelease says there are 12 cubic feet of them, & many growing in the Shaw Garden. We will go to St. Louis presently and study them. After we are thro' there, I think we will be in condition to call in the other collections & do something with them. I have a large number of Col[orad]o & Arizona *cacti* in bloom in my rooms, sent in by my correspondents. *Echinocactus Simpsoni* is in one mass of bloom before me.

Years passed before Coulter's studies were completed but, when ready, were published as *Contributions from the United States National Herbarium*. Although his "Revision of North American Cornaceae"[8] and his determinations of Smith's Guatemala collection,[9] "New or Noteworthy *Compositae* from Guatemala," were published by the *Botanical Gazette*, most of Coulter's other taxonomic works were made known as *Contributions* from the National Herbarium: "Upon a Collection of Plants made by Mr. G. C. Nealley, in the Region of the Rio Grande, in Texas, from Brazos, Santiago, to El Paso County" (1890);[10] "Preliminary Revision of the North American Species of Cactus, Anhalonium, and Lophophora" (1894);[11] "Report on Mexican Umbelliferae, Mostly from the State of Oaxaca, Recently Collected by C. G. Pringle and E. W. Nelson" (1895);[12] "Preliminary Revision of the North American Species of Echinocactus, Cereus, and Opuntia" (1896);[13] "Hesperogenia, a New Genus of Umbelliferae from Mount Rainier" (1899);[14] "Monograph of the North American Umbelliferae" (1900);[15] and others.

On August 24, 1891, from the president's office of the University of Indiana at Bloomington, Dr. Coulter wrote Dr. Watson:

You see by the heading of the letter that I have moved. The new position will give me greater advantages for botanical work. I have transferred my herbarium & books, from Crawfordsville, & the Board of Trustees has given me $2000 to spend for plants this first year, and $1500 for books. I think that is well enough for a start in addition to what I had already accumulated.

---

[8] *Botanical Gazette*, XV (1890), Number 2, pp. 30-38, 86-97, with Walter H. Evans.

[9] On July 19, 1890, Coulter told Watson: "I have a lot of Donnell Smith's new Guatemala material that I am coming to Cambridge with this fall to study. . . . The 'Flora of W. Texas' that the Depart. of Agric. has engaged me to prepare is a much larger thing than I at first imagined, & as new & Mexican species are being found there all the time it will be exasperatingly incomplete when completed." The former, in part, was published in *Botanical Gazette*, XVI (1891), pp. 95-102.

[10] I (1890), pp. 29-65.

[11] III (1894), Number 2, pp. 91-132.

[12] III (1895), Number 5, pp. 289-309.

[13] III (1896), Number 7, pp. 355-462.

[14] V (1899), pp. 2-3 (with Joseph Nelson Rose).

[15] VII (1900), pp. 1-256.

Coulter had written Robinson at Harvard on April 20, telling him: "You are just the sort of man that I would like to see helping develop the botany of the State University." Coulter wanted "to appoint a full professor of morphological & cryptogamic botany, retaining for myself only what might be called a chair of systematic botany. The former chair will be to your taste, I think," he told Robinson. Dr. Coulter had been offered a chair in botany and a few years later was made the president of the university. Although an able administrator and honored as successor to David Starr Jordan, Coulter was not a university president by nature but a botanist with genius as a teacher of the subject and remarkable vision for its future.

In 1893 he became president of Lake Forest University, located near Chicago and the north shore of Lake Michigan in Illinois. During 1896 he resigned this position to become head of the department of botany where in the course of thirty years he directed original and creative research in botany, including morphology, cytology, advanced taxonomy, and all other branches. At first a visiting lecturer, when made head of the department, Coulter soon summoned Charles R. Barnes for plant physiology and placed Henry Chandler Cowles to develop further the comparatively new subject of ecology—a study having its real origins in Europe and developed in America at the University of Nebraska and University of Minnesota a short time before its more elaborate cultivation at Chicago. Coulter and Bessey became the great teachers of morphology and under their guidance and vision the scope of botanic investigation increased incalculably.

Before the section of biology and as vice-president of Section F of the American Association for the Advancement of Science, Coulter described what he regarded as "The Future of Systematic Botany."[16] The systematist, he said, must not only collect and describe plants, studying also plant life histories, but must construct a natural system in the plant kingdom:

... the last and highest expression of systematic work is the construction of a natural system based upon the accumulations of those who collect and describe, and those who study life-histories; that this work involves the completest command of literature and the highest powers of generalization; that it is essential to progress for a natural system to be attempted with every advance in knowledge; that all the known facts of affinity, thus brought within reach, should be expressed in all systematic literature,

was now evident and were the objects and ends in view for systematic

[16] A biography of Coulter, with bibliography, has been prepared by this author. References, therefore, in this work are not cited.

work. Cytological study—the "minute tracing, cell by cell, from the primitive cell to the mature plant"—was by this time conceded able to "reveal more of the deep secrets of affinity" than perhaps any other method.

Coulter aided strongly in freeing morphology of the weight of its older and rigid doctrine of types. Coulter aided in preparing the way for a new science of plant anatomy. Mature *and developing* organs and tissues were studied, aids to, and yet separate from, taxonomy. Nascent organs were more closely examined. Coulter aided strongly in freeing the science from exaggerated theological influences—from concepts such as predestination which had worked its way into belief that plant organs, cells, cell groups, etc. develop with inevitable necessity and purpose as predetermined. He aided in substituting for the teleological or personification explanation of adaptation the concept of plant responses to varying stimuli, of mechanical causation as a functioning factor in the processes of growth and reproduction. At least, said Coulter and others, in showing a great need for research experimentation in botany, proof of adaptation must be confirmed not only by field observation but reproduced under controlled conditions in the laboratory.

In other words, Coulter, like Bessey and many others, brought about a renewed concentration of attention on the plants themselves; shifting attention from the mature plant organ to the organ in its development; from knowledge not only of external structure but interior also; from an understanding not only of reproductive organs but of the vegetative structures. The study of organic functions, of plants generally in all their relationships, of all plants—vascular and nonvascular—seed plants to algae—widened the botanical horizon immeasurably. Paleobotanical researches were joined to ever increasing study searching out affinities rather than differences and the great basic evolutionary studies soon presented a staggering story of life development in the plant kingdom. Coulter said at the turn of the century:

. . . there was developed for the first time what may be called a philosophy of the plant kingdom, organizing the details of morphology into one coherent whole about such facts as alternation of generations and heterospory. Study of the metamorphoses of plant organs was replaced by a study of their development and of "life histories," and the earliest stages of gametophyte and sporophyte, and reproductive organs were scrutinized and recorded in the greatest detail in the search for relationships. . . . No longer was the flower of highly organized angiosperms read down into the structure of the lower groups; but from the simplest beginnings structures were traced through increasing complexity and seen to end in the flower, explaining what it is.

Before immense possibilities in study of phylogenies, botanists saw a new science. The problem of the origin of species was placed on an experimental basis. Experimentation studied the behavior of plants and the behavior of protoplasm. Natural selection and modification of species as explanations of evolutionary development and as Darwin explained them would be added to by DeVries in his study of mutation. "Progressive evolution"—the study of combining forms among the great groups—would not only tabulate the history of the groups but aim to show their relationships.

The history of botany, beginning with taxonomy, has been a history that began with the tips of the branches and has proceeded in converging lines towards the common trunk. The fundamental unity of the whole science, in fact, of biological science, however numerous the branches may be, is becoming more and more conspicuous.

So concluded Coulter in 1904. In 1912 he would say:

The recent development of our knowledge of the structures of fossil plants is familiar to botanists, constituting as it does one of the most remarkable chapters in the history of our science. This has been due not only to the elaboration of a technique for sectioning petrifaction, but also to the inclusion of the vascular system among the morphological material that is recognized to be significant in conclusions concerning phylogeny. . . . [I]t is clear that paleobotany must learn to recognize the relationships of fossil plants, or there would be no reliable taxonomy or phylogeny. So long as paleobotany depended upon the form resemblances of detached organs, there could be no taxonomy in the real sense. It was merely a cataloguing of material. But when it learned to uncover structure, it began to establish a real taxonomy.

The great work of not only Torrey, Gray, and other great systematizers, but also of Leo Lesquereux and Newberry in paleobotany underwent enlargement and modification under the guiding work of Penhallow, Wieland, Ward, Jeffrey, Campbell, Bessey, Coulter, Knowlton, David White, Edward W. Berry, and others. Taxonomy remained a classifier and cataloguer of material. But its system, its points of emphasis, changed, remaining withal in large part the same. Always, as Coulter said, there must be the collector and describer of plants, and in this work Nathaniel Lord Britton and many others carried on.

Under Britton's skillful direction, the Columbia College Herbarium, composed of Torrey's, Meisner's, Chapman's, Austin's, Newberry's, and other herbaria, was arranged geographically and according to Bentham and Hooker's *Genera Plantarum*. With a botanical library of more than 2,000 volumes and an equal number of pamphlets, the herbarium was combined by a cooperative agreement with the great New York Botanical Garden during the last years of the century and the institution, combining research and graduate work in Columbia University, soon became one of the great botanic institutions of the United States and the world. The *Bulletin of the Torrey Botanical Club*, published under its auspices, became a botanical journal of the very highest authority, similarly as did the Club's *Memoirs* commenced in 1889 with a volume containing "Studies of Some Types of the Genus Carex" by L. H. Bailey, Jr., which demonstrated his scholarship even as *Talks Afield about Plants and the Science of Plants* showed his ability "to present the truths of science correctly to the great mass of people." After the Garden's work had gotten effectively under way, its *Bulletin* and later *Journal* appeared. In 1901 *Torreya*, another Torrey Botanical Club publication, began issues. *Contributions from the Columbia College Herbarium, Mycologia, Addisonia*, and eminently notable products of a "well-rounded research program, not confined to the taxonomy of the flowering plants and cellular cryptogams . . . including plant physiology, plant pathology, palaeobotany, and popular education," established the Garden as one of the world's great botanic centers. For direction and consummation of these accomplishments, too much credit cannot be given Britton. Acquisition of the DeCandolle library, the Ellis and Everhart collection of fungi, and the Mitten collection of mosses were achievements of supreme importance to the Garden and American botany generally. Development of a largely planned research program in paleobotanic study, one of the most effective in the United States, added prestige; as did one of the most progressive experimental laboratory series of studies to be found in America. Today the Garden's library is one of the most complete and fully equipped centers for botanic and bibliographical, as well as biographical, study of American materials to be found anywhere. The founding of the New York Botanical Garden was Britton's great work.

However, for much else Britton's name is justifiably memorable. One of the greatest programs of botanical exploration ever instituted by a single institution of world botany was initiated under Britton's directorship. Beginning principally with Rusby's South American collections which extended over a long period of years and during the study of which Britton went to England in 1888—added to also by Thomas Morong's collections in Paraguay and Miguel Bang's collections in Bolivia—Britton's great interest in South Atlantic materials was aroused. After the Spanish American War when much new land came under ownership or protection of the United States, the Garden, and other great taxonomic botanic centers, instituted investigations, many being initial ones, of floras of many South Atlantic islands or mainlands. Studies went along the entire Atlantic seaboard as far south as northern

South America and gradually extending, covered vast areas in all Mexico, Central and South America, proceeding in the early 1900's to the Pacific to include American explorations of the Philippine Islands, China, Africa—indeed tabulation of all regions would require much space. Decades have transpired and still the tasks remain uncompleted. Nor does this mean that northern exploration ceased. Nor continued exploration in interior United States. Britton himself collected many North American plants, notably in New York, New Jersey, Virginia, and southeastern Kentucky. He also enumerated collections from Texas, Arizona, Colorado, and other regions. However, his studies of plant life of Florida and the Bahamas, of Cuba, of Bermuda, of Porto Rico and the Virgin Islands, of Jamaica, of Trinidad, and such, are most important, especially that of the West Indies. Systematizations of numerous genera and species came from him; and his *Manual of the Flora of the Northern States and Canada*, published in 1901, became a real competitor of the historic and famous Gray's *Manual*. His three volume work, prepared with Addison Brown, *An Illustrated Flora of the Northern United States, Canada, and the British Possessions*, became known and used by every North American botanist and, when first published, represented "the first fully illustrated 'Flora' on any part of North America." With John Adolph Shafer he published *North American Trees,* being descriptions and illustrations of trees growing independently of cultivation in North America, north of Mexico and the West Indies. His monograph, *The Cactaceae*, prepared with Joseph Nelson Rose, stands as one of the great works of American botany.[17]

Britton did not persist arguing nomenclatural matters. In 1889 he concluded the Torrey *Bulletin* should have only so much nomenclatural material. After 1891, he was too busy for much controversy. On June 1, he told Watson, "This big Botanic Garden project has used all my time, but it looks as though it would be carried out after a while." He served on committees reporting on nomenclature. Although he argued for a scientific taxonomy, after adoption of serviceable Codes, he was more contented.

At Jamaica Plain, Massachusetts, there flourished another institution speedily becoming one of the great botanic establishments of North America—the Arnold Arboretum. About 1870, James Arnold of Providence, Rhode Island, had devised to three trustees a sum of $100,000 for the improvement of agriculture or horticulture. These trustees had determined to dedicate an institution for purposes of study of forestry

---

[17] For an excellent biographical memoir of Britton, see that prepared by E. D. Merrill, with bibliography by John Hendley Barnhart, *Nat. Acad. Sci.*, XIX, 5th memoir (1938), pp. 147-202.

and dendrology. On November 26, 1873, Gray had written DeCandolle:

I am going this morning to witness the nuptials of my colleague and friend Professor Sargent and a charming young lady of Boston; and, on the chance of their having a day in Geneva, I wish to introduce the happy couple to you and Madame De Candolle. . . .

Professor Sargent is given to horticulture and arboriculture. He not only takes charge of the university Botanic Garden, but also of a recent and noble foundation for an arboretum, from which much may in due time be expected. . . .

When Sargent accepted directorship of the Arboretum he found himself confronted with a large task. A "worn out farm, partly covered with natural plantations of native trees nearly ruined by excessive pasturage [had] to be developed into a scientific garden with less than $3000 available for that purpose, without equipment or the support and encouragement of the general public who then knew nothing about an arboretum or what it [was] expected to accomplish." There were, however, groves of trees, a picturesque hill with cliffs, a beautiful stretch of hemlocks, and a meandering stream. In the course of years, Sargent began working with Frederick Law Olmstead when the latter was planning and constructing a park system for the city of Boston and in the early 1880's an agreement was effected whereby the city took the land and leased it back to the college. In the adjoining Bussey Institution greenhouses, plants were raised and propagated but large scale planting of the grounds was not begun until 1886 since the city was slow in building roads and gravel paths.

On December 25, 1881, Gray had written Hooker:

Sargent has got his arboretum at length on to the hands of the city of Boston to make the roads for, to repair and to light and police. He seems to have made a mark in his Census forestry work. He has developed not only a power of doing work, but of getting work done for him by other people, and so can accomplish something.

Not many years had passed before the institution became the most important of its kind in the United States and one of the most valuable of the world. In April 1887 the *Botanical Gazette* announced that Dr. Sargent had reported that 70,000 trees and shrubs had been planted at the arboretum during the past year. Equipped with a large tract of land, there was by 1896 a planted area of 160 acres. Its museum was of growing importance. And from its able personnel issued the prominent scientific journal, *Garden and Forest*. Despite some unfavorable criticism, largely controversial in nature, Sargent's *Silva of North America* was received as one of the most significant publications of a generation of North American botany. Nathaniel Britton characterized it as "great," as did the great majority of American botanists. Sargent was severely criticized for his treatment of collectors such as Pringle and others but, as Gray said, his productions justified his insistence on swiftness of accomplishment and efficiency.[18]

The influence of Greene, Gray, Watson, and then, Britton, is shown in the work of Henry Hurd Rusby. In 1895 appeared Rusby's *Essentials of Vegetable Pharmacognosy*,[19] Part I devoted to the gross structure of plants and Part II to the minute structures, and in 1899 was published his *Morphology and Histology of Plants*, designed as a guide to plant analysis and classification and as an introduction to pharmacognosy and vegetable physiology.[20] Rusby, an explorer, professor, and reformer, became an authority on American pharmacopoeia. He continued South American exploration and systematization of plants from there, writing many articles such as "Floral Features of the Amazon Valley,"[21] "Botanical Collecting in the Tropical Andes,"[22] and "Concerning Exploration upon the Lower Orinoco."[23] He enumerated Miguel Bang's[24] and R. S. Williams's[25] Bolivian collections, and collections from the Republic of Colombia,[26] in addition to describing new genera and species from his own collections with the Orinoco Exploration and Colonization Company in 1896 and the Mulford Biological Exploration of the Amazon Valley in 1921-1922.[27] His *Jungle Memories* published in New York and London in 1933 is descriptive of his 1885-1886 botanical exploration extending in great part North American searches to South America. With Britton he aided in the determination of Arizona and New Mexican collections of Dr. E. A. Mearns.[28] What a great echo of authority these have as compared with that of the young man in New Jersey of the early 1880's about to begin exploration in New Mexico! On August 16, 1880, George Thurber had written Gray:

[18] See *Letters of Asa Gray*, II, pp. 645, 729; *Botanical Gazette*, XII, Number 4 (April 1887), p. 94; N. L. Britton, "Botanic Gardens—Origin and Development," *Science*, n.s., IV, Number 88 (September 4, 1896), pp. 284-293. Also Alfred Rehder's biographical account of Sargent, published by the Arboretum.

[19] *A Treatise on Structural Botany*—designed especially for pharmaceutical and medical students, pharmacists and physicians (New York). 149 pp.

[20] Part I. *The Morphology of Plants*. Part II. *Plant Histology* (New York). 378 pp. Both works illustrated.

[21] *New England Druggist*, I (1889), pp. 14-15, 18-19.

[22] *Bulletin of Phar.*, April 1891.

[23] *Alumni Journal*, III (1896), pp. 185-191.

[24] *Memoir Torr. Bot. Club*, III, IV, and VI (1893).

[25] *Bull. N.Y. Bot. Gard.*, VI (1910), pp. 487-528.

[26] *Mem. Torr. Bot. Club*, XVII (1918), pp. 39-47.

[27] *Mem. N.Y. Bot. Gard.*, VII (1927), pp. 205-387. See *Journal of the New York Botanical Garden*, XXIII (August 1922), pp. 101-112.

[28] New York, 1888-1889. Note: No effort has been made to make this brief list of Rusby's writings complete, nor has the list been brought complete to any one date.

Rusby's grasses came, preceded by your "Who is Rusby—anyhow?"—He is a young schoolmaster who belongs at Franklin, about 6 miles from here. If by "Who," the conundrum had been "*What* is R," I should have quoted an old uncle, who in his classification of human kind, had one set that he called "The devil's unaccountables"—the *Logoniaceae*, as it were, of humanity, into which went whoever didn't fit elsewhere. R[usby] is provokingly good natured and well meaning —would run his toes off to oblige another, and thinks everyone else willing to do the same for him. [He] has a smattering of various things and [is] rather spoiled by the deference of those who know nothing. He is one of those chaps who grasp at problems but has not the least definite idea of structure. He sends me a grass with the florets all gone. I blow him up for first collecting such stuff, and secondly for bothering me with it, tell him if he has any more such to burn 'em. He writes back, "I like abnormal curiosities as well as specimens." I don't know what started him off to New Mexico. I told him it would be folly to make such specimens as he did at home, as people would not buy or take them as a gift. He is just one of those unpractical and impracticable chaps that hang around. He reminds me much of Torrey's "Old Holton." Only H[olton] knew lots, but always had it when he couldn't use it. But why the chap should be bothering you?—Now I have the grasses, I have no word from [him] as to what shall be done with them. I hoped that going off and roughing it for a while might knock some of the nonsense out of him. By the way he is the President of the "Botanical Society (I am not sure that [it] isn't Club) of Northern New Jersey"—Ah ha.

Thurber was an editor, not a teacher. Never once did Gray turn Rusby away so long as Rusby maintained his earnest, conscientious interest in botany. In establishing relations between botany and pharmaceutical science, Rusby had much to do with extending applications of medical botany. It is true that Gray and Rusby had very little dealings, mainly because Rusby's most important work began as Gray's work ended. However, it was Gray who sent a letter to Rusby as he sailed for his adventurous exploration of Bolivia. And when Rusby, on his return, turned to Britton, he made certain Gray approved. Had Engelmann lived, his interest would doubtless have been as manifest as Gray's. In fact, Engelmann watched Rusby's youthful botanical development in New Mexico with interest. Rusby went to Greene because Greene knew New Mexican plants. Rusby was a typical representative of the next generation of North American botanists. In large part, the entire branch of botany known as pharmacognosy has since developed.

Points of view would still further develop in taxonomy and leaders such as Britton, Coulter, Robinson, and others, would discuss for several decades the issues involved; enacting during the period codes applicable to American needs but culminating finally in a great international instrument. Morphological research, dominated by great teachers and investigators such as Coulter, Bessey, Charles J. Chamberlain, Edward C. Jeffrey, Trelease, and a score of other able students, would continue, developing phylogeny, cytology, anatomy, and allied subjects. Ecology and physiology would come forward, training up a host of American students—some trained by Gray, such as Charles Reid Barnes, and others schooled by Gray's, Goodale's, and Farlow's students. The list is too lengthy for naming here. Suffice it to say that the one great period of North American botany dominated by one man—Gray—and his associates produced some of the ablest botanists this continent has known.

After Gray came organization in North American botany. When in August 1892 the Botanical Club of the American Association for the Advancement of Science met, on recommendation of Dr. Arthur, a committee was appointed to consider entertaining an International Botanical Congress to meet at the time of the World's Columbian Exposition of 1893. The committee reported negatively. Nevertheless, after Dr. Lucien M. Underwood returned from Genoa and told how the "Rochester resolutions" on American nomenclature had favorably impressed a world assemblage of botanists gathered there, the proposal for an International Congress to be held near Chicago was revived. Madison, Wisconsin, was selected and in August 1893 the "Madison Botanical Congress" under leadership of Edward Lee Greene met at Science Hall of the University of Wisconsin. The American members were disappointed at the fewness of European botanists present and so internationally important issues in nomenclature were for the most part not settled. However, one accomplishment of great significance was effected. A committee was directed to organize an American society of botanists and on August 15, 1894, at Brooklyn, New York, the Botanical Society of America was established with Dr. Trelease the first president. Impressive, wise, and tactful leadership to form a "pure science" society of American botanists—from introduction of the original resolution to the naming of charter members—had been assumed by Liberty Hyde Bailey. He did not participate in every proceeding toward the society's establishment but the force of his genius for organization, working officially in committee or botanical club meetings and by correspondence, energized the movement which led to the Society's establishment. Only botanists qualified in research and authors of works of recognized merit were to be eligible. Today this organization flourishes as a most potent influence in the American botanical domain. And organization of separate branches of the science have followed. All are living symbols of work begun in the first great transition period of American botany, stalks grown from earlier seed.

What shall we say of the transition period as a whole? Certainly it was a period of search and accomplishment—a search that has come to fuller truth and may yet expand in realms of undiscovered knowledge.

ACKNOWLEDGMENTS

MOST of the material contained in this book has been quoted direct—or obtained from unpublished correspondence which has never before been generally available. For information that has been previously published the most authoritative sources have been made use of and references to these have been incorporated as footnotes. For valuable assistance in securing this material and for helpful criticism afforded the author in writing this book, he wishes to express his indebtedness to the following persons and institutions.

To Dr. Liberty Hyde Bailey, formerly Dean of the New York State College of Agriculture and Director of the Cornell University Agricultural Experiment Station, and now Director of the Bailey Hortorium of Cornell University, who has read the whole of the manuscript, giving much helpful aid and advice, suggesting corrections, and other forms of improvement. Dr. Bailey's years of acquaintance with Asa Gray made his aid and advice especially valuable. Those portions, relating to taxonomy—in fact, the entire manuscript in first writing—have been read by Dr. C. A. Weatherby, Research Associate of the Gray Herbarium, Harvard University, Cambridge, Massachusetts.

To Dr. Edward W. Berry, Professor Emeritus of Paleobotany of Johns Hopkins University, Baltimore, Maryland, who has read all parts of the book relating to paleobotany, for much helpful aid, corrections, and criticism.

To the Gray Herbarium for permission to photostat a large amount of correspondence with Asa Gray from Engelmann, Lesquereux, Parry, Rothrock, Palmer, Coulter, Britton, Greene, Tuckerman, Rusby, Smith, Fendler, Chapman, Dana, Bolander, Vasey, Eaton, Macoun, Cleveland, Muir, Lemmon, and others. To the Missouri Botanical Garden for permission to photostat a large amount of correspondence with George Engelmann from Watson, Parry, Palmer, Lesquereux, and others. To the Archives of the Smithsonian Institution for permission to photostat and copy letters written by Lesquereux to Spencer Baird and other officers of the Institution. To the American Philosophical Society for permission to copy correspondence of Lesquereux with J. Peter Lesley. To Dr. Ernst A. Bessey, Professor of Botany, Michigan State College of Agriculture and Applied Science, for permission to copy correspondence with Charles E. Bessey maintained by the number of botanists referred to in this book. To the Farlow Herbarium and Reference Library of Cryptogamic Botany, Harvard University, for permission to copy diverse letters and materials. To Iowa State College for permission to use the letter of Gray to Parry dated October 30, 1887.

To many libraries, too numerous to mention all, but especially libraries of the Gray Herbarium, the Missouri Botanical Garden, the Ohio State University Botanical and Zoological Library, the University of California at Los Angeles, the Academy of Natural Sciences of Philadelphia, and the Smithsonian Institution. For material on John Strong Newberry, indebtedness to the New York Botanical Garden is acknowledged.

To the American botanists who have aided the author by conference and consultation on many various points of doubt, and especially members of the Department of Botany of Ohio State University before whom the author delivered a lecture at its Seminar incorporating the most important materials concerned with Lesquereux's early work in North American paleobotany. Members of the department have all willingly and helpfully aided the author. Others whom the author especially wishes to thank are Dr. John Hendley Barnhart, Dr. Francis W. Pennell, Dr. William R. Maxon, Dr. J. M. Greenman, Dr. Carl C. Epling, Dr. Rogers McVaugh, and many others. The interest and encouragement of Dr. Samuel Wood Geiser of Southern Methodist University is greatly appreciated.

To the author's mother and father, Mr. and Mrs. Andrew Denny Rodgers, for much aid and encouragement, together with many suggestions for improvement.

INDEX

* Since the name of Asa Gray appears on almost every page of this book, his name has been omitted from this index.